*Rudolf Dvorak and*
*Christoph Lhotka*

**Celestial Dynamics**

## *Related Titles*

Kopeikin, S., Efroimsky, M., Kaplan, G.

**Relativistic Celestial Mechanics of the Solar System**
**892 pages with 65 figures and 6 tables**

2011
ISBN: 978-3-527-40856-6

Barnes, R. (ed.)

**Formation and Evolution of Exoplanets**
**320 pages with 136 figures and 5 tables**

2010
ISBN: 978-3-527-40896-2

Szebehely, V. G., Mark, H.

**Adventures in Celestial Mechanics**
**320 pages with 86 figures**

1998
ISBN: 978-0-471-13317-9

*Rudolf Dvorak and Christoph Lhotka*

# Celestial Dynamics

Chaoticity and Dynamics of Celestial Systems

WILEY-VCH Verlag GmbH & Co. KGaA

**The Authors**

**Prof. Dr. Rudolf Dvorak**
University of Vienna
Department of Astronomy
Tuerkenschanzstrasse 17
1180 Vienna
Austria

**Dr. Christoph Lhotka**
University of Namur
Department of Mathematics
Rempart de la Vierge 8
5000 Namur
Belgium



**Library of Congress Card No.:**
applied for

**British Library Cataloguing-in-Publication Data:**
A catalogue record for this book is available from the British Library.

**Bibliographic information published by the Deutsche Nationalbibliothek**
The Deutsche Nationalbibliothek lists this publication in the Deutsche Nationalbibliografie; detailed bibliographic data are available on the Internet at http://dnb.d-nb.de.



**Print ISBN** 978-3-527-40977-8
**ePDF ISBN** 978-3-527-65188-7
**ePub ISBN** 978-3-527-65187-0
**mobi ISBN** 978-3-527-65186-3
**oBook ISBN** 978-3-527-65185-6

**Cover Design** Grafik-Design Schulz, Fußgönheim
**Typesetting** le-tex publishing services GmbH, Leipzig
**Printing and Binding** Markono Print Media Pte Ltd, Singapore

Printed in Singapore
Printed on acid-free paper

*Dedicated to Caroline, Sophia, Stephanie, and Barbara*

# Contents

# Preface

The idea of writing a book about astrodynamics primarily for students and colleagues wishing to learn about and perhaps to work in this interesting field of astronomy came into my mind (RD) already some twenty years ago. In those days my concept was a different one than the book in your hands: based on a very special topic equally interesting for nonlinear dynamics and astrodynamics (see the book of J. Moser[1]) the so-called *Sitnikov Problem* I wanted to present all the classical ideas of perturbation theory and connect them to the modern tools of chaos theory. Years went by without such a book being written! But then, some ten years ago, I was fortunate enough to have a brilliant young student (CL) working especially on modern development in nonlinear dynamics who extensively made use of the tools of computer algebra. After his PhD he then moved as a postdoc to two famous Mathematics Departments (Rome and Namur under the direction of two world known colleagues, Profs. Alessandra Celletti, respectively Anne Lemaitre). Immediately after his PhD, still in Vienna, I (RD) got an offer by Wiley to write a book on the subject I have been working on for 40 years. Being aware of this opportunity I invited my former student to write a book with me which, on one hand, deals with the modern tools of nonlinear dynamics and, on the other hand, introduces to the classical methods used since more than two centuries with great success. We realize that this is a difficult task because recently two excellent books concerning this subject appeared written by Prof. S. Ferraz-Mello[2] and by Prof. A. Celletti[3]. Nevertheless we hope to combine with our book different aspects of Celestial Mechanics in an understandable way for interested students and colleagues in Physics and Astronomy, and to succeed in infecting the reader with our enthusiasm for the subject.

Vienne and Namur, December 2012 *Rudolf Dvorak and Christoph Lhotka*

1) Stable and Random Motion in Dynamical Systems.
2) Canonical Perturbation Theories.
3) Stability and Chaos in Celestial Mechanics.

# 1
# Introduction: the Challenge of Science

Science is one of the main challenges of mankind: it is not only to follow the traces of nature and to try to understand why things are there and how they 'function' – it is also to help humans to live in better conditions. Nevertheless we need to be cautious since our history shows that the technical development is gradually destroying our mother planet Earth. One may understand science as one side of a coin, where art is on the other side, a wonderful possibility of using the human spirit to produce things which are not present in nature and are created just for the sake of beauty. Both are the outcome of a development of nature from the primitive forms of life to human beings, the only creatures being able to think about their own existence. However we begin to discover more and more 'intellectual abilities' even in the 'animal kingdom', for example elephants recognizing themselves in a mirror, different species using tools for preparing their food and their place to live. Still, there seems to be a development in the human brain, far beyond the capacity of any animal or supercomputer, that allows us to be creative in a dual sense, in art AND science.

Tracing back the development of astronomy we need to go back thousands of years into the past, when our ancestors were looking into the sky during daytime following the path of the sun and during night time when the scintillating stars were visible in the dark sky. The Moon is the only 'body' sometimes visible during daytime and sometimes during the night, its shape changed while it apparently was also 'eating' the Sun for up to several minutes. Although the view of the world changed with time – and even now it is changing with new exciting discoveries like the dark matter and dark energy – the main point of view was a geocentric picture of the universe. In principle it did not change because mankind today still continuously follows an aspect of being in the center of the universe and the only intelligence capable of questioning their own existence.

Definitely the Sun, the Moon and the stars were subjects of religious worship, unapproachable and unalterable, but at the same time priests (and astronomers) tried to predict the celestial phenomena and to use them for measuring time and for making a calendar for these civilized nations in China, India, Central America, in Babylon and in Egypt and later also in Greece. Relatively early they could dis-

*Celestial Dynamics*, First Edition. R. Dvorak and C. Lhotka.

tinguish between the planets, moving in the sky in somewhat interesting 'orbits' and the stars, which never change their positions with respect to each other. Of special importance was to predict eclipses of the Sun and the Moon, which were often regarded to be able to alter the destiny of mankind. It seems that the highest development reached was by the Babylonians with big influence on the Greek astronomy and later on the Arabic astronomers which then was the basis for the European view of the world up to the middle ages, which was not always favorable[1] for the progress.

Even in these early days of the dawn of science the description of the motion of the Sun, the Moon and the planets asked for a knowledge of determining positions on the sky relative to a never changing background of the stars. In this sense measurements of positions of celestial objects – astrometry – was born; their results serve as basis for the topic of the book on astrodynamics. In the old civilizations the motions of the Sun, the Moon and the planets were necessary to determine with as high precision as possible for the purpose of computing a calendar based on periodic events. The day is automatically the shortest measure for time, but the problem is that the length of the day from sunrise to sunset is changing. But again the long period where the length of the day repeats can well be determined and defines the year. Between the day and the year the motion of the Moon defines another time period, the month, which was also determined with quite a good precision from the ancient astronomers.

The intention in this introduction is not to give a course on the – very interesting – history of astronomy[2], but to describe the big steps forward in the development of the description of the motion of celestial bodies caused by outstanding discoveries and of technical developments. We thus skip the developments of the astronomical knowledge up to the discoveries of the noncircular orbits of the planets around the Sun by Johannes Kepler[3], the discoveries of Galileo Galilei[4] and the brilliant discovery of the universal law of gravitation by Isaac Newton[5].

Up to the early twentieth century the idea was that all events in nature may be described properly because we are living in a fully deterministic world as it was the

1) Think of the dragging influence of Aristotle with respect to science: he had in his thinking no concepts of mass and force. He had a basic conception of speed and temperature, but no quantitative understanding of them, which was partly due to the absence of basic experimental devices, like clocks and thermometers.
2) There are excellent books on the subject on the market like 'Early Physics and Astronomy: A Historical Introduction' by Olaf Pedersen [1].
3) (1571–1630) derived the surprising fact that Mars is in an elliptic orbit based on the excellent observations by Tycho Brahe (1546–1601). And he then formulated his three laws concerning the motion of the two body problem (see details in Chapter 5).
4) (1564–1642) who observed with a telescope the motion of the 4 large Moons of Jupiter, the Galilean satellites, which form a 'planetary system' themselves on a smaller scale.
5) (1642–1717) in his *magnum opus* 'Philosophia Naturalis Principia Mathematica' (1678) where he described the fundamental law of gravitation.

thinking of Laplace[6)]

> "Une intelligence qui, à un instant donné, connâtrait toutes les forces dont la nature est animé et la situation respective des êtres qui la compose embrasserait dans la même formule les mouvements des plus grands corps de l'univers et ceux du plus léger atome; rien ne serait incertain pour elle, et l'avenir, comme le passé, serait présent à ses yeux."
>
> *Pierre Simon de Laplace,*
> Essai philosophique sur les probabilités, Bacheliers, Paris 1840.

In the concept of determinism it is believed that it is possible to describe nature with the basic laws of physics and the aid of appropriate mathematical tools (here we need to mention excellent mathematicians and astronomers of these periods like Leonard Euler (1707–1783), Lagrange (1736–1813), d'Alembert (1717–1783) and Carl Friedrich Gauss (1777–1855) without being complete in this list).

And just at the end of the nineteenth century and beginning of the twentieth century the work of Henri Poincaré (1854–1912) led to a new understanding of describing nature: in his *Méthodes Nouvelles de la Mécanique céleste* he already described a phenomenon which some 50 years later entered into the scientific mind as *CHAOS*; in contrary to the idea of Laplace it is NOT possible to describe nature in a deterministic way. About this time also from the theories of specific and general relativity by Albert Einstein (1879–1955) and quantum theory[7)] we learned that we have to accept the fact that nature can be described only statistically.

Three breakthroughs in the last century originated in new challenges for science in general and for astronomy and astrodynamics in particular: the rôle of the chaos described above, the fast development of computers opened up new frontiers, and the launch of the first artificial satellites by the Russians in 1957.[8)]

6) (1749–1827), "We may regard the present state of the universe as the effect of its past and the cause of its future. An intellect which at a certain moment would know all forces that set nature in motion, and all positions of all items of which nature is composed, if this intellect were also vast enough to submit these data to analysis, it would embrace in a single formula the movements of the greatest bodies of the universe and those of the tiniest atom; for such an intellect nothing would be uncertain and the future just like the past would be present before its eyes." [Laplace, A Philosophical Essay, New York, 1902, p. 4]

7) Here we need to mention besides Einstein Max Planck (1858–1947), Erwin Schrödinger (1887–1961), Niels Bohr (1885–1962), Wolfgang Pauli (1900–1958) and Werner Heisenberg (1901–1976) as some of the main founders of quantum physics.

8) In this year on the fourth of October the Soviet Union Sputnik 1 was the first man made satellite in orbit around the Earth. Only 4 years after the Soviet cosmonaut Yuri Gagarin on board of Vostok 1, was the first human in space orbiting the Earth. In fact already in 1903 the Russian Konstantin Tsiolkovsky in his main work (The Exploration of Cosmic Space by Means of Reaction Devices) established the basis for the later space missions with rockets. The endpoint (beginning) of a race between the Americans and the Soviets ended with the landing on the Moon of the first human in 1969. On board of Apollo 11, which landed on the Moon the American astronauts Neil Armstrong and Edwin Aldrin, Jr., acceded the Moons surface: *That's one small step for man – one – giant leap for mankind.*

Although there had been made extremely interesting discoveries in astronomy during the last decades like the existence of dark matter and the surprisingly 'fact' of an accelerating expansion of our universe due to dark energy, we believe that for the human race the discovery of extrasolar planets is the most exciting one and a real breakthrough for scientific thinking. This first discovery of a planet around a main sequence star some almost twenty years ago was a real positive 'shock' for astronomers.

The idea of other worlds besides our own is an idea which goes back to the Greek philosopher Epicure and later in the middle ages as Giordano Bruno expressed his fantastic idea of other worlds *This space we declare to be infinite... In it are an infinity of worlds of the same kind as our own.* He believed in the Copernican theory that the Earth and other planets orbit the Sun. In addition he was convinced that the stars are similar to our Sun, and that they are accompanied by inhabited planets. Giordano Bruno was burned in 1600 in Rome primarily because of his pantheistic ideas.

Ever since that first discovery of an extrasolar planet astronomers work hard to confirm the existence of extrasolar planets and today more than 700 exoplanets are known and the number of discoveries is growing fast with finer and better tools of detecting them around other stars. Although most of the planets discovered seem to be 'alone', we have to assume that other planets are present in such systems because from the theory of formation of planets it is rather improbable that they form alone. To understand the diversity of the of the planets and especially the architecture of multiplanetary systems is one big task to be answered by astrodynamics.

The concept of our book is the following: In the first chapters we introduce the basic mathematical tools to understand the complexity of the subject we are dealing with in celestial mechanics. The next chapters are dedicated to classical problems in celestial mechanics from the two-body problem to the motion of the Moon. We briefly outline the content of each chapter:

We start by describing Hamiltonian mechanics, which is the basis of all later subjects. We explain the concept of canonical transformations, what are action angle variables and so on. We then move on to describe the different methods of solving the complex problems: we shortly explain the principle of mappings with some simple examples, introduce an efficient tool – the Lie-integrator – for solving differential equations and demonstrate how analytical methods may be used for solutions of problems in nonlinear dynamics in general. A central section is devoted to the problem of the stability of motions where we also present the theories of Arnold–Kolmogorov–Moser and of Nekhoroshev.

Only now we discuss the two-body problem with the Kepler laws and the famous transcendental Kepler-equation and some methods of solving this simple looking equation. We also make use of a complex notation which makes the two-body problem even in three dimensions easy to describe in closed form. An important step from the integrable Kepler-problem to the problem where 3 bodies (one of them regarded as massless) are involved is described in the chapter of the restricted three-body problem. The equations of motion, in a rotating coordinate system, are well adapted to find a concept of qualitative description of stable motion via the zero-velocity curves. Because the basics in the motion of the Moon can be described in a dynamical system with a massless Earth companion we then turn to give insights into the Lunar motion.

A kind of intermezzo is dedicated to the Sitnikov problem where we both (RD and CL) worked a lot during the last years. Analytical approaches are presented together with results of extensive numerical integrations. The richness of motion in its different adaption from the circular and the elliptic problem (two equally massive bodies and a massless third body) is shown in many details up to the self-similarity in the representation of the Poincaré surfaces of section.

We move on in the next part to develop the main ideas of the planetary theories, where, after giving a simple example, we deal with first and second-order perturbations, and the role of small divisors. Mean motion resonances, secular resonances and three-body resonances are the topic of the next chapter, where we also describe many recent results about the interesting motions of a group of asteroids, the Trojans. Then we give an introduction into the Lunar theory based on the former results of the restricted three-body problem and discuss the main 'inequalities' of the motion of the Moon based on the perturbing function.

Finally, to conclude we list books dealing with Celestial Mechanics from the time of Poincaré on and briefly say some words about their content. This very personal choice is far from being complete and we are aware of the fact that many important contributions are not mentioned; we apologize for that.

*Enjoy our book!*

# 2
# Hamiltonian Mechanics

In this chapter we introduce the basic concepts and formalism of Hamiltonian mechanics which are needed to present the contents of this book. The following short exposition is based on [2], see [3, 4] for more details on this interesting subject.

A dynamical system given in $\mathbb{R}^3$ consisting of $N$ particles and with $k$ holonomic constraints has $n = 3N - k$ degrees of freedom. The motion of the remaining free variables is described by the Lagrangian equations of the form:

$$\frac{d}{dt}\frac{\partial L}{\partial \dot{q}_j} - \frac{\partial L}{\partial q_j} = 0 \tag{2.1}$$

with $j = 1, \ldots, n$ and where $(q_j, \dot{q}_j)$ are the positions and velocities of the $j$-th body, respectively. Here, the function $L$ is the Lagrangian function,

$$L = L(q_1, \ldots, q_n, \dot{q}_1, \ldots, \dot{q}_n)$$

The notation was introduced by William Rowan Hamilton, and named after the French(-Italian) mathematician Joseph Louis Lagrange. In classical or Lagrangian mechanics the function $L$ is defined as the difference

$$L = T - V$$

where $T = T(\dot{q}_1, \ldots, \dot{q}_n)$ is the kinetic and $V = V(q_1, \ldots, q_n)$ is the potential energy of the system. Lagrangian equations of the form (2.1) are valid only for holonomic constraints and for forces which can be derived from a potential. It can be shown that they are equivalent to Newton's classical equations of motion. To obtain (2.1) one takes $T - V$ in $3N$ generalized coordinates $(q_j, \dot{q}_j)$, with $j = 1, \ldots, 3N$, and expresses the $3N$ coordinates through $3N - k$ using the $k$ additional constraints. The equations of motion follow from (2.1) and define a set of $2n$ ordinary equations of motion of second order. They can also be expressed by $2n$ ordinary equations of first order by introducing the generalized momenta

$$p_j = \frac{\partial L}{\partial \dot{q}_j} \quad \text{with} \quad j = 1, \ldots, n$$

which will replace the $\dot{q}_j$ (note, that $p_j$ defines the mechanical momenta only if the potential $V$ does not depend on the velocities). In this setting the motion of

*Celestial Dynamics*, First Edition. R. Dvorak and C. Lhotka.

the $N$ particles takes place in a $2n$-dimensional phase space spanned by $(q_j, p_j)$, with $j = 1, \ldots, n$. We obtain the corresponding ordinary differential equations of first order by means of a Legendre transformation of (2.1). We summarize, briefly, the main steps of the transformation as follows: the Legendre transformation of a function $f(x, y)$ gives another function $g(u, y)$, where $u := \partial f/\partial x$. From the total derivative

$$df = \frac{\partial f}{\partial x}dx + \frac{\partial f}{\partial y}dy = udx + vdy$$

with

$$u = \frac{\partial f}{\partial x}, \quad v = \frac{\partial f}{\partial y}$$

we get a new function $g = f - ux$ and also

$$dg = df - udx - xdu = vdx - xdu$$

Taking the function $g = g(u, y)$ we have

$$dg = \frac{\partial g}{\partial u}du + \frac{\partial g}{\partial y}dy$$

and from comparison with the above relation we get

$$v = \frac{\partial g}{\partial y}, \quad x = -\frac{\partial g}{\partial u}$$

We repeat a similar calculation for the Lagrangian function $L$ taking into account the case that $L$ may also depends on time, that is, where $L = L(q, \dot{q}, t)$. We define another function $H$, which is called the Hamiltonian function, according too:

$$H(p, q, t) = \sum_{i=1}^{n} \dot{q}_i p_i - L(q, \dot{q}, t)$$

and we calculate the total derivative of $H$, with respect to time, which gives:

$$dH = \sum_{i=1}^{n}\left(\dot{q}_i dp_i + p_i d\dot{q}_i - \frac{\partial L}{\partial q_i}dq_i - \frac{\partial L}{\partial \dot{q}_i}d\dot{q}_i\right) - \frac{\partial L}{\partial t}dt$$

The total derivative of a generic function $H(p, q, t)$ is

$$dH = \sum_{i=1}^{n}\left(\frac{\partial H}{\partial q_i}dq_i + \frac{\partial H}{\partial p_i}dp_i\right) + \frac{\partial H}{\partial t}dt$$

and the comparison of the related coefficients of the two equations above gives:

$$\begin{aligned}\dot{q}_i &= \frac{\partial H}{\partial p_i}\\ \dot{p}_i &= -\frac{\partial H}{\partial q_i}\\ \frac{\partial H}{\partial t} &= -\frac{\partial L}{\partial t}\end{aligned} \tag{2.2}$$

with $i = 1, \ldots, n$. The relation that involves the preceding derivatives $\dot{q}$, $\dot{p}$ are called the Hamilton equations and correspond to the system of $2n$ ordinary differential equations of first order. They are completely equivalent to the Lagrangian equations given in (2.1). For given initial conditions $(q_i(0), p_i(0))$ they determine the motion in the space spanned by $(q_i, p_i)$ with $i = 1, \ldots, n$. Due to the special symmetry of (2.2) the set is also called the canonical system of equations of motion, $(q_i, p_i)$ are called canonical variables, $p_i$ the canonical (or conjugated) momentum to the generalized variable $q_i$ (with $i = 1, \ldots, n$).

Note, that the calculation of $H$ can be simplified if the holonomic constraints are time independent and the forces can be derived from a potential. In that case the Hamiltonian function reads

$$H = T + V$$

and thus the Hamiltonian is the sum of kinetic and potential energy.

Remark 2.1

The Hamilton function $H(p, q, t)$ is defined as a function of the time and the canonical variables only. It is mandatory to express the velocities of the original Lagrangian system just in terms of them. Thus to obtain $H$ for given $L$ one needs to write $\dot{q}_i = \dot{q}_i(p, q, t)$ to obtain

$$H(p, q, t) = \sum_{i=1}^{n} \dot{q}_i(p, q, t) p_i - L(q, \dot{q}_i(p, q, t), t)$$

which is then a function of $(p, q, t)$ only.

In the next step we investigate the total time derivative of $H$. From

$$\frac{dH}{dt} = \sum_{i=1}^{n} \left( \frac{\partial H}{\partial q_i} \dot{q}_i + \frac{\partial H}{\partial p_i} \dot{p}_i \right) + \frac{\partial H}{\partial t}$$

which also gives

$$\frac{dH}{dt} = \sum_{i=1}^{n} \left( \frac{\partial H}{\partial q_i} \frac{\partial H}{\partial p_i} + \frac{\partial H}{\partial p_i} \frac{\partial H}{\partial q_i} \right) + \frac{\partial H}{\partial t}$$

we find

$$\frac{dH}{dt} = \frac{\partial H}{\partial t} = -\frac{\partial L}{\partial t}$$

Thus, the Hamilton function is a constant of motion if and only if it does not depend on time explicitly. The identification of $H$ with the total energy $E$ of the system is a subcase of the condition

$$\frac{\partial H}{\partial t} = 0$$

which can also be true for $H \neq E$, that is, when the Hamiltonian is not equal to the total mechanical energy of the system.

## 2.1 Hamilton's Equations from Hamiltonian Principle

Another way to demonstrate that the Hamilton equations are equivalent to the Lagrange equations is to show that both can be derived from Hamilton's principle of stationary action. The action of a dynamical system, between time $t_1$ and $t_2$, is defined as

$$S = \int_{t_1}^{t_2} L(q, \dot{q}, t) dt$$

which we have to express in terms of $(q, p, t)$, that is, compatible to the phase space variables. From

$$S = \int_{t_1}^{t_2} L dt = \int_{t_1}^{t_2} \left[ \sum_{i=1}^{n} \dot{q}_i p_i - H(p, q, t) \right] dt$$

and the variation principle

$$\delta q_i(t_1) = \delta q_i(t_2) = \delta p_i(t_1) = \delta p_i(t_2) = 0$$

we find

$$\delta S = \int_{t_1}^{t_2} \sum_{i=1}^{n} \left( \dot{q}_i \delta p_i + p_i \delta \dot{q}_i - \frac{\partial H}{\partial q_i} \delta q_i - \frac{\partial H}{\partial p_i} \delta p_i \right) dt = 0$$

Since the variations vanish at $t_1$ and $t_2$ a partial integration gives

$$\int_{t_1}^{t_2} p_i \dot{q}_i dt = - \int_{t_1}^{t_2} \dot{p}_i d q_i dt$$

and thus

$$\delta S = \int_{t_1}^{t_2} \sum_{i=1}^{n} \left[ \left( \dot{q}_i - \frac{\partial H}{\partial p_i} \right) \delta p_i - \left( \dot{p}_i + \frac{\partial H}{\partial q_i} \right) \delta q_i \right] dt = 0$$

Since the individual integrands of the preceding relation have to vanish term by term (the $\delta q_i$ and $\delta p_i$ are arbitrary and independent from each other) we obtain the equations (2.2) from $\delta = 0$.

Summarized briefly, the Hamilton formalism is equivalent to the Lagrangian one since it follows from the Legendre transformation of the Lagrangian function. Moreover, the Lagrangian function follows from Hamilton's principle by means of a variational problem. Both lead to the canonical equations of motion. The finding of the solutions in terms of the new equations is usually not easier than in terms of the Lagrangian equations of motion. The benefit of the Hamiltonian formalism lies in the fact, that further concepts can be introduced quite easily. Moreover, it is possible to interpret the solution by geometrical means in the phase space and, as we will see, to transform the system to different coordinates, which usually admit a better understanding of the underlying dynamical problem. It has to be stressed, that the Hamiltonian corresponds to the total energy of the system only for time independent holonomic constraints and where the forces can be derived from a potential - the system being conservative. If it does not depend on time it is a conserved quantity.

## 2.2
## Poisson Brackets

A differentiable function $f(p, q, t)$ of variables $(p(t), q(t), t)$ is called a dynamic variable (or observable). The derivative with respect to time becomes

$$\frac{df}{dt} = \sum_{i=1}^{n} \left( \frac{\partial f}{\partial q_i} \dot{q}_i + \frac{\partial f}{\partial p_i} \dot{p}_i \right) + \frac{\partial f}{\partial t}$$

Taking into account the definition of the canonical equations (2.2) we also get:

$$\frac{df}{dt} = \sum_{i=1}^{n} \left( \frac{\partial f}{\partial q_i} \frac{\partial H}{\partial p_i} - \frac{\partial f}{\partial p_i} \frac{\partial H}{\partial q_i} \right) + \frac{\partial f}{\partial t}$$

We define the Poisson-bracket of two dynamic variables $f(p, q, t)$ and $g(p, q, t)$ as

$$\{f, g\} = \sum_{i=1}^{n} \left( \frac{\partial f}{\partial q_i} \frac{\partial g}{\partial p_i} - \frac{\partial f}{\partial p_i} \frac{\partial g}{\partial q_i} \right)$$

and note, that the time derivative of a function $f(p, q, t)$ can be written as

$$\frac{df}{dt} = \{f, H\} + \frac{\partial f}{\partial t}$$

We are now able to define a condition whereby a function $f(p, q, t)$ is a constant (or an integral of motion) along the flow $H$ as

$$\{f, H\} = 0$$

Since

$$\frac{\partial f}{\partial q_i} = -\{p_i, f\}\,, \quad \frac{\partial f}{\partial p_i} = \{q_i, f\}$$

we are also able to write the Hamilton equations of motion as

$$\dot{q}_i = \frac{\partial H}{\partial p_i} = \{q_i, H\}\,, \quad \dot{p}_i = -\frac{\partial H}{\partial q_i} = \{p_i, H\}$$

The properties of the Poisson bracket are:

1. linearity:

$$\{c_1 f + c_2 g, h\} = c_1\{f, h\} + c_2\{g, h\}$$

where $c_1, c_2$ are constants and $f, g, h$ are dynamical variables of $(p, q, t)$.
2. antisymmetry:

$$\{f, g\} = -\{g, f\}$$

where $f, g$ are dynamical variables of $(p, q, t)$.
3. existence of a zero element:

$$\{c, f\} = 0$$

where $c$ is a constant and $f = f(p, q, t)$.
4. associativity:

$$\{fg, h\} = f\{g, h\} + \{f, h\}g$$

where $f, g, h$ are dynamical variables of $(p, q, t)$.
5. Jacobi's identity:

$$\{f, \{g, h\}\} + \{g, \{h, f\}\} + \{h, \{f, g\}\} = 0$$

where $f, g, h$ are dynamical variables.
6. fundamental Poisson brackets:

$$\{q_i, q_j\} = \{p_i, p_j\} = 0\,, \quad \{q_i, p_j\} = \delta_{ij}$$

where $i, j = 1, \ldots, n$ and $\delta_{ij}$ is the Kronecker-delta.

We call a phase space spanned by $(p, q)$ which obtains the property 6. a phase space with Poisson structure related to the canonical nature of the equations of motion. The inverse of the Poisson brackets are called the Lagrange brackets $[f, g]$ which are defined as

$$[u_k, u_l] = \sum_{i=1}^{n} \left( \frac{\partial q_i}{\partial u_k} \frac{\partial p_i}{\partial u_l} - \frac{\partial q_i}{\partial u_l} \frac{\partial p_i}{\partial u_k} \right)$$

with $k, l = 1, \ldots, 2n$ and where $u_k = u_k(p, q, t)$ are functions of $p, q$. We will not use them in the further discussion.

The Poisson brackets are useful to test integrals of motion and to define the so-called Poisson structure of the phase space. They can also be used to write the canonical equations of motion in a different form and to check that a transformation of variables preserves the canonical structure of the phase space.

## 2.3
## Canonical Transformations

In many problems, found in celestial mechanics, the change of the coordinate system already allows to solve the dynamical problems. If it does not solve the problem, the change may still simplify the dynamical system and/or reduce it to a form where the most interesting dynamics become more visible. As already noted above not all transformations from one system to another preserve the fundamental properties of the system. So it is desirable to preserve the Poisson structure of the phase space when dealing with Hamiltonian vector fields. We are looking for the class of transformations which preserve the canonical equations of motion, which we denote by canonical transformations in short.

First we investigate the kind of transformations of the form $Q_i = Q_i(q)$, where $q = (q_1, \ldots, q_n)$ and $Q_i$ denotes a new generalized coordinate which should replace $q_i$. They are denoted by point transformations since they only change the generalized coordinates within the phase space. An easily generalization allows them to depend on time also, in such a way that $Q_i = Q_i(q, t)$. The new Lagrangian $L'(Q, \dot{Q}, t)$ may be obtained from $L(q, \dot{q}, t)$ by expressing $q$ by $Q$:

$$L'(Q, \dot{Q}, t) = L(q(Q, \dot{Q}, t), \dot{q}(Q, \dot{Q}, t), t)$$

The Lagrange equations of motion remain the same:

$$\frac{d}{dt}\frac{\partial L'}{\partial \dot{Q}_i} - \frac{\partial L'}{\partial Q_i} = 0$$

for $i = 1, \ldots, n$ although the explicit form may change completely. To transform the generalized momenta we use

$$p_i = \frac{\partial L}{\partial \dot{q}_i} \rightarrow P_i = \frac{\partial L'}{\partial \dot{Q}_i} = \sum_{j=1}^{n} \frac{\partial L}{\partial \dot{q}_j}\frac{\partial \dot{q}_j}{\partial \dot{Q}_i}$$

which gives

$$P_i = \sum_{j=1}^{n} p_j \frac{\partial q_j}{\partial Q_i} = \sum_{j=1}^{n} a_{ij}(q, t) p_j$$

for $i = 1, \ldots, n$. Here $(a_{ij})$ denotes a rectangular matrix whose elements depend on time $t$ and $(q_1, \ldots, q_n)$, respectively. We obtain the Hamiltonian $H'$

$$H'(P, Q, t) = \sum_{i=1}^{n} P_i \dot{Q}_i(P, Q, t) - L'\left(Q, \dot{Q}(P, Q, t), t\right)$$

and note that the canonical equations remain the form

$$\dot{Q}_i = \frac{\partial H'}{\partial P_i}, \quad \dot{P}_i = -\frac{\partial H'}{\partial Q_i} \quad \text{with} \quad i = 1, \ldots, n$$

Thus transformations of the form $Q_i = Q_i(q, t)$ preserve also the Poisson structure of the phase space and point transformations are canonical.

We investigate transformations of the form

$$\begin{aligned} q_i &\to Q_i(p, q, t) \\ p_i &\to P_i(p, q, t) \end{aligned}$$

where $i = 1, \ldots, n$ which gives $2n$ relations between $(q_i, p_i)$ and $(Q_i, P_i)$ the difference to above is that we allow also the transformations of the momenta and that the transformation of the coordinates may also involve the old momenta. Again we are interested in transformations that retain the Poisson structure of the phase space. The idea behind is to show the existence of a Hamiltonian $H'(P, Q, t)$ such that for

$$\dot{q}_i = \frac{\partial H}{\partial p_i}, \quad \dot{p}_i = -\frac{\partial H}{\partial q_i}$$

we also find

$$\dot{Q}_i = \frac{\partial H'}{\partial P_i}, \quad \dot{P}_i = -\frac{\partial H'}{\partial Q_i}$$

with $i = 1, \ldots, n$. Note, that we no longer require $H'$ to represent anything special (like the total energy of the system); on the contrary we just require the existence of $H'$ for given transformation $(q, p) \to (Q, P)$.

Formally, we make use of the fact that the canonical equations follow from the Hamilton's principle. For a given Hamiltonian $H$ and

$$\delta \int_{t_1}^{t_2} \left[ \sum_{i=1}^{n} p_i \dot{q}_i - H(p, q, t) \right] dt = 0$$

we require that the Hamiltonian in new variables $H'$ fulfills

$$\delta \int_{t_1}^{t_2} \left[ \sum_{i=1}^{n} P_i \dot{Q}_i - H'(P, Q, t) \right] dt = 0$$

This leads to the requirement that the two integrands differ only by a constant $c \neq 0$ and the total derivative of any arbitrary function $F(p, q, P, Q, t)$:

$$\sum_{i=1}^{n} p_i \dot{q}_i - H(p, q, t) = c \left[ \sum_{i=1}^{n} P_i \dot{Q}_i - H'(P, Q, t) \right] + \frac{d}{dt} F(p, q, P, Q, t)$$

In other words, a transformation $(q, p) \to (Q, P)$ is canonical if and only if there exists a constant $c \neq 0$ and a function $F(p, q, P, Q, t)$ which fulfills the preceding relation. Without loss of generality we assume for the subsequent discussion that

$c = 1$, which gives

$$\left[\sum_{i=1}^{n} p_i \dot{q}_i - H(p, q, t)\right] - \left[\sum_{i=1}^{n} P_i \dot{Q}_i - H'(P, Q, t)\right] = \frac{d}{dt} F(p, q, P, Q, t) \tag{2.3}$$

The functions $F$ depends on $4n$ variables but since the transformation $(q, p) \rightarrow (Q, P)$ relates $2n$ of them it can only depend on time and the remaining $2n$ variables. The most important cases are of the form

$$F_1(q, Q, t)\,, \quad F_2(q, P, t)\,, \quad F_3(p, Q, t)\,, \quad F_4(p, P, t)$$

and arbitrary combinations of them. We investigate in more detail the cases $F_1$–$F_4$ which we denote by generating functions (of the transformations) in short.

**Case $F_1$** The generating function has the form $F_1(q, Q, t)$ and we get the total derivative:

$$\frac{dF_1}{dt} = \sum_{i=1}^{n} \left(\frac{\partial F_1}{\partial q_i} \dot{q}_i + \frac{\partial F_1}{\partial Q_i} \dot{Q}_i\right) + \frac{\partial F_1}{\partial t}$$

From (2.3) we deduce

$$\sum_{i=1}^{n} p_i \dot{q}_i - H(p, q, t) = \sum_{i=1}^{n} \left(P_i \dot{Q}_i + \frac{\partial F_1}{\partial q_i} \dot{q}_i + \frac{\partial F_1}{\partial Q_i} \dot{Q}_i\right) - H'(P, Q, t) + \frac{\partial F_1}{\partial t}$$

Since we require the old $(q)$ and new $(Q)$ coordinates to be independent each summand above must vanish independently and thus we obtain:

$$\begin{aligned} p_i &= \frac{\partial F_1(q, Q, t)}{\partial q_i} \\ P_i &= -\frac{\partial F_1(q, Q, t)}{\partial Q_i} \\ H' &= H(p(P, Q, t), q(P, Q, t), t) + \frac{\partial F_1(q(P, Q, t), Q, t)}{\partial t} \end{aligned}$$

In conclusion, any function $F_1$ that fulfills the first two of the preceding equations generates a canonical transformation; the new Hamiltonian is defined by the third equation. If $F_1$ is known, the transformation is defined in the following way: first one solves the system $Q = Q(p, q, t)$ to get $q_i \rightarrow Q_i(p, q, t)$; in the next step one solves for the new generalized momenta $P_i$ in terms of $q_i$ and $Q_i(p, q, t)$ to get $p_i \rightarrow P_i(p, q, t)$. On the contrary for given transformation $p_i(q, Q, t)$ and $P_i(q, Q, t)$ one easily calculates the partial derivatives $\partial F_1/\partial q_i$ and $\partial F_1/\partial Q_i$ to determine $F_1(q, Q, t)$, which is given up to a function $g(t)$ just depending on time.

**Case $F_2$** From $F_2(q, P, t)$ we get the total derivative

$$\sum_{i=1}^{n}(p_i \dot{q}_i - P_i \dot{Q}_i) - H + H' = \frac{d}{dt}\hat{F}_2(q, P, t)$$

where $\hat{F}_2$ will be related to $F_2$ in the following calculation. From (2.3) we have

$$\sum_{i=1}^{n}\left[p_i \dot{q}_i - P_i \sum_{j=1}^{n}\left(\frac{\partial Q_j}{\partial q_j}\dot{q}_j + \frac{\partial Q_i}{\partial P_j}\dot{P}_j\right) - P_i \frac{\partial Q_i}{\partial t}\right] - H + H'$$
$$= \sum_{i=1}^{n}\left(\frac{\partial \hat{F}_2}{\partial q_i}\dot{q}_i + \frac{\partial \hat{F}_2}{\partial P_i}\dot{P}_i\right) + \frac{\partial \hat{F}_2}{\partial t}$$

The comparison of the coefficients gives:

$$p_i = \sum_{j=1}^{n} P_j \frac{\partial Q_j}{\partial q_i} + \frac{\partial \hat{F}_2}{\partial q_i} = \frac{\partial}{\partial q_i}\left(\hat{F}_2 + \sum_{j=1}^{n} P_j Q_j\right)$$

$$Q_i = \sum_{j=1}^{n} P_j \frac{\partial Q_j}{\partial P_i} + \frac{\partial \hat{F}_2}{\partial P_i} = \frac{\partial}{\partial P_i}\left(\hat{F}_2 + \sum_{j=1}^{n} P_j Q_j\right)$$

$$H' = H + \sum_{i=1}^{n} P_i \frac{\partial Q_i}{\partial t} + \frac{\partial \hat{F}_2}{\partial t} = H + \frac{\partial}{\partial t}\left(\hat{F}_2 + \sum_{j=1}^{n} P_j Q_j\right)$$

It simplifies the calculations to define

$$F_2(q, P, t) = \hat{F}_2(q, P, t) + \sum_{j=1}^{n} P_j Q_j(q, P, t)$$

to obtain the simple set of determining equations:

$$p_i = \frac{\partial F_2(q, P, t)}{\partial q_i}$$
$$Q_i = \frac{\partial F_2(q, P, t)}{\partial P_i}$$
$$H' = H + \frac{\partial F_2}{\partial t}$$

Again, any transformation which leaves the $2n$ variables $(q_i, P_i)$ independent is canonical if there exists a generating function $F_2(q, P, t)$ that fulfills the above equations. From the first of the foregoing equations one obtains the new generalized momenta $P_i = P_i(q, p, t)$ and from the second $Q_i(q, p, t)$.

**Case $F_3$** For $F_3(p, Q, t)$ we find in a similar way as before the equations:

$$q_i = -\frac{\partial F_3(p, Q, t)}{\partial p_i}$$
$$P_i = -\frac{\partial F_3(p, Q, t)}{\partial Q_i}$$
$$H' = H + \frac{\partial F_3}{\partial t}$$

for $i = 1, \ldots, n$ and any $F_3(p, Q, t)$ fulfilling the preceding equations generates a canonical equation.

**Case $F_4$** We conclude the discussion with the generating function of the form $F_4(p, P, t)$ together with the equations

$$q_i = -\frac{\partial F_4(p, P, t)}{\partial p_i}$$
$$Q_i = \frac{\partial F_4(p, P, t)}{\partial P_i}$$
$$H' = H + \frac{\partial F_4}{\partial t}$$

with $i = 1, \ldots, n$.

In all cases the new Hamiltonian function $H'(P, Q, t)$ is defined as the old Hamiltonian function $H$ plus the partial derivative of the generating function $\partial F_k / \partial t$:

$$H'(P, Q, t) = H + \frac{\partial F_k}{\partial t}$$

(for $k = 1, \ldots, 4$). We discuss the case for which $\partial F_k / \partial t = 0$ and $H' = H$. For time independent transformations of the form:

$$q_i \to Q_i(p, q)$$
$$p_i \to P_i(p, q)$$

the condition (2.3) becomes

$$\sum_{j=1}^{n} (p_i \dot{q}_i - P_i \dot{Q}_i) = \frac{dF}{dt}$$

which means that the left hand side of the preceding equations is a total differential $dF$:

$$\sum_{j=1}^{n} (p_i d\dot{q}_i - P_i d\dot{Q}_i) = dF$$

We conclude the discussion on canonical transformations with the example of the harmonic oscillator. Setting the parameters equal to unity we find in terms of Cartesian like variables

$$H(p,q) = \frac{p^2}{2} + \omega^2 \frac{q^2}{2}$$

We aim to write the Hamiltonian in polar like coordinates

$$p \propto P\cos(Q)\,, \quad q \propto P\sin(Q)$$

A possible generating function of the first kind is given by:

$$F_1(q,Q) = \omega \frac{q^2}{2}\cot(Q)$$

which gives (see Case $F_1$):

$$p = \frac{\partial F_1}{\partial q} = \omega q \cot(Q)$$
$$P = -\frac{\partial F_1}{\partial Q} = \omega \frac{q^2}{2\sin^2(Q)}$$

We solve for $(p,q)$ as

$$q = \sqrt{\frac{2P}{\omega}}\sin(Q)\,, \quad p = \sqrt{2\omega P}\cos(Q)$$

and get from $H' = H(p(P,Q), q(P,Q))$ the transformed Hamiltonian

$$H'(P,Q) = \omega P$$

Since $Q$ is an ignorable coordinate we find the very simple form of the equations of the canonical equations:

$$\dot{Q} = \omega\,, \quad \dot{P} = 0$$

which admits the trivial solution

$$Q(t) = \omega t + Q(0)$$
$$P(t) = P(0)$$

where $P(0)$, $Q(0)$ are the initial conditions and where from $P = H'/\omega$ we find $P(0)$ for given $(H,\omega)$. From the inverse of the transformation $P(p,q)$ and $Q(p,q)$ we find the solution in terms of the old variables

$$q(t) = \sqrt{2H\omega^{-2}}\sin(\omega t + q(0))$$

Usually it is as difficult to find the generating function and transformation as it is to find the solution of the problem in terms of the old variables. The importance of the canonical transformations lies in their ability to simplify dynamical systems and to introduce advanced concepts like the Hamilton–Jacobi theory or the notion of action-angle variables.

## 2.4
## Hamilton–Jacobi Theory

We see in the simple example of the harmonic oscillator that the transformation into new variables leads to an ignorable variable $Q$ and a trivial solution for $P(t) = P(0)$. It would therefore be desirable, in the generic case, $q = (q_1, \ldots, q_n)$, $p = (p_1, \ldots, p_n)$ and $Q = (Q_1, \ldots, Q_n)$, $P = (P_1, \ldots, P_n)$, to be able to find a change of coordinates $(p_i, q_i) \rightarrow (P_i, Q_i)$ such that the canonical equations become:

$$\dot{P}_i = -\frac{\partial H'}{\partial Q_i} = 0$$

and in addition

$$\dot{Q}_i = \frac{\partial H'}{\partial P_i} = 0$$

In the best case the solution of the preceding equations becomes:

$$P(t) = P(0) = \text{const}\,, \qquad Q(t) = Q(0) = \text{const}$$

and the system is trivially solved. In the following paragraph we investigated the possibility to find a generating function $S$ of type $F_2(q, P, t)$ which transforms $H(p, q, t)$ into $H'(P, Q, t) = 0$. We start with variables of the form

$$\begin{aligned} Q_i &= Q_i(q(t), p(t), t) \\ P_i &= P_i(q(t), p(t), t) \end{aligned}$$

with $i = 1, \ldots, n$. A canonical transformation

$$q_i(t) \rightarrow Q_i(q(t), p(t), t) = \text{const} \tag{2.4}$$

$$p_i(t) \rightarrow P_i(q(t), p(t), t) = \text{const} \tag{2.5}$$

exists if $H' = $ const. and thus

$$\dot{Q}_i = \frac{\partial H'}{\partial P} = \dot{P}_i = -\frac{\partial H'}{\partial Q} = 0$$

Since for any $S$ the new Hamiltonian is given by $H' = H + \partial S/\partial t$ we look for a generating function which is determined by:

$$H(q, p, t) + \frac{\partial S(q, P, t)}{\partial t} = 0$$

Since $2n$ of the above $q_i, p_i, P_i$, with $i = 1, \ldots, n$, are independent from each other we express the $p_i$ in $H$:

$$p_i = \frac{\partial S(q, P, t)}{\partial q_i}$$

with $i = 1, \dots, n$, which gives

$$H\left(q_1, \dots, q_n, \frac{\partial S}{\partial q_1}, \dots, \frac{\partial S}{\partial q_n}, t\right) + \frac{\partial S}{\partial t} = 0 \tag{2.6}$$

This is the so-called Hamilton–Jacobi equation. It is a nonlinear partial differential equation of an unknown function

$$S(q_1, \dots, q_n, t)$$

If it is possible to find $S$ we are able to generate the transformation (2.6) and the problem is in principle solved. The solution in terms of the old variables $(q_i(t), p_i(t))$ with $i = 1, \dots, n$ can then be obtained from:

$$p_i = \frac{\partial S(q, P, t)}{\partial q_i}$$
$$Q_i = \frac{\partial S(q, P, t)}{\partial P_i}$$

with $i = 1, \dots, n$. First one solves the second of above for $q_i = q_i(Q, P, t)$ and rewrites the first to get $p_i(Q, P, t)$. From the theory of partial differential equations the solution for $S$ admits $n$ independent constants $\alpha_i$ with $i = 1, \dots, n$. For this reason the generating function has also the form:

$$S = S(q_1, \dots, q_n; \alpha_1, \dots, \alpha_n, t)$$

where the $\alpha_i$ can be identified either with i) $\alpha_i = Q_i$ or with ii) $\alpha_i = P_i$ since $Q_i = P_i = \text{const}$. In the last step we identify

$$p_i = \frac{\partial S(q, \alpha, t)}{\partial q_i}$$
$$\beta_i = \frac{\partial S(q, \alpha, t)}{\partial \alpha_i}$$

with $i = 1, \dots, n$, and where $\beta_i = -P_i$ or $\beta_i = Q_i$ depending on the choice i) or ii) and find the constants $\alpha_i, \beta_i$ from the initial conditions $(q_i(0), p_i(0))$ from:

$$\alpha_i = \alpha_i(q(0), p(0)), \quad \beta_i = \beta_i(q(0), p(0))$$

This, in principle, concludes the calculation to find the solution also in terms of the original variables. Using two examples, we demonstrate how to solve a Hamiltonian system by solving the Hamilton–Jacobi equation.

Example 1

The Harmonic Oscillator

The simplest formulation of the harmonic oscillator takes the form:

$$H = \frac{p^2}{2} + \frac{q^2}{2}$$

where $(p, q) \in \mathbb{R}^2$. We express the momentum $p$ by

$$p = \frac{\partial S}{\partial q}$$

and add

$$\frac{\partial S}{\partial t}$$

to the original Hamiltonian. We get the Hamilton–Jacobi equation:

$$\frac{1}{2}\left(\frac{\partial S}{\partial q}\right)^2 + \frac{q^2}{2} + \frac{\partial S}{\partial t} = 0$$

To separate the time from the generating function we use the following form for it:

$$S(q, \alpha, t) = f(q, \alpha) - \alpha t$$

If we insert it into the above equation we get for $f$:

$$\frac{1}{2}\left(\frac{\partial f}{\partial q}\right)^2 + \frac{q^2}{2} = \alpha$$

Since the left hand side of the preceding equation defines the original Hamiltonian (note, that $p = \partial S/\partial q = \partial f/\partial q$) we identify the energy $E \equiv H = \alpha$. Integration with respect to $q$ gives:

$$S(q, \alpha) = f(q, \alpha) - Et = \int \sqrt{2E - q^2}\, dq$$

We do not give the integration limits since we do not need to solve the integral but rather find the derivative with $\partial S/\partial \alpha$:

$$\frac{\partial S}{\partial \alpha} = \frac{\partial S}{\partial E} = \int \frac{dq}{\sqrt{2E - q^2}} - t = \beta$$

The integral gives:

$$t + \beta = \arcsin\left(\sqrt{\frac{1}{2E}}\, q\right)$$

which results in

$$q(t) = \sqrt{2E}\sin(t + \beta)$$

For given $E$ and $q(0) = q_0$ we get the phase $\beta$.

Example 2

Motion in the Central Force Field

The Hamiltonian for a particle moving in the plane in a central force field $V(r)$ is given by:

$$H = \frac{1}{2}\left(p_r{}^2 + \frac{p_\varphi{}^2}{r^2}\right) + V(r)$$

where $(r, p_r) \in \mathbb{R}^2$ and $(\varphi, p_\varphi) \in \mathbb{T}^2$ are the polar coordinates with their conjugated momenta, respectively. We express the momenta by the derivatives of a generating function in connection with their respective coordinates:

$$p_r = \frac{\partial S}{\partial r}, \quad p_\varphi = \frac{\partial S}{\partial \varphi}$$

and get the Hamilton–Jacobi equation:

$$\frac{1}{2}\left[\left(\frac{\partial S}{\partial r}\right)^2 + \frac{1}{r^2}\left(\frac{\partial S}{\partial \varphi}\right)^2\right] + \frac{\partial S}{\partial t} + V(r) = 0$$

Again, we use a form to separate the time from the unknown $S$:

$$S(r, \varphi, \alpha_1, \alpha_2, t) = f(r, \varphi, \alpha_1, \alpha_2) - \alpha_1 t$$

and get for $f$:

$$\frac{1}{2}\left[\left(\frac{\partial f}{\partial r}\right)^2 + \frac{1}{r^2}\left(\frac{\partial f}{\partial \varphi}\right)^2\right] + V(r) = \alpha_1$$

For the same argument as in the example of the harmonic oscillator we identify the energy $E$ with $E \equiv H = \alpha_1$. In the next step we try to separate the unknown function $f$ as:

$$f(r, \varphi, E, \alpha_2) = f_1(r, E, \alpha_2) + f_2(\varphi, E, \alpha_2)$$

which yields the equation:

$$\left(\frac{\partial f_2}{\partial \varphi}\right)^2 = 2r^2[E - V(r)] - r^2\left(\frac{\partial f_1}{\partial r}\right)^2$$

Since the left-hand side depends only on $\varphi$ and the right-hand side only on $r$ we deduce that the left- and the right-hand sides are constant equal to $\alpha_2$ and we get:

$$f_1(r, E, \alpha_2) = \int \sqrt{2[E - V(r)] - \frac{\alpha_2}{r^2}}\, dr$$

$$f_2(\varphi, E, \alpha_2) = \sqrt{\alpha_2}\varphi$$

On the basis of

$$\frac{\partial S}{\partial \varphi} = p_\varphi = \frac{\partial f_2}{\partial \varphi} = \sqrt{\alpha_2}$$

we find from $S = f_1 + f_2 - Et$

$$S(r, \varphi, E, \alpha_2) = \int \sqrt{2[E - V(r)] - \frac{p_\varphi{}^2}{r^2}}\, dr + p_\varphi \varphi - Et$$

which also gives:

$$\frac{\partial S}{\partial \alpha_1} = \frac{\partial S}{\partial E} = \int \frac{dr}{\sqrt{2[E - V(r)] - \frac{p_\varphi{}^2}{r^2}}} - t = \beta_1$$

$$\frac{\partial S}{\partial \alpha_2} = \frac{\partial S}{\partial p_\varphi} = -\int \frac{p_\varphi\, dr}{r^2\sqrt{2[E - V(r)] - \frac{p_\varphi{}^2}{r^2}}} + \varphi = \beta_2$$

The Hamilton–Jacobi theory does not imply a solution to a given Hamiltonian system it is a reformulation of the dynamical problem in terms of a generating function and the Hamilton–Jacobi equation. The importance of the theory lies in the fact that for separable systems (where the generating function can be separated) the problem can be transformed into the problem to solve $n$ ordinary integrals rather than $n$ coupled ordinary differential equations as in the Lagrange problem or $2n$ coupled ordinary differential equations of first order. Moreover, the Hamilton–Jacobi theory is the basis for advanced concepts in perturbation theory and the concept of the so-called action angle-variables.

## 2.5
## Action-Angle Variables

In the final part of this chapter we introduce the concept of action-angle variables for periodic and completely separable systems. In one dimension, two types of periodicity can be defined: libration and rotation. In terms of phase space geometry it is easy to define the two different kinds of motion. We first define periodicity in one dimension and generalize the concept to arbitrary dimension based on it.

If the curve $(p_i, q_i)$ is closed we speak of librational motion (see Figure 2.1a). Both $p_i, q_i$ are bounded within a given domain which we denote by the librational regime of motion. In the one dimensional case it is possible to define $p_i(q_i)$ from $H(p_i, q_i) \equiv E$. The curve is usually symmetric with respect to the $q_i$-axis. Typical examples are the one of the harmonic oscillator or the librational regime of motion of the pendulum. The second case concerns the rotational motion (see Figure 2.1b). The curve $(p_i, q_i)$ is not closed and one of the variables, say $q_i$ depends linearly on time. It is possible to define a period $T$ such that $p_i(q_i + T) = p_i(q_i)$.

If the dimension is higher such that $p = (p_1, \ldots, p_n)$ and $q = (q_1, \ldots, q_n)$ the system is called periodic if it is periodic in the foregoing sense in each of the $(p_j, q_j)$-planes, with $j = 1, \ldots, n$. The system is still called periodic if it is composed of librational and rotational subdynamics and we do not require the fre-

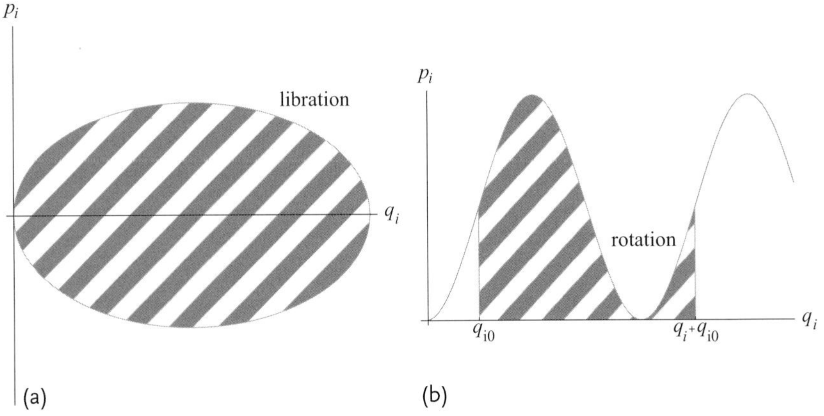


**Figure 2.1** Action-angle variables in the hyperplane $(p_i, q_i)$. (a) action integral for the librational motion. (b) action integral for the rotational motion.

quencies (respective periods) to be the same or rationally dependent. Note, that as consequence of the multidimensionality the orbit itself does not need to be closed even if $p_j(q_j)$ is periodic and/or $(p_j, q_j)$ are closed.

In the following discussion we require systems which are separable in all their arguments $q_i, t$ with $i = 1, \ldots, n$ and where the Hamiltonian does not depend explicitly on time. With this we are able to introduce the concept of action angle variables by means of Hamilton–Jacobi theory. We define the action variable $J_i$ for the subspace defined by $(p_i, q_i)$ in terms of:

$$J_i = \oint p_i dq_i = \oint \frac{\partial S_i(q_i, \alpha_1, \ldots, \alpha_n)}{\partial q_i} dq_i$$

where the integral has to be evaluated along a complete librational boundary and over a complete rotational period, respectively. The integral over $q_i$ implies that $J_i$ depend only on the constants $\alpha_1, \ldots, \alpha_n$:

$$J_i = J_i(\alpha_1, \ldots, \alpha_n)$$

with $i = 1, \ldots, n$. The $J_i$ are therefore integrals of motion, that is, constant. The inversion of the above relations gives:

$$\alpha_i = \alpha_i(J_1, \ldots, J_n)$$

with $i = 1, \ldots, n$, the $J_i$ are independent functions of the $\alpha_i$ and can be used to introduce new generalized momenta. In terms of Hamilton–Jacobi theory the generating function $S$ takes the form:

$$\begin{aligned} S &= S\left[q_1, \ldots, q_n, \alpha_1(J_1, \ldots, J_n), \ldots, \alpha_n(J_1, \ldots, J_n)\right] \\ &= S(q_1, \ldots, q_n, J_1, \ldots, J_n) \end{aligned}$$

Since we assume that the system is separable we may write

$$S(q_1,\ldots,q_n,J_1,\ldots,J_n)=\sum_{i=1}^{n} S_i(q_i,J_1,\ldots,J_n)$$

To complete the transformation we use

$$p_i=\frac{\partial S_i(q_i,\alpha)}{\partial q_i}=\frac{\partial S_i(q_i,J)}{\partial q_i}$$
$$\varphi_i=-\frac{\partial S(q,J)}{\partial J_i}$$

where $\alpha=(\alpha_1,\ldots,\alpha_n)$, $J=(J_1,\ldots,J_n)$ and $\varphi=(\varphi_1,\ldots,\varphi_n)$ are the generalized coordinates conjugated to $J$. We label $(J,\varphi)$ action-angle variables, where $\varphi$ is called the angle. The transformation from original to action-angle variables:

$$(p,q)\rightarrow(J,\varphi)$$

is defined in terms of the generating function $S$. The Hamilton function in terms of the new variables is given by:

$$H'(J,\varphi)=H(p(J,\varphi),q(J,\varphi))$$

which is

$$H'=H+\frac{\partial S}{\partial t}$$

Since $S$ does not depend on $t$ we conclude $H'=H$ and since the Hamilton–Jacobi equation reads:

$$H\left(q_1,\ldots,q_n,\frac{\partial S_1}{\partial q_1},\ldots,\frac{\partial S_1}{\partial q_n}\right)=\alpha_1$$

we find

$$H'=H=\alpha_1(J_1,\ldots,J_n)$$

In conclusion the new Hamiltonian only depends on the action $J$. The canonical equations are of the form:

$$\dot{J}_i=-\frac{\partial H'}{\partial \varphi_i}=0$$
$$\dot{\varphi}_i=\frac{\partial H'}{\partial J_i}=\omega(J)$$

with $i=1,\ldots,n$. In terms of $(J,\varphi)$ we are thus able to solve the system easily:

$$J_i(t)=J_{i0}$$
$$\varphi_i(t)=\omega_i(J_{i0})t+\varphi_{i0}$$

where $J_{i0} = J_i(0)$ and $\varphi_{i0} = \varphi_i(0)$, respectively. We call $\omega_i$ the fundamental frequencies of the system. They are called fundamental since it can be shown that in terms of the old variables $q_i = q_i(J, \varphi)$ the solution in Fourier series form becomes:

$$q_i(J, \varphi) = \sum_{k \in \mathbb{Z}^n} a_k^i(J) e^{i(k_1\varphi_1 + \cdots + k_n\varphi_n)}$$

Since $\varphi_i = \varphi_i(t)$ and $J_i = \text{const}$ the solution $q_i = q_i(t)$ is periodic or quasi-periodic in time. We clearly see that the concept of action angle variables is strongly related to the concept of periodicity in the system.

Hamiltonian mechanics allows to introduce the concept of normal forms (introduced in Section 3.4) and to formulate the so-called stability problem in a much easier way, which is done in Chapter 4. To this end we will use it to derive and easily transform the equations of motion of celestial dynamical systems as done with the restricted three-body problem in Section 6.1.

# 3
# Numerical and Analytical Tools

In this chapter we briefly discuss the main tools which are needed to investigate dynamical systems. First we introduce the reader to the concepts and definitions by means of simple mapping equations. We then present an efficient numerical integration code the Lie-series integration method. The next section is dedicated to present different kinds of chaos indicators and finally we speak about analytical tools based on perturbation theory.

## 3.1
## Mappings

A discrete system of dimension $d$ is given by the set of equations of the form

$$x_{n+1} = f(x_n, n) \tag{3.1}$$

where $x_n = (x_n^{(1)}, \dots, x_n^{(2)})$, $f = (f^{(1)}, \dots, f^{(d)})$ and $n \in \mathbb{Z}$ labels the discrete time $n\Delta t$, where $\Delta t$ is the time step that passes by between the state of the system at a given time $(n)$ and $(n+1)$. If $f$ does not depend explicitly on discrete time $n$ it is called autonomous like in the continuous case. The map is called symplectic if it is a diffeomorphism that preserves the symplectic structure. In this case it is possible to find local coordinates such that

$$(D f^{-1})(D f) = J \tag{3.2}$$

where $f^{-1}$ is the inverse of $f$, $(D f)$ denotes the Jacobian matrix and $J$ is the skew symmetric matrix given by:

$$J = \begin{pmatrix} 0_d & I_d \\ -I_d & 0_d \end{pmatrix}$$

with $0_d$, $I_d$ the $d$-dimensional zero and identity matrix, respectively. The relation (3.2) implies that symplectic mappings also conserve the orientation and the volume in phase space, that is, in two dimensions it is area preserving and thus, $\det(D f) = 1$. Symplectic mappings are sometimes called symplectomorphisms

*Celestial Dynamics*, First Edition. R. Dvorak and C. Lhotka.

and can also be identified with canonical transformations, that is, with the concepts of canonical transformations stemming from Hamiltonian dynamics. Mappings of the form (3.1) have important applications in different fields of science. For review papers on mappings found in physics, astronomy and that is, celestial mechanics see, for example [5, 6]. They were used to investigate the problem of asteroidal resonances in [7–12] as well as in [13, 14].

### 3.1.1 Simple Examples

In this section we briefly introduce the main concepts of dynamical system theory. We aim not to be technical in the definitions but rather to give examples in terms of simple mappings to provide a 'visual' idea of important 'objects' found when dealing with the subject. Definitions and further information will be provided later on if needed. The examples are those which can also be found on various resources of the world wide web, for example Scholarpedia [15] or MathWorld [16].

#### 3.1.1.1 Twist Map

A twist map is a class of area preserving maps of the form:

$$x_{n+1} = x_n + 2\pi\alpha(y_n)$$
$$y_{n+1} = y_n$$

where $(y, x)$ are a kind of action-angle variables and $\alpha(y)$ defines the strength of the twist. The mapping is integrable and serves as the integrable approximation to many dynamical systems. Assuming $x(t + 2\pi) = x$ and $\alpha(y_0) = 2\pi p/q$ the trajectory $\{(y_i, x_i), i \in \mathbb{Z}\}$ is a periodic orbit with the period given by $q$ while $p$ defines the number of iterations in the interval $(0, 2\pi)$ before the system returns to its initial state (see Figure 3.1a). If $\alpha(y_0)$ is irrational the motion is called quasiperiodic and the orbit $\{(y_i, x_i), i \in \mathbb{Z}\}$ fills the line $y_i = y_0$ densely. The motion will come arbitrary close to its initial state $(y_0, x_0)$ but will never reach it (see Figure 3.1b). The two kind of motions lead to the concept of the rotation number $\omega$ which is defined by:

$$\omega = \lim_{j\to\infty} \frac{x_j - x_0}{j}$$

which tends to $p/q$ for periodic orbits and to $\omega \in \mathbb{R}\backslash\mathbb{Q}$ otherwise. The phase portraits of twist maps are spanned by circles which are mapped into circles under the twist condition. A special case of a perturbed twist map is the standard map, which becomes a twist map for $K = 0$. The twist map can also be related to a Hamiltonian system of the form:

$$H(y, x) = h_0(y)$$

where $\partial h_0/\partial y$ defines the frequency of the motion. Perturbed twist maps have become an important subject of research by its own. See [17] for further information.

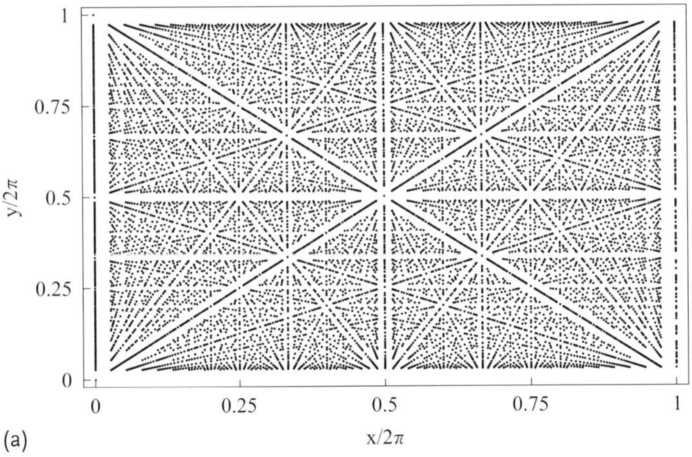


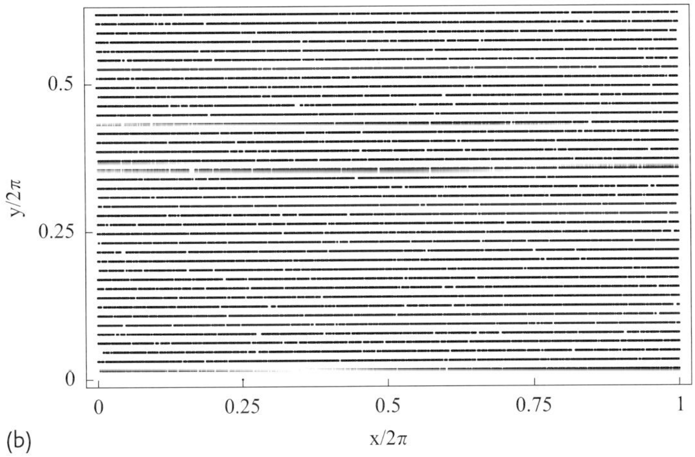


**Figure 3.1** Phase portrait of the twist mapping. (a) iteration of rational initial conditions lead to periodic orbits. (b) iteration of irrational initial conditions leads to quasiperiodic orbits.

#### 3.1.1.2 Logistic Map

The logistic map is given by the quadratic equation

$$x_{n+1} = rx_n(1 - x_n)$$

Here, the variable $x$ is defined on the interval $(0, 1)$ and defines the ratio of an existing population to the maximum population at year $n$. The parameter $r \in (0, 4)$ defines the combined rate of reproduction and starvation of the population also known as the biotic potential. A period $p$-fixed point of the map exists under the condition $x_*^{(p)} = x_{n+p} = x_n$. For a period 1 fixed point ($p = 1$) we have:

$$rx_*^{(1)}\left(1 - x_*^{(1)}\right) - x_*^{(1)} = rx_*^{(1)}\left[x_*^{(1)} - (1 - r^{-1})\right] = 0$$

with the solution $x_{*,1}^{(1)} = 0$ and $x_{*,2}^{(1)} = 1 - r^{-1}$. Higher-order fixed points can be obtained in a similar way, where one makes use of previous solutions $x_*^{(<p)}$ to obtain $x_*^{(p)}$. The investigation of their domain of existence allows to develop the concept of period doubling or bifurcation, for example for $p = 2$ we find

$$x_{*,1,2}^{(2)} = \frac{1}{2}\left[(1 + r^{-1}) \pm \sqrt{(r - 3)(r + 1)}\right]$$

and thus the solution is real if and only if $r \geq 3$ which implies period doubling at $r = 3$. Since the map is contractive the fixed points also correspond to periodic attractors and thus the presence of more and more attractors would imply chaotic motion also (small changes in the initial conditions imply that the orbit falls on different attractors). This can be seen if we plot the locations of the fixed points (found numerically) versus the parameter $r$ and we get a bifurcation diagram as it is shown in Figure 3.2 As we can see with increasing $r$ more and more fixed points solutions exist while in specific domains of $r$ the number of attractors decreases again, an effect which is known as intermittency in chaotic dynamics. We plot typical orbits in the so-called web-diagram in Figure 3.3. Here the parabola represents the $f(x_n)$ in $x_{n+1} = f(x_n)$, the dashed line corresponds to the fixed-point condition $x_{n+1} = x_n$. We start with a point $x_0$ on the curve $(x_0, f(x_0))$ and plot the line to the successive point of the iteration on the median $(f(x_0), f(x_0))$. We complete the iteration and plot a line to the point $(f(x_0), f(f(x_0)))$. If the orbit tends to an attractor it clearly lies on the parabola. An orbit tending to the period 1 fixed point at $r = 2.8$ is shown in Figure 3.3a, an orbit tending to the period 2 fixed point is shown in Figure 3.3b. In the chaotic regime, the orbit does not tend to a specific point, the complete diagram is filled with lines (see Figure 3.3c). We also show the case where the fixed point is unstable and becomes a repeller in Figure 3.3d: starting close to the fixed point location the orbit spirals outwards until it is attracted

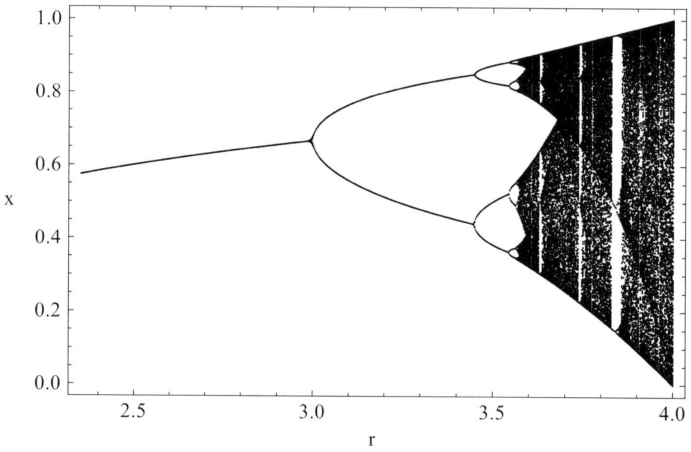


**Figure 3.2** Bifurcation diagram of the logistic map. Period doubling occurs at specific values of $r$. Chaos is maximal for $r \rightarrow 4$. In the chaotic sea windows of stability open up – a phenomenon found in chaos theory called intermittency.

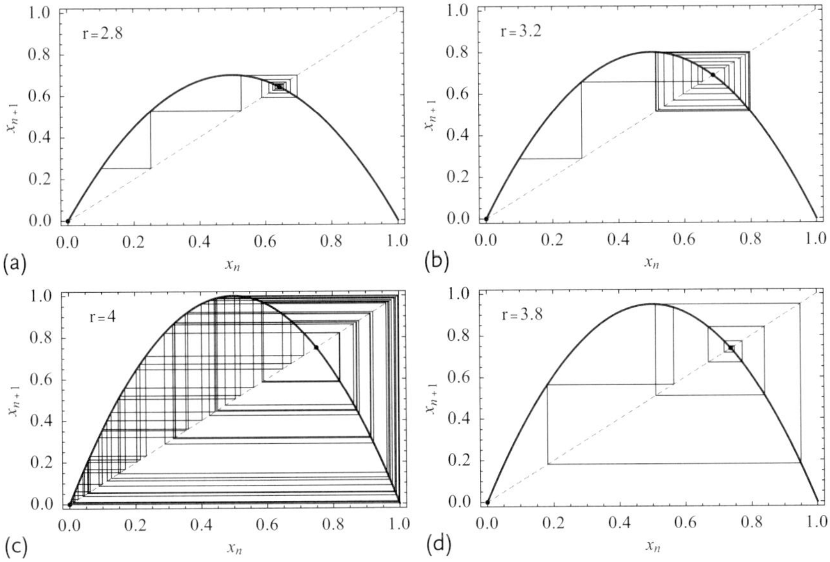


**Figure 3.3** The evolution of orbits in the web diagram. Attraction to a period 1 fixed point (a); attraction to a period 2 fixed point (b); chaotic trajectory (c); unstable period 1 fixed point (d).

by another fixed point (if it exists). The stability of a generic fixed point may be obtained from the Jacobian of the map which is given by:

$$J(r, x) = \frac{dx_{n+1}}{dx_n} = r(1 - 2x_n)$$

and from the condition that $|J(r, x_*)| < 1$.

### 3.1.1.3 Arnold Cat Map

The Arnold cat map, which is named after Vladimir Arnold, is a very simple chaotic map of the form:

$$x_{n+1} = 2x_n + y_n \pmod 1$$
$$y_{n+1} = x_n + y_n \pmod 1$$

The map is invertible, ergodic and mixing and not symplectic but area preserving. It has an unique hyperbolic fixed point at the origin. If one writes the mapping in matrix notation as

$$\begin{pmatrix} x_{n+1} \\ y_{n+1} \end{pmatrix} = \begin{pmatrix} 2 & 1 \\ 1 & 1 \end{pmatrix} \begin{pmatrix} x_n \\ y_n \end{pmatrix}$$

one easily finds the eigenvalues by

$$\begin{pmatrix} 1 - \lambda & 1 \\ 1 & 2 - \lambda \end{pmatrix} = \lambda^2 - 3\lambda + 1 = 0$$

and thus

$$\lambda_\pm = \frac{1}{2}\left(3 \pm \sqrt{5}\right)$$

The eigenvectors are found by the matrix equation

$$\begin{pmatrix} 1-\lambda_\pm & 1 \\ 1 & 2-\lambda_\pm \end{pmatrix} \begin{pmatrix} x \\ y \end{pmatrix} = \begin{pmatrix} 0 \\ 0 \end{pmatrix}$$

which gives for $\lambda_+$:

$$y = \phi x$$

with the golden ratio $\phi$ defined by:

$$\phi = \frac{1}{2}\left(1 + \sqrt{5}\right)$$

For the solution $\lambda_-$ we find

$$y = -\phi^{-1} x$$

so the eigenvectors are given by:

$$\begin{pmatrix} 1 \\ \frac{1}{2}\left(1 + \sqrt{5}\right) \end{pmatrix}, \begin{pmatrix} 1 \\ \frac{1}{2}\left(1 - \sqrt{5}\right) \end{pmatrix}$$

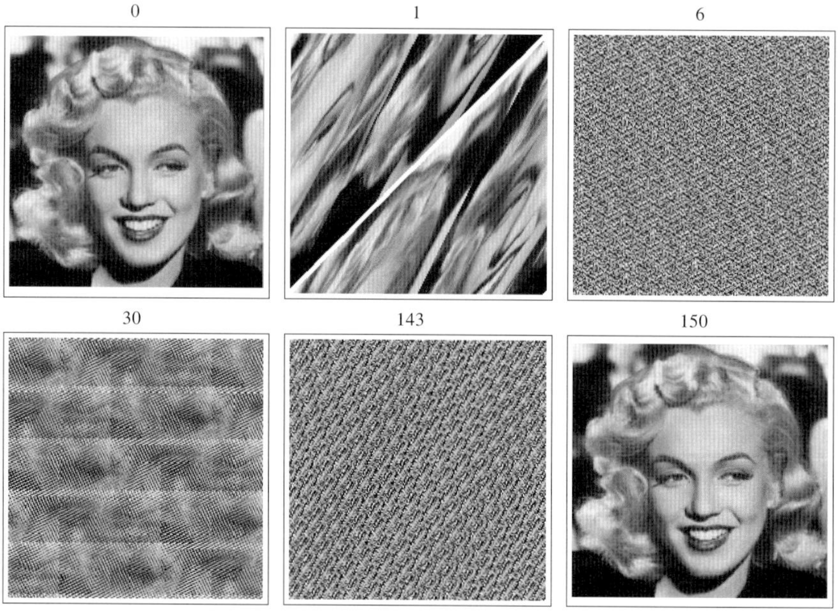


**Figure 3.4** Iteration of the Arnold cat map of a rasterized picture of Marilyn Monroe. The original picture is randomized under the application of the mapping rule but reappears after 150 iterations of the mapping.

respectively. The notion cat map comes from the possible application of the map to pictures (originally a cat) that are mapped to the unit square: if one defines the map over the set of integers modulo $N$ and identifies $(x, y) \in (0, N)\times(0, N)$ as the matrix element of a rasterized picture of dimension $N \times N$ the application of the mapping to each entry of the rasterized image defines a transformation $T : (n) \to (n + 1)$ from a picture given at state $(n)$ to a picture to the state $(n + 1)$ also. From the Poincaré recurrence theorem it follows that after a certain number of iterations the 'orbits' of the map should return arbitrarily close to their initial conditions. It follows that the application of the mapping to the picture should reproduce the picture after a while. This is shown in Figure 3.4 where the picture of Marylin Monroe was iterated 150 times. While after some steps the picture seems to be randomized ($n = 6$) the picture returns after 150 iterations to its initial state.

#### 3.1.1.4 **Standard Map**

The standard- (also called Chirikov- or Taylor-) map has the form (see [18])

$$\begin{aligned} y_{n+1} &= y_n + K\sin(x_n) \\ x_{n+1} &= x_n + y_{n+1} \end{aligned}$$

It is related to the time dependent Hamiltonian

$$H(p, q, t) = \frac{p^2}{2} + K\cos(q)\delta_1(t)$$

where $\delta_1(t)$ is a periodic delta function with period 1. The dynamics can be interpreted by a sequence of free propagations interleaved with periodic kicks. It has found application in particle physics, astrodynamics as well as in the theory of dynamical systems. The fixed points are located at $(x_*, y_*) = (0, 0)$ and $(x_*, y_*) = (0, \pi)$. They are derived from the conditions

$$\begin{aligned} y_* &= y_{n+1} = y_n \\ x_* &= x_{n+1} = x_n \end{aligned}$$

From the first we get $K\sin(x_*) = 0$ and from $\sin(x_*) = 0$ we get $x_* = 0, \pi$. From the second requirement one gets $y_* = y_* + K\sin(x_*) = 0$ which yields $y_* = 0$. The linear stability of the fixed points can be derived from

$$\begin{aligned} \delta y_{n+1} &= \delta y_n + K\cos(x_n)\delta x_n \\ \delta x_{n+1} &= \delta y_n + \left[1 + K\cos(x_n)\right]\delta x_n \end{aligned}$$

which gives in matrix form

$$\begin{pmatrix} \delta y_{n+1} \\ \delta x_{n+1} \end{pmatrix} = \begin{pmatrix} 1 & K\cos(x_n) \\ 1 & 1 + K\cos(x_n) \end{pmatrix} \begin{pmatrix} \delta y_n \\ \delta x_n \end{pmatrix}$$

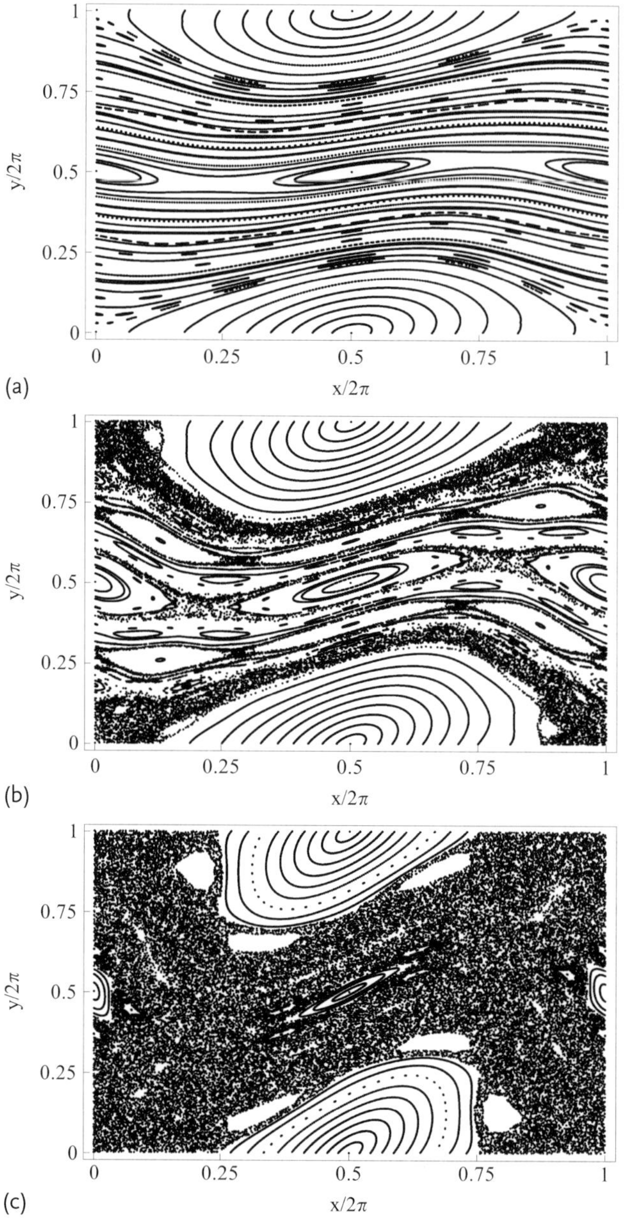


**Figure 3.5** Phase portraits of the standard map. Orbits for the parameter $K = 0.5$ (a), $K = K_c$ (b), and $K = \phi$ (c), respectively. With increasing $K$ more and more quasi-periodic invariant curves are distorted and replaced by larger and larger chaotic regimes.

From the eigenvalues obtained from the characteristic polynomial,

$$\left|\begin{pmatrix} 1-\lambda & K\cos(x_n) \\ 1 & 1+K\cos(x_n)-\lambda \end{pmatrix}\right| = 0$$

which is

$$\lambda^2 - \lambda\left[K\cos(x_n)+2\right] + 1 = 0$$

one has the solution

$$\lambda_\pm = \frac{1}{2}\left\{K\cos(x_n) + 2 \pm \sqrt{\left[K\cos(x_n)+2\right]^2 - 4}\right\}$$

For $(x_*, y_*) = (\pi, 0)$ one therefore finds

$$\lambda_\pm = \frac{1}{2}\left(2 - K \pm \sqrt{K^2 - 4K}\right)$$

and the fixed point will be stable for $K \in (0, 4($. If we repeat the above calculation for $(x_*, y_*) = (0, 0)$ we get

$$\lambda_\pm = \frac{1}{2}\left(2 + K \pm \sqrt{K^2 + 4K}\right)$$

From the larger eigenvalue $\lambda_+$ we find the condition for stability

$$\frac{1}{2}\left|2 + K + \sqrt{K^2 + 4K}\right| < 1$$

which gives

$$-4 - K < \sqrt{K^2 + 4K} < -K$$

Since $K > 0$ the above inequality cannot be fulfilled and we conclude that the fixed point located at (0, 0) is unstable. We plot typical phase portraits in Figure 3.5. For $K = 0$ and initial conditions $x_0 = 0$, $y_0/2\pi = p/q$ with $p, q$ integer we get Figure 3.1a. For $K = 0$ and $x_0 = 0$, $y_0/2\pi = r$ with $r$ irrational we get Figure 3.1b. For $K = 0.5$ and initial conditions $(x_0, y_0) \in (0, 2\pi) \times (0, 2\pi)$ we get Figure 3.5a with $K = 0.971$ close to the critical value we get Figure 3.5b and for $K = \phi$ (golden ratio) we get Figure 3.5c. The critical parameter $K_c$ was estimated to be $K_c > 0.029$, $K_c > 0.838$ [19], $K_c \simeq 0.9716\ldots$ (using Greene's method [20]), $K_c < 0.984$ [21] and $K_c < 1.333\ldots$ (see [22]). Open questions, like the proof that the mapping is ergodic are still open.

#### 3.1.1.5 The Circle Map

The circle map is related to the standard map by setting $\theta_n = x_n$, fixing $y_n = \Omega$ and taking the second equation of the standard map to get:

$$\theta_{n+1} = \theta_n + \Omega + \frac{K}{2\pi}\sin(2\pi\theta_n)$$

It defines a one-dimensional map which maps the circle onto itself. The Jacobian gives:

$$\frac{\partial \theta_{n+1}}{\partial \theta_n} = 1 - K\cos(2\pi\theta_n)$$

so the map is not area preserving. The unperturbed circle map has the form

$$\theta_{n+1} = \theta_n + \Omega$$

If $\Omega = p/q$ is rational, it defines the so-called map winding number; if $\Omega = r$ is irrational the motion is quasi-periodic. The domain close to rational $\Omega$ is known as an Arnold tongue. Such domains are known to be phase locked regions related to the rational value of $\Omega = p/q$. At $K = 0$ they correspond to isolated sets of measure zero. The map originated with Andrey Kolmogorov. We plot the Arnold tongues in the parameters space $(K, \Omega)$ versus the recurrence time $N(\epsilon)$ in Figure 3.6. The recurrence time is defined as the smallest integer for which:

$$\left|\left(\theta_{n+N(\epsilon)} - \theta_n\right) \bmod 1\right| < \epsilon$$

where $\epsilon$ is a small quantity tending to zero. For $N(\epsilon) \to N$ for $\epsilon \to 0$ it is sometimes called the Poincaré recurrence time related to the periodicity in the system.

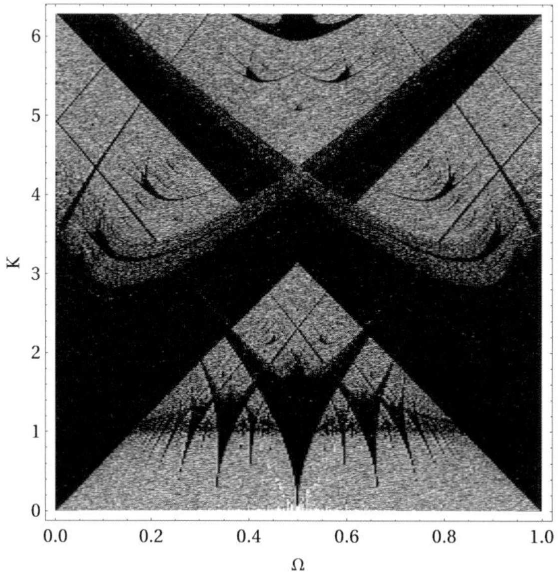


**Figure 3.6** The circle map. The figure shows the average discrete time $N$, which is needed to return to a randomly chosen initial condition $\theta_0$ in the parameter space $\Omega$ (x-axis) versus $K$ (y-axis). The quantity $N$, which is normalized between 0 (black) and 1 (white), gives a measure of the so-called Poincaré recurrence time. Arnold tongues are indicated in black and originate from specific $\Omega$ for $K = 0$.

#### 3.1.1.6 The Dissipative Standard Map

The dissipative standard map is of the form

$$\begin{aligned} y_{n+1} &= b y_n + c + K f(x_n) \\ x_{n+1} &= x_n + y_{n+1} \end{aligned}$$

where $f$ is a periodic function, $y \in \mathbb{R}$ and $x \in \mathbb{T}$ with $b, c, K$ are constants. The quantities $b, c \in \mathbb{R}_+$ are called the dissipative and drift parameters, respectively. For $b = 1$ and $c = 0$ one recovers with $f(x) = \sin(x)$ the classical conservative standard map. The mapping is contractive since the Jacobian becomes

$$\det \begin{pmatrix} b & \dfrac{K\partial f}{\partial x} \\ b & 1 + \dfrac{K\partial f}{\partial x} \end{pmatrix} = b$$

For $K = 0$ the mapping reduces to

$$\begin{aligned} y_{n+1} &= b y_n + c \\ x_{n+1} &= x_n + y_{n+1} \end{aligned}$$

and one introduces the quantity

$$\alpha = \frac{c}{1-b}$$

since $\{y = \alpha\} \times \mathbb{T}$ is an invariant object. From the first of the preceding equations one gets:

$$\alpha = b\alpha + c$$

and therefore also the relation $c = \alpha(1 - b)$. The dynamics of the dissipative standard map is foliated by attracting periodic orbits, invariant quasi-periodic attractors and strange attractors. They can be seen as analogous to the periodic, quasi-periodic and chaotic orbits of the classical standard map, respectively. It is known that

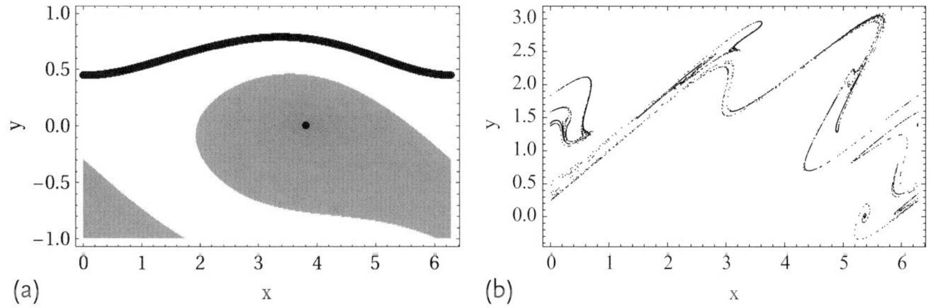


**Figure 3.7** Phase portraits of the dissipative standard map. (a) coexistence of a periodic and quasiperiodic attractor with corresponding basin of attraction. (b) strange attractor for $f(x) = \sin(x) + \sin(3x)$ (see text).

strange attractors have a fractal noninteger dimension. We plot the coexistence of a periodic and quasi-periodic orbit in Figure 3.7a. The basin of attraction, which is the set of initial conditions which fall on a given attractor, is shown in light gray for the quasi-periodic, and in dark for the periodic attractor. A strange attractor (for $f(x) = \sin(x) + \sin(3x)$) is shown in Figure 3.7b.

### 3.1.2
### Hadjidemetriou Mapping

The examples of the previous sections allow to provide simple phase portraits that nevertheless already include the rich phenomenology of dynamical systems theory and found in more complex systems. We would like to conclude the mapping section with a method that makes it possible to construct symplectic mappings to Hamiltonian flows by means of averaging theory. This enables us to investigate specific problems using the tools of discrete systems, and moreover, provides a method to derive a mapping from a continuous flow that recovers most of the essential dynamics of the original system. The mapping that we are going to present is based on averaging theory. It thus will not be able to reproduce the individual orbits with all their frequencies but rather make it possible to investigate the phase space structure from a qualitative point of view. The idea of using mappings to understand 'real-world problem' in astronomy dates back to the work of [7, 23]. The method presented here is the one which was introduced in [8].

We assume a Hamiltonian function $H = H(p, q)$ of the form

$$H(p, q) = h(p) + h_\varepsilon(p, q; \varepsilon)$$

with $p = (p_1, \ldots, p_d)$ and $q = (q_1, \ldots, q_d)$, action-angle variables, where $\varepsilon$ is a small parameter and $d$ is the positive dimension $d \in \mathbb{N}$. It is possible to approximate the dynamics in terms of a symplectic mapping in the following way: without loss of generality we assume that $q_d(t + T_d) = q_d(t)$ is periodic with period $T_d$. We discretize the state variables $(p, q)$ as

$$q_n = \left(q_n^{(1)}, \ldots, q_n^{(d)}\right)$$
$$p_{n+1} = \left(p_{n+1}^{(1)}, \ldots, p_{n+1}^{(d)}\right)$$

where $n$ defines multiples of time $t = nT_d$ with $n \in \mathbb{Z}$ from now on. The idea is to replace the evolution in time of $(p(t), q(t))$ by a transformation in phase space which maps $(p(nT_d), q(nT_d))$ into $(p((n + 1)T_d), q((n + 1)T_d))$. In other words we aim to find a transformation from $(p_n, q_n) \rightarrow (p_{n+1}, q_{n+1})$ which follows the solution $(p(t), q(t))$ at discrete time steps $nT_d$. Since the phase space is symplectic the transformation needs to be canonical too. From a theoretical point of view we need not distinguish between the transformation in time and the transformation of the state variables. We thus may use the standard theory of canonical transformations and define the transformation of variables $(p_n, q_n)$ to the 'new' variables

$(p_{n+1}, q_{n+1})$ in terms of a generating function $S = S(p_{n+1}, q_n)$ by

$$q_{n+1}^{(i)} = \frac{\partial S(p_{n*1}, q_n)}{\partial p_{n+1}^{(i)}}$$

$$p_n^{(i)} = \frac{\partial S(p_{n+1}, q_n)}{\partial q_n^{(i)}} \quad \text{with} \quad i = 1, \ldots, d$$

The question remains how to relate $S$ with $H$. It was shown in [8] that a good choice is to define the generating function in the following form:

$$S(p_{n+1}, q_n)$$
$$= p_{n+1} \cdot q_n + T_d \bar{H}\left(p_{n+1}^{(1)}, \ldots, p_{n+1}^{(d-1)}, q_n^{(1)}, \ldots, q_n^{(d-1)}; p_{n+1}^{(d)}\right)$$

where '·' denotes the dot product, and $\bar{H} = \bar{H}(p_{n+1}, q_n)$ is the averaged Hamiltonian $H$ over the variable $q_n^{(d)}$:

$$\bar{H}(p_{n+1}, q_n) = \frac{1}{T_d} \int_0^{T_d} H(p_{n+1}, q_n) dq_d \tag{3.3}$$

While the term

$$p_{n+1} \cdot q_n = p_{n+1}^{(1)} q_n^{(1)} + \ldots p_{n+1}^{(d)} q_n^{(d)}$$

defines the identity transformation

$$q_{n+1}^{(i)} = q_n^{(i)}$$
$$p_{n+1}^{(i)} = p_n^{(i)}$$

the contribution $T_d \bar{H}$ replaces the time-evolution by its average over the period $T_d$. Averaging theory tells that for the long-term dynamics the short periodic effects can usually be neglected. If one chooses $T_d$ such that $T_d \ll T_k$ with $k = 1, \ldots, d$, then the error of the approximation becomes minimal. It can be estimated by replacing the integral (3.3) itself by a canonical transformation:

$$(p, q) \rightarrow (P, Q)$$

where $P = (P_1, \ldots, P_d)$ and $Q = (Q_1, \ldots, Q_d)$ are the new variables and the Hamiltonian takes the form:

$$H(P, Q) = h(P) + \varepsilon h_1(P, Q') + \cdots + \varepsilon^r h_r(P, Q') + \varepsilon^{r+1} h_{r+1}(P, Q)$$

Here $Q' = (Q'_1, \ldots, Q'_{d-1})$ and thus $h_k = h_k(P, Q')$ with $k \leq r$ are independent of the angle $Q_d$. This can always be done if $H$ is analytic and can be expanded into a Taylor series around $\varepsilon = 0$. One implements the transformation by means of a generating function, say $\chi$, or by means of a perturbative approach similar to the Lie-series transformation. Since the transformation is known and its inverse can be constructed in a suitable domain, we may obtain the solution $(p(t), q(t))$ from $(P(t), Q(t))$ also. If we moreover neglect the term $h_{r+1}$ in $H(P, Q)$ the error becomes

$$\varepsilon^{r+1}\|h_{r+1}\| \leq C\varepsilon^{r+1}$$

where $C$ is a constant bounding the norm $\|h_{r+1}\|$. With this it can be shown that (3.3) corresponds to a first-order approximation in the small parameter $\varepsilon$. A typical application of a Hadjidemetriou mapping found in astrodynamics is given by the $1:1$ mean motion resonance of an asteroid with Jupiter. In a simplified restricted three body model the Hamiltonian takes the form

$$H(a, e, \omega, \lambda, \lambda') = -\frac{1}{2a^2} + \mu R(a, e, \omega, \lambda, \lambda')$$

where $(a, e, \omega, \lambda)$ are the orbital elements – which describe the motion of the two-body problem[1] – and $\lambda'$ denotes the orbital longitude of Jupiter. The term proportional to $a^{-2}$ defines the unperturbed two-body motion, while $R$ defines the potential due to the Jupiter. If one is interested just in the dynamics close to the $1:1$ resonance one may replace $R$ by its average over the mean orbital longitude of $\lambda'$ and construct a mapping in the above way. With this it is possible to investigate the dynamics related to the ratio of the semimajor axis and the difference of orbital longitudes as well as the eccentricity and argument of the perihelion of the

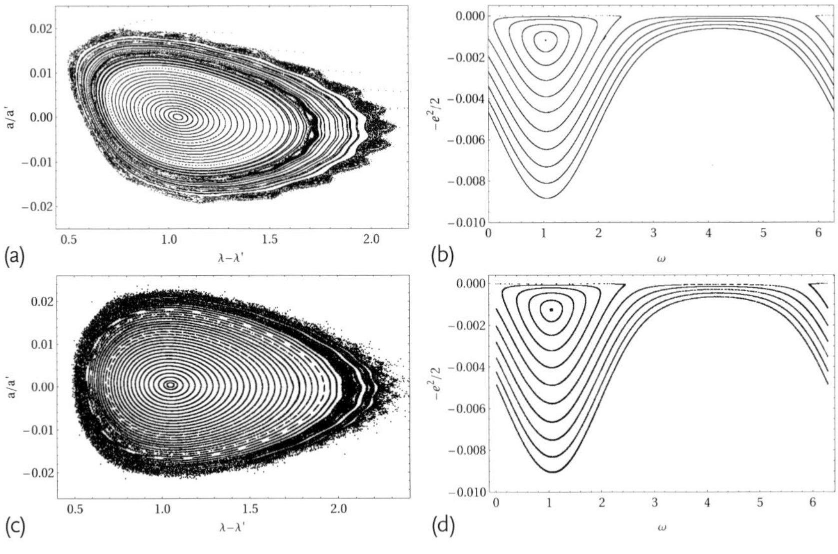


**Figure 3.8** Comparison of a Hadjidemetriou mapping (a,b) with direct numerical integration (c,d). (a,c) projections of the phase portrait to the $(a/a', \lambda - \lambda')$-plane. (b,d) projection of the phase portrait to the $(-e^2/2, \omega)$-plane.

1) See definition in Chapter 6. A detailed description of the mapping for the elliptic restricted three-body problem can be found, for example in [14].

asteroid. Typical projections of the phase portrait are shown in Figure 3.8 where we compare the dynamics obtained from the mapping equations (Figure 3.8a,b) with the dynamics obtained from numerical integration of the unaveraged problem (Figure 3.8c,d).

## 3.2 Lie-Series Numerical Integration

Ordinary differential equations that need to be solved in problems of celestial mechanics can be appropriately solved by numerical methods, where the solution for special initial conditions is derived step by step. There exists a large body of literature on this topic, and many different integrators are available (Bulirsch–Stoer, Runge–Kutta, Adams–Moulton–Bashforth to name a few). For an extensive review we refer the reader to the recently published comparison of the most commonly used integrations methods in [24]. In this section we introduce a method based on the Lie-series, named after the mathematician Sophus Lie, that turned out to be very efficient and precise for many of our problems. In celestial mechanics it was successfully introduced by [25] for the gravitational N-body problem; for the restricted three-body problem in rotating coordinates it was used by [26] and for the problem of bodies with variable masses by [27]. Additional broad applications for it were found for the stability of Trojans in the Solar system [28–30], and generally for planets in extrasolar planetary systems (e.g. [31–34]).

The general properties of the Lie-series are well described in the books [35] and also [36], where the whole theory is developed in details. In Gröbner's book the Lie-operator $D$ is defined:

$$D = \theta_1(z)\frac{\partial}{\partial z_1} + \theta_2(z)\frac{\partial}{\partial z_2} + \cdots + \theta_n(z)\frac{\partial}{\partial z_n} \tag{3.4}$$

where $D$ is an arbitrary linear differential operator. A point $z = (z_1, z_2, \ldots, z_n)$ lies in the $n$-dimensional $z$-space and the functions $\theta_i(z)$ are holomorphic within a certain domain $G$ which means that they can be expanded in converging power series. For a holomorphic function $f(z)$ in the same region as $\theta_i(z)$ the operator $D$ can be applied to it:

$$Df = \theta_1(z)\frac{\partial f}{\partial z_1} + \theta_2(z)\frac{\partial f}{\partial z_2} + \cdots + \theta_n(z)\frac{\partial f}{\partial z_n} \tag{3.5}$$

If we proceed applying $D$ to $f$ we get

$$\begin{aligned} D^2 f &= D(Df) \\ &\vdots \\ D^n f &= D(D^{n-1} f) \end{aligned}$$

We now define the *Lie-series* operator acting on a function *f*:

$$L(z,t) = \sum_{\nu=0}^{\infty} \frac{t^\nu}{\nu!} D^\nu f(z) = f(z) + t D f(z) + \frac{t^2}{2!} D^2 f(z) + \dots$$

Since the Taylor-series expansion of the exponential reads

$$e^{tD} f = 1 + tD^1 + \frac{t^2}{2!} D^2 + \frac{t^3}{3!} D^3 + \dots \tag{3.6}$$

we can write the Lie-series $L(z,t)$ operator also in a symbolic form

$$L(z,t) = e^{tD} f(z) \tag{3.7}$$

For the lengthy proof of convergence of $L(z,t)$ we refer to the book of Gröbner. An interesting property of Lie-series that turns out to be most useful is the *Vertauschungssatz*:

Let $F(z)$ be a holomorphic function in the neighborhood of $(z_1, z_2, \dots, z_n)$. If the corresponding power series expansion converges at the point $(Z_1, Z_2, \dots, Z_n)$ then we have:

$$F(Z) = \sum_{\nu=0}^{\infty} \frac{t^\nu}{\nu!} D^\nu F(Z) \tag{3.8}$$

which we can write also in the form

$$F(e^{tD})z = e^{tD} F(z) \tag{3.9}$$

It can be used to show how Lie-series are well adapted to solve differential equations. For such a system

$$\frac{dz_i}{dt} = \theta_i(z) \tag{3.10}$$

for $(z_1, z_2, \dots, z_n)$ it is possible to write immediately the solution with the aid of Lie-series

$$z_i = e^{tD} \xi_i \tag{3.11}$$

*D* is the Lie-operator and the $\xi_i$ are the initial conditions $z_i(t = 0)$. The proof is straightforward. We need to differentiate (3.10) with respect to time *t*:

$$\frac{dz_i}{dt} = D e^{tD} \xi_i = e^{tD} D \xi_i \tag{3.12}$$

Because of

$$D\xi_i = \theta_i(\xi_i) \tag{3.13}$$

we obtain the following result, which turns (3.10) into:

$$\frac{dz_i}{dt} = e^{tD} \theta_i(\xi_i) = \theta_i(e^{tD} \xi_i) = \theta_i(z_i) \tag{3.14}$$

### 3.2.1
### A Simple Example

We demonstrate the principle of the Lie-integration on the simple example of an unperturbed harmonic oscillator

$$\frac{d^2x}{dt^2} + \alpha^2 x = 0 \tag{3.15}$$

To begin we need to separate the equation into two first-order differential equations

$$\begin{aligned}\frac{dx}{dt} &= y = \theta_1(x, y)\\ \frac{dy}{dt} &= -\alpha^2 x = \theta_2(x, y)\end{aligned} \tag{3.16}$$

with initial conditions $z(t = 0) = \xi$ and $y(t = 0) = \eta$. The Lie-operator reads

$$D = \theta_1 \frac{\partial}{\partial \xi} + \theta_2 \frac{\partial}{\partial \eta} = \eta \frac{\partial}{\partial \xi} - \alpha^2 \xi \frac{\partial}{\partial \eta} \tag{3.17}$$

Immediately the solution can now be written in form of the Lie-series

$$x - e^{\tau D}\xi \quad \text{and} \quad y = e^{\tau D}\eta \tag{3.18}$$

Again we use the symbolic development of $e^{\tau D}$ and compute

$$\begin{aligned}D^1\xi &= \eta = \theta_1\\ D^2\xi &= D\eta = -\alpha^2\xi = \theta_2\\ D^3\xi &= -\alpha^2 D\xi = -\alpha^2\eta\\ D^4\xi &= -\alpha^2 D\eta = \alpha^4\xi\\ D^5\xi &= \alpha^4 D\xi = \alpha^4\eta\\ D^6\xi &= \alpha^4 D\eta = -\alpha^6\xi\\ &\vdots\end{aligned}$$

For the different Lie-terms (even and odd) we find

$$D^{2n}\xi = (-1)^n\alpha^{2n}\xi\,, \quad D^{2n+1}\xi = (-1)^n\alpha^{2n}\eta$$

so that the solution for $z$ reads

$$z = \xi + \tau\eta - \frac{\tau^2}{2!}\alpha^2\xi - \frac{\tau^3}{3!}\alpha^2\eta + \frac{\tau^4}{4!}\alpha^4\xi \ldots$$

After the factorization of $\xi$ and of $\eta$ we find

$$\begin{aligned}z = {}& \xi\left(1 - \frac{\tau^2}{2!}\alpha^2 + \frac{\tau^4}{4!}\alpha^4 - \frac{\tau^6}{6!}\alpha^6 + \ldots\right)\\ &+ \frac{\eta}{\alpha}\left(\tau\alpha - \frac{\tau^3}{3!}\alpha^3 + \frac{\tau^5}{5!}\alpha^5 - \frac{\tau^7}{7!}\alpha^7 + \ldots\right)\end{aligned}$$

that is the Taylor series expansion of the solution of the harmonic oscillator:

$$z(t) = \xi \cos \alpha\tau + \frac{\eta}{\alpha} \sin \alpha\tau$$

A more difficult example, that is the 'heart of celestial mechanics', is to find a precise solution of an N-body system like our own planetary system. We show in the following how the Lie-integrator needs to be constructed such that is keeps its strength of being precise and fast. We start with the equations of motion for the planetary problem (see Chapter 8)

$$m_\nu \ddot{\boldsymbol{q}}_\nu = k^2 \sum_{\mu=1,\nu\neq\mu}^{N} \frac{m_\nu m_\mu (\vec{r}_\mu - \vec{r}_\nu)}{\|\vec{r}_\mu - \vec{r}_\nu\|^3} \tag{3.19}$$

where we denote by $m_\nu$ the corresponding masses. We divide (3.19) by $m_\nu$ and split these (three) second-order differential equations into 6 first-order differential equations like we have done it before in the simple example ($k = 1$)

$$\dot{\boldsymbol{q}}_\nu = \vec{v}_\nu \tag{3.20}$$

$$\dot{\boldsymbol{v}}_\nu = k^2 \sum_{\mu=1,\nu\neq\mu}^{N} \frac{m_\mu (\boldsymbol{q}_\mu - \boldsymbol{q}_\nu)}{\|\boldsymbol{q}_\mu - \boldsymbol{q}_\nu\|^3} \tag{3.21}$$

For the step-size $\tau$ the solution after a time-step $\tau$ reads

$$\boldsymbol{q}_\nu(\tau) = e^{\tau D} \boldsymbol{q}_\nu(0) \tag{3.22}$$

$$\bar{\boldsymbol{q}}_\nu(\tau) = e^{\tau D} \bar{\boldsymbol{q}}_\nu(0) \tag{3.23}$$

where the Lie-Operator $D$ for the N-body problem is

$$D = \sum_{i=1}^{3} \sum_{\nu=1}^{N} \left( v_\nu^i \frac{\partial}{\partial r_\nu^i} + \sum_{\mu=1,\nu\neq\mu}^{N} m_\mu r_{\mu\nu}^i \rho_{\nu\mu}^{-3} \frac{\partial}{\partial v_\nu^i} \right)$$

We used in this equation the index $i$ for the $i$th component of the respective vectors for the position and the velocity. For the sake of a better understanding we define the Lie-Operator to act on one argument only:

$$D\boldsymbol{q}_\nu = D\boldsymbol{q}_\nu, D\bar{\boldsymbol{q}}_\nu = \bar{\boldsymbol{q}}_\nu, \quad \text{and} \quad D^n \boldsymbol{q}_\nu = D\left(D(D \ldots \boldsymbol{q}_\nu)\right) \tag{3.24}$$

We introduce the following quantities for the connecting position vector of particles $\nu$ and $\mu \boldsymbol{q}_{\mu\nu}$ its norm $\boldsymbol{r}_{\nu\mu}$ which stands for the distance between the masses $\nu$ and $\mu$ respectively $\bar{\boldsymbol{q}}_{\nu\mu}$) for their mutual velocity $\bar{\boldsymbol{q}}_{\nu\mu}$. It is evident that the following identity holds

$$\begin{aligned}
\boldsymbol{q}_{\nu\mu} &= \boldsymbol{q}_\mu - \boldsymbol{q}_\nu = -\boldsymbol{q}_{\mu\nu} \\
\rho_{\nu\mu} &= \|\vec{r}_\mu - \vec{r}_\nu\| = \|\vec{r}_{\nu\mu}\| \\
\bar{\boldsymbol{q}}_{\nu\mu} &= \bar{\boldsymbol{q}}_\mu - \bar{\boldsymbol{q}}_\nu = -\bar{\boldsymbol{q}}_{\mu\nu}
\end{aligned}$$

As next step we perform a series expansion which leads to

$$\begin{aligned} \boldsymbol{q}_\nu(\tau) &= e^{\tau D}\boldsymbol{q}_\nu(0) = \left[\sum_{n=0}^{\infty}\frac{(\tau D)^n}{n!}\right]\boldsymbol{q}_\nu(0) \\ &= \left[1+\tau D+\frac{(\tau)^2}{2!}D^2+\frac{(\tau)^3}{3!}D^3+\cdots+O\left(\frac{(\tau)^n}{n!}D^n\right)\right]\boldsymbol{q}_\nu(0) \qquad n\in\mathbb{N}_0 \end{aligned}$$

At the core of the integration algorithm lies the computation of the Lie series operator, that is the calculation of the orders[2] $D^n\boldsymbol{q}(0)$.

$$\begin{aligned} D^0\boldsymbol{q}_\nu &= \boldsymbol{q}_\nu \qquad \text{(position)} \\ D^1\boldsymbol{q}_\nu &= \bar{\boldsymbol{q}}_\nu \qquad \text{(velocities)} \\ D^2\boldsymbol{q}_\nu &= \sum_{\mu=1,\nu\neq\mu}^{N} m_\mu\left(\boldsymbol{q}_{\mu\nu}\rho_{\nu\mu}^{-3}\right) \qquad \text{(acceleration)} \\ D^3\boldsymbol{q}_\nu &= \sum_{\mu=1,\nu\neq\mu}^{N} m_\mu\left(D\boldsymbol{q}_{\mu\nu}\rho_{\nu\mu}^{-3}+\boldsymbol{q}_{\mu\nu}D\rho_{\nu\mu}^{-3}\right) \\ &= \sum_{\mu=1,\nu\neq\mu}^{N} m_\mu\left(D\boldsymbol{q}_{\mu\nu}\rho_{\nu\mu}^{-3}-3\boldsymbol{q}_{\mu\nu}\rho_{\nu\mu}^{-4}D\rho_{\nu\mu}\right) \\ &= \sum_{\mu=1,\nu\neq\mu}^{N} m_\mu\left[D\boldsymbol{q}_{\mu\nu}\rho_{\nu\mu}^{-3}-3\boldsymbol{q}_{\mu\nu}\rho_{\nu\mu}^{-4}\rho_{\nu\mu}^{-1}\left(\boldsymbol{q}_{\mu\nu}\cdot\bar{\boldsymbol{q}}_{\mu\nu}\right)\right] \\ &= \sum_{\mu=1,\nu\neq\mu}^{N} m_\mu\left[D\boldsymbol{q}_{\mu\nu}\rho_{\nu\mu}^{-3}-3\boldsymbol{q}_{\mu\nu}\rho_{\nu\mu}^{-5}\left(\boldsymbol{q}_{\mu\nu}\cdot\bar{\boldsymbol{q}}_{\mu\nu}\right)\right] \end{aligned}$$

At higher and higher orders this leads to quite complicated expressions and therefore we introduce new variables $\phi$ and $\Lambda$ [37].

$$\begin{aligned} \phi_{\nu\mu} &= \rho_{\nu\mu}^{-3} \\ \Lambda_{\mu\nu} &= \boldsymbol{q}_{\mu\nu}\cdot\bar{\boldsymbol{q}}_{\mu\nu} = \boldsymbol{q}_{\mu\nu}\cdot D\boldsymbol{q}_{\mu\nu} \end{aligned}$$

so that the third-order operator can be written as follows

$$\begin{aligned} D^3\boldsymbol{q}_\nu &= \sum_{\mu=1,\nu\neq\mu}^{N} m_\mu\left(D\boldsymbol{q}_{\mu\nu}\rho_{\nu\mu}^{-3}+\boldsymbol{q}_{\mu\nu}D\rho_{\nu\mu}^{-3}\right) \\ &= \sum_{\mu=1,\nu\neq\mu}^{N} m_\mu\left(\phi_{\nu\mu}D\boldsymbol{q}_{\mu\nu}+D\phi_{\nu\mu}\boldsymbol{q}_{\mu\nu}\right) \\ &= \sum_{\mu=1,\nu\neq\mu}^{N} m_\mu\left(D\boldsymbol{q}_{\mu\nu}\phi_{\nu\mu}-3\boldsymbol{q}_{\mu\nu}\rho_{\nu\mu}^{-2}\phi_{\nu\mu}\Lambda_{\mu\nu}\right) \end{aligned}$$

2) Note that in the 'real' uses for an efficient computer code we need orders up to $n = 15$.

where we can use $D\phi_{\nu\mu} = (-3)\rho_{\nu\mu}^{-2}\phi_{\nu\mu}\Lambda_{\mu\nu}$. This expression is not simpler than the previous one, but it is possible to find recurrence relations for higher orders with the aid of $\phi$ and $\Lambda$. We emphasize that the order $n$ of the derivative $D^n$ is proportional to the order of the time-step $\tau$ and in principle such a recurrence relation allows one, theoretically, to compute the positions and velocities to any order of $\tau$ very efficiently[3]:

$$D^3\mathbf{q}_\nu = \sum_{\mu=1,\nu\neq\mu}^{N} m_\mu \left(\phi_{\nu\mu} D\mathbf{q}_{\mu\nu} + D\phi_{\nu\mu}\mathbf{q}_{\mu\nu}\right)$$

$$D^4\mathbf{q}_\nu = \sum_{\mu=1,\nu\neq\mu}^{N} m_\mu \left(\phi_{\nu\mu} D^2\mathbf{q}_{\mu\nu} + 2D\phi_{\nu\mu} D\mathbf{q}_{\mu\nu} + D^2\phi_{\nu\mu}\mathbf{q}_{\mu\nu}\right)$$

$$D^5\mathbf{q}_\nu = \sum_{\mu=1,\nu\neq\mu}^{N} m_\mu \left(\phi_{\nu\mu} D^3\mathbf{q}_{\mu\nu} + 3D\phi_{\nu\mu} D^2\mathbf{q}_{\mu\nu} + 3D^2\phi_{\nu\mu} D\mathbf{q}_{\mu\nu} + D^3\phi_{\nu\mu}\mathbf{q}_{\mu\nu}\right)$$

$$D^6\mathbf{q}_\nu = \sum_{\mu=1,\nu\neq\mu}^{N} m_\mu \left(\phi_{\nu\mu} D^4\mathbf{q}_{\mu\nu} + 4D\phi_{\nu\mu} D^3\mathbf{q}_{\mu\nu} + 6D^2\phi_{\nu\mu} D^2\mathbf{q}_{\mu\nu} \right.$$
$$\left. + 4D^3\phi_{\nu\mu} D\mathbf{q}_{\mu\nu} + D^4\phi_{\nu\mu}\mathbf{q}_{\mu\nu}\right)$$

The 'kernel' of the Lie-series integration method is summarized as follows:

$$D^n\mathbf{q}_\nu = \sum_{k=1,l\neq k}^{N} m_\mu \sum_{l=0}^{n-2} \binom{n-2}{l} D^l\phi_{\nu\mu} D^{n-2-l}\mathbf{q}_{\mu\nu} \tag{3.25}$$

where we also need to compute $D$ acting on $\phi_{\mu\nu}$

$$D^n\phi_{\nu\mu} = \rho_{\nu\mu}^{-2}\left(\sum_{l=0}^{n-1} a_{n,l+1} D^{n-1-l}\phi_{\nu\mu} D^l\Lambda_{\mu\nu}\right) \tag{3.26}$$

with the respective coefficients:

$$a_{n,n} = -3\,, \quad n \geq 0$$
$$a_{n,1} = a_{n-1,1} - 2\,, \quad n \geq 1$$
$$a_{n,l} = a_{n-1,l-1} + a_{n-1,l}\,, \quad 1 \leq l < n$$

Another important quantity we need to compute $D$ acting on $\Lambda_{\mu\nu}$

$$D^n\Lambda_{\mu\nu} = \sum_{i=1}^{3}\sum_{l=0}^{\mathrm{nint}(n/2)} b_{n,l} D^l\mathbf{q}_{\mu\nu} D^{n+1-l}\mathbf{q}_{\mu\nu} \tag{3.27}$$

3) This is clear from the fact that in the series expansion of the exponential, factorial numbers must be computed. Their rapid growth, as well as the weak convergence behavior at high orders, constitutes the major limitations of the accuracy of Lie-Series expansions.

where in addition we need to compute the coefficients[4)]

$$
\begin{aligned}
b_{n,0} &= 1\,, & n &\geq 0\\
b_{n,l} &= b_{n-1,l-1} + b_{n-1,l}\,, & 1 < l &< \operatorname{nint}\left(\frac{n}{2}\right)\\
b_{n,\operatorname{nint}(n/2)} &= b_{n-1,\operatorname{nint}(n/2)-1}\,, & n &\text{ uneven}\\
b_{n,\operatorname{nint}(n/2)} &= 2b_{n-1,\operatorname{nint}(n/2)} + b_{n-1,\operatorname{nint}(n/2)-1}\,, & n &\text{ even}
\end{aligned}
$$

The final $n$th-order approximation to the solution of equations (3.20) and (3.21) for one time-step $\tau$ are given by:

$$q_\nu(\tau) = \left[1 + \tau D + \frac{(\tau)^2}{2!} D^2 + \frac{(\tau)^3}{3!} D^3 + \cdots + O\left(\frac{(\tau)^n}{n!} D^n\right)\right] q_\nu(0) \quad (3.28)$$

$$
\begin{aligned}
\bar{q}_\nu(\tau) &= \left[1 + \tau D + \frac{(\tau)^2}{2!} D^2 + \frac{(\tau)^3}{3!} D^3 + \cdots + O\left(\frac{(\tau)^n}{n!} D^n\right)\right] \bar{q}_\nu(0)\\
&= \left[D + \tau D^2 + \frac{(\tau)^2}{2!} D^3 + \frac{(\tau)^3}{3!} D^4 + \cdots + O\left(\frac{(\tau)^n}{n!} D^{(n+1)}\right)\right] q_\nu(0)
\end{aligned}
\quad (3.29)
$$

Let us state the advantages of the Lie-series integrator:

1. The use of recurrence formulas from order to order makes the evaluation of higher terms very efficient.
2. The distances between the acting bodies have to be evaluated *only once,* namely for $D^2 q_\nu$.
3. The stepsize can be changed from one step to the other.

and also the disadvantages:

1. The integrator is not symplectic the energy is not conserved (see [24]).
2. For every problem one needs to find out the most effective scheme for the recurrence formulas.
3. For complicated right hand sides of the differential equations the finding of recurrence formulae for $D$ becomes more and more complicated.

All in all, for most problems in celestial mechanics it is an efficient tool, well designed to deal with close encounters between celestial bodies, for example near-Earth asteroids, or comets suffering from transformed orbits due to close encounters to Jupiter. Even nongravitational force can be included (like the Yarkovsky effect for asteroids) but this requires special care and it is not straightforward to include relativity (e.g. the relativistic restricted three-body problem [38]) easily.

4) The function nint($x$) stands for the nearest integer close to a real number (e.g. nint(2.345) = 2, nint(2.5) = 3, nint(2.876) = 3).

## 3.3
## Chaos Indicators

Different kinds of chaos indicators exist – their full treatment is beyond the scope of this book. They have in common that they allow to separate regular and chaotic orbits. On the one hand, mathematically speaking, regular orbits are related to regular functions, which are well defined in a given domain. In the regular case we are able to describe the motion as Fourier series in terms of amplitudes, frequencies and phases. Once these quantities (and their possible evolution in time) are known, the signal can, in principle, be reconstructed by standard arguments of Fourier theory. On the other hand, chaotic orbits are known to depend sensitively on their initial conditions and may produce patterns of complex structure. Their existence has been known since the beginning of the twentieth century, but it took some more decades to see them in the context of a complete new field of research in the area of dynamical systems: chaos theory. With the invention of computers and the rise of numerical 'computer' experiments, the need for additional tools to identify chaotic orbits led to the concept of chaos indicators. We briefly discuss some of them, their definition, and fundamental properties and outline their possible use in the fields related to celestial mechanics.

### 3.3.1
### Lyapunov Characteristic Exponent

Lyapunov Characteristic Exponents (LCE), Largest Lyapunov Characteristic Exponents, and the Maximum Lyapunov Numbers are different notions for the concept of stability after Lyapunov which is based on the concept of linearization proposed in the PhD thesis of Alexandr Lyapunov in 1892. The linearized system of a given dynamical system

$$\dot{x} = F(x) \tag{3.30}$$

where $x = (x_1, \ldots, x_n)$, $F = (F_1, \ldots, F_n)$ is given by:

$$\dot{y} = J(x_*)y \tag{3.31}$$

where $y = (y_1, \ldots, y_n)$ are defined as $y = x - x_*$. In case $x_*$ is an equilibrium, then $F(x_*) = 0$ and the linear stability of the fixed point is defined by the eigenvalues of $J$, say $\mu_1, \ldots, \mu_n$. In this simple case the LCE are defined as the real part of the $\mu_i$, with $i = 1, \ldots, n$, and give a measure of the rate of change of infinitesimal small perturbations. The idea now is to generalize the concept to arbitrary orbits $x = x(t)$, which do not necessarily define an equilibrium of (3.30). Let us assume that $x(t + T) = x(t)$ defines a periodic orbit with period $T$ and replace (3.31) by

$$\dot{y} = J(x(t))y \tag{3.32}$$

If we integrate the equation from $t = 0$ up to $t = T$ we can still define a matrix $M$ which replaces the time evolution of $y$ by:

$$y_{t+T} = M y_t$$

The eigenvalues $\mu'_1, \ldots, \mu'_n$ of the matrix $M$, which is also known as the time operator conjugated to (3.30), can easily be found and define the so-called Floquet exponents

$$\alpha_i = \frac{\ln \mu'_i}{T} \quad \text{with} \quad i = 1, \ldots, n$$

Again the LCE are defined as the real parts of them. In the last step one generalizes the concept further to arbitrary (also nonperiodic) orbits and replaces the matrix $M$ by $R = M \cdot M^{\mathrm{T}}$, where $M^{\mathrm{T}}$ is the transposed of $M$. Since in general $M = M(t)$ the Floquet exponents $\mu''_1, \ldots, \mu''_n$ of $R$ themselves become functions of time, that is, $\mu''_i = \mu''_i(t)$. In an analogous way the Floquet exponents can be defined as:

$$\alpha_i = \alpha_i(t) = \frac{\ln \mu''_i}{2t}$$

with $i = 1, \ldots, n$. In the final step we consider the time limit $t \to \infty$ and define the LCE as

$$\lambda_i = \lim_{t \to \infty} \left( \sup_t (\lambda_i(t)) \right)$$

where the $\sup_t[\lambda_i(t)]$ accounts for the worst possible fluctuations of $\lambda_i$ with $i = 1, \ldots, n$. It has been shown (Oseledec-theorem, [39]) that the LCEs are independent of the initial conditions. Note that the LCEs are also defined as real numbers only.

From a geometrical point of view the set $(\lambda_1, \ldots, \lambda_n)$ defines a volume which tends to expand or shrink. The rate of change of the $n$-dimensional volume is strongly related to the stability of the system and measures the effect of $n$ independent infinitesimal perturbations over infinite times. The complete set is also called a Lyapunov spectrum. Another way to see their meaning is as follows: let us define the distance of two close-by orbits by $\delta y = (\delta y_1, \ldots, \delta y_n)$. The linearized evolution in time $\delta y(t)$ according to (3.32) is approximately given by:

$$\delta y(t) = e^{\lambda t} \delta y(0)$$

where $\delta y(0) = (\delta y_1(0), \ldots, \delta y_n(0))$ is the initial distance. The LCE $\lambda_i$ ($i = 1, \ldots, n$) therefore define the exponential growth rate of the separation with time. It can be bounded by the largest LCE number which is defined as:

$$\lambda_{\max} = \max_{i=1,\ldots,n} \lambda_i$$

With this we are able to state the following result (after Lyapunov):

If a linearized system is composed of analytic (regular) functions and the largest LCE is negative then the solution of the original system is asymptotically Lyapunov stable. This result provides a mathematical background for many chaos indicators used nowadays in literature. See for example [39, 40] and [41] for further reading.

### 3.3.2 Fast Lyapunov Indicator

The concept of the divergence of nearby orbits can be used directly to detect chaotic trajectories in higher-dimensional phase space since in the definition of chaos the sensitivity of the final state of a system on its initial conditions plays a crucial role. Unfortunately, the Lyapunov spectrum can not be computed directly from its definition for most dynamical systems, and one has to calculate the LCE by numerically means. Standard numerical techniques require several finite time approximations of the LCEs which can be used to find the limiting $\lambda_{\max}$. In addition rescaling is necessary to avoid numerical instabilities in the calculations. Since the techniques are time consuming additional attempts have been made to make the theory a powerful tool (Fast Lyapunov Indicator – FLI, Relative Lyapunov Indicator, Dynamical Lyapunov Indicator, and so on). Today the commonly used definition of the Fast Lyapunov Indicator (FLI) is given by:

$$\mathrm{FLI}(t) = \ln \|v(t)\| = \ln \delta(t)$$

where $v = (v_1, \dots, v_n)$ is a basis of deviation vectors of initially unitary length. We refer to the works of [42–44] as well as [41, 45] (and references within) for further reading.

### 3.3.3 Mean Exponential Growth Factor of Nearby Orbits

The MEGNO (Mean Exponential Growth of Nearby Orbits) was first introduced in [46, 47]. It can be related to the FLI in the following way. First one takes the time average in the interval $(0, t)$ defined by:

$$\overline{\mathrm{FLI}}(t) = \frac{1}{t}\int_0^t \mathrm{FLI}(s)ds = \frac{1}{t}\int_0^t \ln \delta(s)ds$$

With this the MEGNO indicator is defined as

$$Y(t) = \frac{2}{t}\int_0^t \frac{\dot{\delta}(s)}{\delta(s)} s ds$$

The formula was derived on the basis of fundamental properties of Hamiltonian dynamics and defines twice the time-weighted average of the relative divergence of orbits. The sensitivity of the indicator could be improved and by definition the distinction of chaotic, regular and resonant orbits was made possible. The indicator was also generalized to the possible application to symplectic mappings. See [47, 48] and [49] for further reading.

### 3.3.4 Smaller Alignment Index

The SALI (Smaller Alignment Index) was introduced in [50, 51] and later generalized to GALI (Generalized Alignment Index) [52]. It has been shown to be an efficient tool for numerically detecting chaos in dynamical systems. The GALI method keeps the normalized directions in the tangent space defined by the eigenvectors $(e_1, \dots, e_n)$ of the linearized system. The $\text{GALI}_k$ index is then defined as the volume of the $k$-parallelepiped spanned by the unitary deviation vectors $e_i$ with $i = 1, \dots, n$. Regular motion on a $m$-dimensional torus are found for special values of $\text{GALI}_k(t)$ while for chaotic orbits $\text{GALI}_k$ tends to zero according to the growth law:

$$\text{GALI}_k(t) \to e^{-[(\lambda_1-\lambda_2)+(\lambda_1-\lambda_3)+\dots+(\lambda_1-\lambda_k)]t}$$

where $(\lambda_1, \dots, \lambda_k)$ are the first $k$ largest Lyapunov characteristic exponents. See [52, 53] and [54] for further reading.

### 3.3.5 Spectral Analysis Method

The spectral analysis method (SAM, after [55]) defines regular orbits as quasiperiodic trajectories which occupy invariant tori and those which are not regular as irregular. Since the regular orbits are quasi-periodic any function $f = f(p, q)$ has the form

$$f(t) = f(p(t), q(t)) = \sum_{k \in \mathbb{Z}^n} \hat{f}_k \exp(ik\omega t)$$

where $\omega = (\omega_1, \dots, \omega_n)$ are the fundamental frequencies of motion. It is known that in the preceding sum all angular frequencies of the type $\Omega = k \cdot \omega$ with arbitrary $k$ may appear but for smooth functions $f = f(t)$ the size of the Fourier coefficients $|\hat{f}_k|$ decreases exponentially with increasing $|k|$ . If we label by $\Omega_r$ the angular frequencies with respect to $r$ such that $\Omega_{|r|+1} < \Omega_r$ and $r = \dots -2, -1, 0, 1, 2, \dots$ the sum is dominated only by a few finite number of terms:

$$f(t) = \sum_r \hat{f}_r \exp(i\Omega_r t)$$

The Fourier transform of the function $f(t)$ is defined as

$$F(\omega) = \int_{-\infty}^{\infty} f(t) \exp(-2\pi i \nu t) d dt$$

where $\nu = \omega/2\pi$. We therefore get:

$$F(\omega) = 2\pi \sum_r \hat{f}_r \delta(\omega - \Omega_r) = \sum_r \hat{f}_r \delta(\nu - \nu_r)$$

with $\nu_r = \Omega_r/2\pi$. In this setting an orbit $(p(t), q(t))$ is called regular if $F = F(\omega)$ is dominated by a countable number of frequency components, that is, by a finite sum over $\delta$-functions. In practice, one computes the orbit $(p(t), q(t))$ for finite times $T$ and numerically finds the Fourier spectrum of the signal $f(p(k\Delta t), q(k\Delta t))$ with $k \in \mathbb{N}$ and $\Delta t$ fixed. Here $f$, in principle, can be any smooth function of $(p, q)$. The definition of the threshold, to distinguish between regular and chaotic orbits, requires some additional work. In practice one defines the threshold of number of peaks in the power spectrum for which the signal is regular by means of experiment (that is by using another chaos indicator outlined previously). See [56, 57] for the application of the method to problems of celestial mechanics.

Further definitions of chaos indicators as well as their discussion and comparison can be found in [58–60] (dynamical stretching number), [61–63] (frequency map analysis) and [64] (comparison of different indicators).

## 3.4 Perturbation Theory

Perturbation methods are useful to find approximate solutions of a dynamical system. They also allow us to obtain the parameter dependence of the solution in terms of simple formulas. Even if the solutions obtained by classical perturbation methods are not as accurate as those obtained by direct numerical integration, they provide us with information that cannot be obtained by any other method. Brilliant books have been written on the topic in recent years. We mention the books [65] and [66] (and references therein) for a further reading. In the following discussion we outline just the basic concepts and are mainly interested in perturbation methods that allow to obtain the so-called normal form of a Hamiltonian flow. Interested readers should also see [67–72]. Let $H$ be a Hamiltonian of the form

$$H = H(J, \varphi) = h_0(J) + \varepsilon h_1(J, \varphi) \tag{3.33}$$

where $h_0, h_1$ are assumed to be analytic in $J = (J_1, \ldots, J_n)$, $\varphi = (\varphi_1, \ldots, \varphi_n)$ and $\varepsilon \in \mathbb{R}$ is a small parameter. We aim to find a transformation $I(J, \varphi)$ and $\psi(J, \varphi)$:

$$(J, \varphi) \to (I, \psi)$$

such that the canonical equations of the transformed Hamiltonian $K(I, \psi) = K(J(I, \psi), \varphi(I, \psi))$ gives the very simple form:

$$\begin{aligned} \dot{I} &= -\frac{\partial K}{\partial \psi} = O(\varepsilon)^{r+1} \\ \dot{\psi} &= \frac{\partial K}{\partial I} = \Omega_r(I) + O(\varepsilon)^{r+1} \end{aligned} \tag{3.34}$$

The system of equations, up to order $O(\varepsilon)^{r+1}$, admits the simple solution:

$$\begin{aligned} I(t) &= I_0 \\ \psi(t) &= \Omega_r(I_0)t + \psi_0 \end{aligned}$$

where $I_0 \equiv I(0)$, $\psi_0 \equiv \psi(0)$. In terms of the transformed variables $(I, \psi)$ the motion therefore takes place on the torus corresponding to the fixed action $I_0$. From (3.34) it is clear that the transformed Hamiltonian $K = H(J(I,\psi), \varphi(I,\psi))$ takes the form:

$$K(I,\psi) = k_0(I) + \varepsilon k_1(I) + \cdots + \varepsilon^r k_r(I) + \varepsilon^{r+1} h_{r+1}[J(I,\psi), \varphi(I,\psi)] \quad (3.35)$$

such that:

$$\frac{\partial h_{r+1}}{\partial \psi} = \frac{\partial h_{r+1}}{\partial I} = O(\varepsilon)^{r+1}$$

We call (3.35) the normal form of the Hamiltonian (3.33), $h_{r+1}$ the remainder and $\Omega_r$ the normalized frequency, respectively. The variables $(I, \psi)$ are usually called normal form coordinates and (3.34) normal form equations.

Different methods exist to transform (3.33) into (3.35). In this section we outline the most basic approach which is based on standard canonical transformations, that is, generating functions of mixed type of variables. Let $S = S(I, \varphi)$ be a function of the form:

$$S(I,\varphi) = \varepsilon S_1(I,\varphi) + \varepsilon^2 S_2(I,\varphi) + \cdots + \varepsilon_r S_r(I,\varphi)$$

The transformation $(J, \varphi) \to (I, \psi)$ is given implicitly through the relations

$$J = I + \frac{\partial S}{\partial \varphi} = I + \varepsilon \frac{\partial S_1}{\partial \varphi} + \varepsilon^2 \frac{\partial S_2}{\partial \varphi} + \cdots + \varepsilon_r \frac{\partial S_r}{\partial \varphi}$$
$$\psi = \varphi + \frac{\partial S}{\partial I} = \varphi + \varepsilon \frac{\partial S_1}{\partial I} + \varepsilon^2 \frac{\partial S_2}{\partial I} + \cdots + \varepsilon_r \frac{\partial S_r}{\partial I}$$

which needs to be inverted to get:

$$\begin{aligned} J &= I + \varepsilon \frac{\partial S}{\partial \psi} + \varepsilon^2 \left( \frac{\partial S_2}{\partial \psi} - \frac{\partial^2 S_1}{\partial \psi^2} \frac{\partial S_1}{\partial I} \right) + \cdots + \varepsilon^r f_r(I,\psi) \\ \varphi &= \psi - \varepsilon \frac{\partial S_1}{\partial I} - \varepsilon^2 \left( \frac{\partial S_2}{\partial I} - \frac{\partial S_1}{\partial I} \frac{\partial^2 S_1}{\partial I \partial \psi} \right) + \cdots + \varepsilon^r g_r(I,\psi) \end{aligned} \quad (3.36)$$

Here $f_r(I,\psi)$, $g_r(I,\psi)$ are known functions depending on $S_j$ and its partial derivatives from $j = 1, \ldots, r-1$. We substitute the above in (3.33) and expand around $\varepsilon = 0$ to get:

$$\begin{aligned} K(I,\psi) &= h_0(I) + \varepsilon \left[ h_1(I,\psi) + \frac{\partial h_0}{\partial I} \frac{\partial S_1}{\partial \psi} \right] \\ &+ \varepsilon^2 \left[ \frac{\partial h_0}{\partial I} \frac{\partial S_2}{\partial \psi} + \frac{1}{2} \frac{\partial^2 h_0}{\partial I^2} \left( \frac{\partial S_1}{\partial \psi} \right)^2 + \frac{\partial S_1}{\partial \psi} \frac{\partial h_1}{\partial I} - \frac{\partial h_0}{\partial I} \frac{\partial^2 S_1}{\partial \psi^2} \frac{\partial S_1}{\partial I} - \frac{\partial h_1}{\partial \psi} \frac{\partial S_1}{\partial I} \right] \\ &+ \cdots + \varepsilon^r \left[ \frac{\partial h_0}{\partial I} \frac{\partial S_2}{\partial \psi} + F_r(I,\psi) \right] \end{aligned} \quad (3.37)$$

where $F_r = F_r(I, \psi)$ are known functions depending on functions of previous orders. From $\omega(I) = \partial h_0/\partial I$ and comparing coefficients order by order with (3.35) we get the system of equations:

$$
\begin{aligned}
\varepsilon^0 &: k_0(I) = h_0(I) \\
\varepsilon^1 &: k_1(I) = h_1(I, \psi) + \omega(I)\frac{\psi S_1}{\partial \psi} \\
\varepsilon^2 &: k_2(I) = F_2(I, \psi) + \omega(I)\frac{\psi S_2}{\partial \psi} \\
\vdots &: \quad \vdots \\
\varepsilon^r &: k_r(I) = F_r(I, \psi) + \omega(I)\frac{\partial S_r}{\partial \psi}
\end{aligned}
\tag{3.38}
$$

where we have identified the known part of the second-order terms proportional to $\varepsilon^r$ by $F_r(I, \psi)$. The system has to be solved recursively for $S_j$ with $j = 1, \dots, r$ to obtain the $k_j$. Let us identify the known part of first order by a function $F_1(I, \psi) \equiv h_1(I, \psi)$ and analyze the equation at order $\varepsilon^j$ with $j \leq r$:

$$k_j(I) = F_j(I, \psi) + \omega(I)\frac{\partial S_j}{\partial \psi} \tag{3.39}$$

We notice that $F_j$ is constructed on the basis of analytic functions, therefore it also has a Fourier series representation of the form:

$$F_j(I, \psi) = \sum_{k \in \mathbb{Z}^n} \hat{f}_k(I) e^{i \cdot (k_1\psi_1 + \dots + k_n\psi_n)} \tag{3.40}$$

which can be written as the sum

$$F_j(I, \psi) = \bar{F}_j(I) + \tilde{F}_j(I, \psi)$$

where $\bar{F}_j$ is called the mean and $\tilde{F}_j$ is called the oscillatory part of $F_j$. With respect to (3.39) we identify the mean part with $k_j$:

$$k_j(I) = \bar{F}_j(I)$$

and subtract it from (3.38) to get:

$$\omega(I)\frac{\partial S_j}{\partial \psi} + \tilde{F}_j(I, \psi) = 0 \tag{3.41}$$

Since $\tilde{F}_j$ contains all Fourier harmonics $k \in \mathbb{Z}^n \backslash \{0\}$ (see (3.40)) the generating function $S_j$ must be of the form:

$$S_j = S_j(I, \psi) = \sum_{k \in \mathbb{Z}^n \backslash \{0\}} \hat{s}_k(I) e^{i \cdot (k_1\psi_1 + \dots + k_n\psi_n)} \tag{3.42}$$

In terms of the coefficients of the Fourier harmonics the solution of the Eq. (3.41) becomes

$$\hat{s}_k = -i \frac{\hat{f}_k}{\omega \cdot k} \tag{3.43}$$

where $k \cdot \omega = \omega_1 k_1 + \cdots + \omega_n k_n$. Once the solutions are known for $j = 1, \ldots, r$ the transformation $(J, \varphi) \to (I, \psi)$ is defined through $K = K(I, \psi)$ from (3.37). However, the method fails if $\omega \cdot k = 0$ since the generating function cannot be defined properly (the so-called zero divisor problem). Since $\omega = (\omega_1, \ldots, \omega_n)$ defines the unperturbed frequencies of the system, that is, for $\varepsilon = 0$:

$$\omega_i = \omega_i(I) = \frac{\partial h_0}{\partial I_i} = \omega_i(J) = \frac{\partial h_0}{\partial J_i}$$

with $i = 1, \ldots, n$, it is the integrable part which defines the domain, say $D$, in $J$ (respective $I$) in which the normal form can be used. The normal form is valid for $J \in D$, where $D$ is small enough to exclude zero divisors from (3.43). The following argument makes clear why $D$ cannot be defined without limiting the domain of Fourier harmonics as well: Let us assume that at normalization order $r$ we could find $D_r$ such that for $J \in D_r$ it is true that $\omega \cdot k \neq 0$ and the normal form can locally be defined. Increasing $r$ we cannot exclude the existence of another Fourier harmonic, say $k^* = (k_1^*, \ldots, k_n^*)$, for which we find $\omega \cdot k^* = 0$. Again we redefine $D_r$ to $D_{r'}$ to ensure that for $J \in D_{r'}$ the normal form is properly defined. The argument can be repeated at every step of the calculation but since for rational $\omega$ and arbitrary $k \in \mathbb{Z}^n$ it is always possible to find $k^*$ such that $\omega \cdot k^* = 0$ we are not able to remove Fourier harmonics of the form

$$\hat{f}_{k^*} e^{ik_* \cdot \omega}$$

from the expansion (3.40) to any normalization order $r$. For this reason we either have to limit $r$ and keep most of the Fourier harmonics in the normal form or the domain $D_r$ for which the normal form would be valid tends to zero as $r \to \infty$. A similar problem, which has not to be confused with the zero divisor problem, is the problem of small divisors: for arbitrary (e.g. irrational) $\omega$ but large $k$ the series may formally converge. However, the accumulation of the so-called small divisors, for which $\omega \cdot k^*$ becomes small (but not zero), still may harm the convergence properties for the practical usage of the series expansions. The only compromise is to try to find $r = r_{\text{opt}}$ for which the effect of the Fourier harmonics that could not be removed is of the same order of magnitude as the effect of the terms which are of order $O(\varepsilon)^{r+1}$.

### 3.4.1 Lie-Transformation Method

A modern approach to obtain a normal form of Hamiltonian systems is based on the Lie-transformation method. We outline the method in short and show how to adapt the method to resonant initial conditions. See for example [70, 73–75] and [76–78] for further information. We consider a $2n$ dimensional phase space in action angle variables $(J, \varphi)$, $J \in \mathbb{R}^n$, $\varphi \in \mathbb{T}^n$, where $J = (J_1, \ldots, J_n)$ are the actions and $\varphi = (\varphi_1, \ldots, \varphi_n)$ are the angles. The Hamiltonian of interest has the same form as in the previous section:

$$H = H(J, \varphi) = h_0(J) + \varepsilon h_1(J, \varphi) \tag{3.44}$$

Again, the aim is to find a transformation of variables to write the Hamiltonian in normal form coordinates, say $(J^{(r)}, \varphi^{(r)})$, which gives the Hamiltonian $H^{(r)}$ normalized at order $r \in \mathbb{N}$:

$$H^{(r)} = H\left(J^{(r)}, \varphi^{(r)}\right)$$

In this notation the normal form equations of motion become:

$$\begin{aligned} \dot{J}^{(r)} &= -\frac{\partial H^{(r)}}{\partial \varphi^{(r)}} = O(\varepsilon)^{r+1} \\ \dot{\varphi}^{(r)} &= \frac{\partial H^{(r)}}{\partial J^{(r)}} = \Omega^{(r)}(J^{(r)}, \varphi^{(r)}) + O(\varepsilon)^{r+1} \end{aligned} \tag{3.45}$$

and we therefore have:

$$H^{(r)} = H^{(r)}(J^{(r)})$$

with $\partial H^{(r)}/\partial \varphi^{(r)} = 0$ up to order $\varepsilon^{r+1}$. From the normal form condition it follows that the normalized Hamiltonian $H^{(r)}$ should be of the form

$$H^{(r)}(J^{(r)}, \varphi^{(r)}) = h_0(J^{(r)}) + \cdots + \varepsilon^r h_r^{(r)}(J^{(r)}) + \varepsilon^{r+1} h^{(r+1)}(J^{(r)}, \varphi^{(r)}) \tag{3.46}$$

In this setting the normal form, say $Z^{(r)}$, is simply

$$Z^{(r)} = Z^{(r)}(J^{(r)}; \varepsilon) \equiv h_0(J^{(r)}) + \cdots + \varepsilon^r h_r^{(r)}(J^{(r)})$$

and the remainder, say $R^{(r+1)}$, is given by

$$R^{(r+1)} = R^{(r+1)}(J^{(r)}, \varphi^{(r)}; \varepsilon) \equiv \varepsilon^{r+1} h^{(r+1)}(J^{(r)}, \varphi^{(r)})$$

and thus:

$$H^{(r)}\left(J^{(r)}, \varphi^{(r)}\right) = Z^{(r)}(J^{(r)}; \varepsilon) + R^{(r+1)}(J^{(r)}, \varphi^{(r)}; \varepsilon)$$

The normalized frequencies $\Omega_r = (\Omega_1^{(r)}, \ldots, \Omega_n^{(r)})$ are given by:

$$\Omega_j^{(r)} = \frac{\partial Z^{(r)}}{\partial J_j^{(r)}}$$

with $j = 1, \ldots, n$ and the solution in terms of $(J^{(r)}, \varphi^{(r)})$ up to order $O(\varepsilon^{r+1})$ is simply given by:

$$J_j^{(r)}(t) = J_{j0}^{(r)}$$
$$\varphi_j^{(r)}(t) = \Omega_j^{(r)}\left(J_{j0}^{(r)}\right) t + \varphi_{j0}$$

where $J_{j0}^{(r)} = J_j^{(r)}(0)$ and $\varphi_{j0}^{(r)} = \varphi_j^{(r)}(0)$ are the initial conditions. In contrast to the previous section we aim to find a sequence of near-identity transformations, say $(J^{(r-1)}, \varphi^{(r-1)}) \to (J^{(r)}, \varphi^{(r)})$, with $r > 1$ of the explicit form:

$$J^{(r)} = J^{(r-1)}(J^{(r)}, \varphi^{(r)})$$
$$\varphi^{(r)} = \varphi^{(r-1)}(J^{(r)}, \varphi^{(r)})$$

which preserves the Hamiltonian structure of the phase space without the usage of mixed-type generating functions. The fundamental idea is to look for transformations which can be expanded close to $\varepsilon = 0$ and which follow a Hamiltonian structure of the form

$$\frac{d\varphi}{d\varepsilon} = \frac{\partial X}{\partial J}$$
$$\frac{dJ}{d\varepsilon} - \frac{\partial X}{\partial \varphi}$$

Here, the transformation is seen in a similar way as the underlying canonical system too. The time flow is replaced by an $\varepsilon$-flow the Hamiltonian is replaced by an arbitrary function $X$. From a formal point of view, the only difference is that $X$ is an arbitrary function and does not need to represent any real physical dynamical system. We define a generating function $X(J_1, \ldots, J_n, \varphi_1, \ldots, \varphi_n)$ in short $X(J, \varphi)$ which is assumed to be analytic in a domain $D_{R,S}$ and can therefore be expanded into a Fourier series of the form

$$X(J, \varphi) = \sum_{k \in \mathbb{Z}^n} \chi_k(J_1, \ldots, J_n; \varepsilon) e^{i(k_1\varphi_1 + \cdots + k_n\varphi_n)}$$

Here $k = (k_1, \ldots, k_n)$ with $k_j \in \mathbb{Z}$, $\varepsilon$ is a small parameter and $\chi_k = \chi_k(J)$ are regular functions in $J$. We define the Lie-derivative along the flow generated by $X$ by making use of the Poisson brackets

$$L_X = \{\#, X\}$$

where # is the argument. The Lie-derivative is an operator which acts on analytic functions $f = f(J, \varphi)$, that is,

$$L_X f = \{f, X\} = \sum_{j=i}^{n} \left( \frac{\partial f}{\partial J_i} \frac{\partial X}{\partial \varphi_i} - \frac{\partial f}{\partial \varphi_i} \frac{\partial X}{\partial J_i} \right)$$

The properties of the Lie-derivative follow from the properties of the Poisson bracket. The properties are similar to that of the standard partial derivative $\partial/\partial t$:

1. linearity in the arguments

$$L_f(g+h) = L_f g + L_f h$$
$$L_f(\alpha g) = \alpha L_f g$$

where $f, g, h$ are analytic functions in $(J, \varphi)$ and $\alpha$ is a constant. The linearity follows from

$$\{f, g+h\} = \{f, g\} + \{f, h\}$$

and

$$\{f, \alpha g\} = \alpha\{f, g\}$$

since

$$\sum_{j=i}^{n}\left(\frac{\partial f}{\partial J_i}\frac{\partial g+h}{\partial \varphi_i} - \frac{\partial f}{\partial \varphi_i}\frac{\partial g+h}{\partial J_i}\right)$$
$$= \sum_{j=i}^{n}\left(\frac{\partial f}{\partial J_i}\frac{\partial g}{\partial \varphi_i} - \frac{\partial f}{\partial \varphi_i}\frac{\partial g}{\partial J_i}\right) + \sum_{j=i}^{n}\left(\frac{\partial f}{\partial J_i}\frac{\partial h}{\partial \varphi_i} - \frac{\partial f}{\partial \varphi_i}\frac{\partial h}{\partial J_i}\right)$$

by rewriting the sums accordingly.

2. anticommutativity:

$$L_f g = -L_g f$$

where $f, g$ are analytic functions in $(J, \varphi)$. The anticommutativity follows from

$$\{f, g\} = -\{g, f\}$$

since

$$\sum_{j=i}^{n}\left(\frac{\partial f}{\partial J_i}\frac{\partial g}{\partial \varphi_i} - \frac{\partial f}{\partial \varphi_i}\frac{\partial g}{\partial J_i}\right) = -\sum_{j=i}^{n}\left(\frac{\partial g}{\partial J_i}\frac{\partial f}{\partial \varphi_i} - \frac{\partial g}{\partial \varphi_i}\frac{\partial f}{\partial J_i}\right)$$
$$= \sum_{j=i}^{n}\left(-\frac{\partial f}{\partial J_i}\frac{\partial g}{\partial \varphi_i} + \frac{\partial f}{\partial \varphi_i}\frac{\partial g}{\partial J_i}\right)$$

3. Jacobi identity

$$L_{(L_h g)} f + L_{(L_f h)} g + L_{(L_g f)} h = 0$$

where $f, g, h$ are analytic functions in $(J, \varphi)$. Jacobi's identity follows from

$$\{f, \{g, h\}\} + \{g, \{h, f\}\} + \{h, \{f, g\}\} = 0$$

To prove the identity one has to calculate all the Poisson brackets and regroup the terms to show the cancelation. First we observe that from the definition of the Poisson bracket the only products that are produced in nested forms of the type $\{\#, \{\#, \#\}\}$ are given by:

$$\frac{\partial^{e_1+e_2} f}{\partial J_i^{e_1} \partial \varphi_i^{e_2}} \frac{\partial^{e_3+e_4} g}{\partial J_i^{e_3} \partial \varphi_i^{e_4}} \frac{\partial^{e_5+e_6} h}{\partial J_i^{e_5} \partial \varphi_i^{e_6}}$$

where $e_1, \ldots, e_6 \in (0, 1, 2)$, that is, we do not produce mixed products (e.g. $(\partial/\partial J_i)(\partial/\partial J_k)$ with $i \neq k$). We introduce the following temporary notation:

$$f_{e_1,e_2} g_{e_3,e_4} h_{e_5,e_6} := \sum_{j=i}^{n} \left( \frac{\partial^{e_1+e_2} f}{\partial J_i^{e_1} \partial \varphi_i^{e_2}} \frac{\partial^{e_3+e_4} g}{\partial J_i^{e_3} \partial \varphi_i^{e_4}} \frac{\partial^{e_5+e_6} h}{\partial J_i^{e_5} \partial \varphi_i^{e_6}} \right)$$

In this setting the term $\{f, \{g, h\}\}$ which we label by T1 gives:

$$\begin{aligned} \mathrm{T1} = {} & f_{1,0} g_{1,1} h_{0,1} - f_{0,1} g_{2,0} h_{0,1} \\ & + f_{1,0} g_{1,0} h_{0,2} - f_{1,0} g_{0,2} h_{1,0} + f_{0,1} g_{1,1} h_{1,0} \\ & - f_{1,0} g_{0,1} h_{1,1} - f_{0,1} g_{1,0} h_{1,1} + f_{0,1} g_{0,1} h_{2,0} \end{aligned}$$

The term $\mathrm{T2} = \{g, \{h, f\}\}$ gives

$$\begin{aligned} \mathrm{T2} = {} & f_{2,0} g_{0,1} h_{0,1} - f_{1,1} g_{1,0} h_{0,1} \\ & - f_{1,0} g_{1,0} h_{0,2} - f_{1,1} g_{0,1} h_{1,0} + f_{0,2} g_{1,0} h_{1,0} \\ & + f_{1,0} g_{0,1} h_{1,1} + f_{0,1} g_{1,0} h_{1,1} - f_{0,1} g_{0,1} h_{2,0} \end{aligned}$$

and the term $\mathrm{T3} = \{h, \{f, g\}\}$ becomes:

$$\begin{aligned} \mathrm{T3} = {} & -f_{2,0} g_{0,1} h_{0,1} + f_{1,1} g_{1,0} h_{0,1} \\ & - f_{1,0} g_{1,1} h_{0,1} + f_{0,1} g_{2,0} h_{0,1} + f_{1,1} g_{0,1} h_{1,0} \\ & + f_{1,0} g_{0,2} h_{1,0} - f_{0,2} g_{1,0} h_{1,0} - f_{0,1} g_{1,1} h_{1,0} \end{aligned}$$

A careful reading shows that any of the $f_{e_1,e_2} g_{e_3,e_4} h_{e_5,e_6}$ is cancelled by another $-f_{e_1,e_2} g_{e_3,e_4} h_{e_5,e_6}$ in T1 + T2 + T3.

4. relation with product:

$$L_f(gh) = h L_f g + g L_f h$$

where $f, g, h$ are analytic functions of $(J, \varphi)$. The property is strongly related to the relation of the partial derivative $\partial/\partial t$. In terms of Poisson bracket notation one has:

$$\{f, gh\} = \{f, g\}h + g\{f, h\}$$

which is

$$
\begin{aligned}
&\sum_{j=i}^{n}\left[\frac{\partial f}{\partial J_i}\frac{\partial(gh)}{\partial \varphi_i}-\frac{\partial f}{\partial \varphi_i}\frac{\partial(gh)}{\partial J_i}\right]\\
&=\sum_{j=i}^{n}\left[\frac{\partial f}{\partial J_i}\left(\frac{\partial g}{\partial \varphi_i}h+g\frac{\partial h}{\partial \varphi_i}\right)-\frac{\partial f}{\partial \varphi_i}\left(\frac{\partial g}{\partial J_i}h+g\frac{\partial h}{\partial J_i}\right)\right]\\
&=\sum_{j=i}^{n}\left(\frac{\partial f}{\partial J_i}\frac{\partial g}{\partial \varphi_i}h+g\frac{\partial f}{\partial J_i}\frac{\partial h}{\partial \varphi_i}-\frac{\partial f}{\partial \varphi_i}\frac{\partial g}{\partial J_i}h-g\frac{\partial f}{\partial \varphi_i}\frac{\partial h}{\partial J_i}\right)\\
&=\sum_{j=i}^{n}\left(\frac{\partial f}{\partial J_i}\frac{\partial g}{\partial \varphi_i}h+g\frac{\partial f}{\partial J_i}\frac{\partial h}{\partial \varphi_i}-\frac{\partial f}{\partial \varphi_i}\frac{\partial g}{\partial J_i}h-g\frac{\partial f}{\partial \varphi_i}\frac{\partial h}{\partial J_i}\right)\\
&=\sum_{j=i}^{n}\left[h\left(\frac{\partial f}{\partial J_i}\frac{\partial g}{\partial \varphi_i}-\frac{\partial f}{\partial \varphi_i}\frac{\partial g}{\partial J_i}\right)+g\left(\frac{\partial f}{\partial J_i}\frac{\partial h}{\partial \varphi_i}-\frac{\partial f}{\partial \varphi_i}\frac{\partial h}{\partial J_i}\right)\right]\\
&=h\sum_{j=i}^{n}\left(\frac{\partial f}{\partial J_i}\frac{\partial g}{\partial \varphi_i}-\frac{\partial f}{\partial \varphi_i}\frac{\partial g}{\partial J_i}\right)+g\sum_{j=i}^{n}\left(\frac{\partial f}{\partial J_i}\frac{\partial h}{\partial \varphi_i}-\frac{\partial f}{\partial \varphi_i}\frac{\partial h}{\partial J_i}\right)
\end{aligned}
$$

To apply the Lie-derivative $L_X$ on a function $f = f(J, \varphi)$ in total $k$-times we write

$$L_X^{(k)} = L_X\left(L_X^{(k-1)}\right)$$

and note that $L_X^{(0)} = \mathrm{Id}$ is the identity operator. With this we define the $k$-th derivative recursively and the zeroth derivative to be the identity as is done with the standard partial derivative $\partial/\partial$. On the basis of the Lie-derivative and the foregoing discussion, we are able to define the Lie-series operator, which is the exponential of the Lie-derivative, in the following way:

$$\exp(L_X) = \sum_{k=0}^{\infty}\frac{1}{k!}L_X^{(k)} = \mathrm{Id} + L_X + \frac{1}{2}L_X^{(2)} + \frac{1}{6}L_X^{(3)} + \dots$$

Note that the infinite sum is an operator and acts due to the linearity of the Lie-derivative term by term on a function $f = f(J, \varphi)$ as an argument:

$$\exp(L_X)f = f + L_X f + \frac{1}{2}L_X^{(2)} f + \frac{1}{6}L_X^{(3)} f + \dots$$

Moreover, the Lie-derivative defines the time-derivative along a Hamiltonian flow defined by $X$. In other words, from

$$
\begin{aligned}
\dot{J}_i &= L_X J_i\\
\dot{\varphi}_i &= L_X \varphi_i
\end{aligned}
$$

and expanding $J(t), \varphi(t)$ in terms of a Taylor series around $t = 0$:

$$J_i(t) = J_i(0) + \frac{dJ_i(0)}{dt} t + \frac{1}{2!} \frac{d^2 J_i(0)}{dt^2} t^2 + \ldots$$
$$\varphi_i(t) = \varphi_i(0) + \frac{d\varphi_i(0)}{dt} t + \frac{1}{2!} \frac{d^2 \varphi_i(0)}{dt^2} t^2 + \ldots$$

we find that the solution $J(t), \varphi(t)$ can also be written in terms of the Lie-series operator

$$J_i(t) = \exp(t L_X) J_i(0)$$
$$\varphi_i(t) = \exp(t L_X) \varphi_i(0)$$

The fundamental idea behind the Lie-series approach is the following: since the time evolution in phase space given by $(J(t), \varphi(t))$ is symplectic and the Lie-series operator gives the solution $(J(t), \varphi(t))$ for given $(J(0), \varphi(0))$ the Lie-series operator can be used to define a symplectic change of coordinates in terms of a generating function $X$. A change of coordinates from old $(J_i^{(\mathrm{old})}, \varphi_i^{(\mathrm{old})})$ to new $(J_i^{(\mathrm{new})}, \varphi_i^{(\mathrm{new})})$ variables which also depends on a small parameter $\varepsilon$ is given by:

$$J_i^{(\mathrm{old})} = \exp(\varepsilon L_X) J_i^{(\mathrm{new})}$$
$$\varphi_i^{(\mathrm{old})} = \exp(\varepsilon L_X) \varphi_i^{(\mathrm{new})}$$

Here, the role of time $t$ was replaced by the role of a small parameter $\varepsilon$ which allows to perform the transformation at order $\varepsilon$. Note, that with the preceding expressions we are also able to define a sequence of transformations $(\chi_1, \ldots, \chi_r)$, where the transformations $(J_i^{(r)}, \varphi_i^{(r)}) \to (J_i^{(r+1)}, \varphi_i^{(r+1)})$ is defined by:

$$J_i^{(r-1)} = \exp(\varepsilon L_{X_r}) \, J_i^{(r)}$$
$$\varphi_i^{(r-1)} = \exp(\varepsilon L_{X_r}) \, \varphi_i^{(r)}$$

where $r > 1$ labels the order of the transformation. Note, that the transformation is given in its explicit form and we can avoid the inversion of the formula by making use of the property of the Lie-derivative and Lie-series operators. Let us introduce the following notation to formalize the idea: let $H^{(0)} = H(J, \varphi)$ be the original Hamiltonian and $(J^{(0)}, \varphi^{(0)}) = (J, \varphi)$ the original variables. We implement the transformation

$$\left(J^{(0)}, \varphi^{(0)}\right) \to \left(J^{(1)}, \varphi^{(1)}\right)$$

to find $H^{(1)} = H^{(1)}(J^{(1)}, \varphi^{(1)})$ which we know is normalized up to order $r = 1$. Iterating the process we may easily find

$$H^{(r)} = H^{(r)}\left(J^{(r)}, \varphi^{(r)}\right)$$

from $H^{(r)} = H^{(r)}(J^{(r)}, \varphi^{(r)})$: we insert the transformation expressed in terms of the Lie-series into the Hamiltonian normalized up to order $r$:

$$H^{(r)} = H^{(r-1)}\left[\exp(\varepsilon L_{X_r}) \, J^{(r)}, \exp(\varepsilon L_{X_r}) \, \varphi^{(r)}\right]$$
$$= h_0\left[\exp(\varepsilon L_{X_r}) \, J^{(r)}\right] + \varepsilon h_1\left[\exp(\varepsilon L_X) J^{(r)}, \exp(\varepsilon L_{X_r}) \, \varphi^{(r)}\right]$$

We make use of the linearity properties of the Lie-series operator, where for any arbitrary function $f = f(J, \varphi)$ we have

$$f\left[\exp(\varepsilon L_{X_r}) J, \exp(\varepsilon L_{X_r}) \varphi\right] = \exp(\varepsilon L_{X_r}) f(J, \varphi)$$

which is called the "Vertauschungssatz" [35]. It follows that

$$\begin{aligned} H^{(r)} &= \exp(\varepsilon L_{X_r}) H^{(r-1)}\left(J^{(r)}, \varphi^{(r)}\right) = \exp(\varepsilon L_{X_r}) H^{(r-1)}\left(J^{(r-1)}, \varphi^{(r-1)}\right) \\ &= \exp(\varepsilon L_{X_r}) h_0\left(J^{(r)}\right) + \varepsilon \exp(\varepsilon L_{X_r}) h_1\left(J^{(r)}, \varphi^{(r)}\right) \end{aligned}$$

Here we just have to replace $(J^{(r-1)}, \varphi^{(r-1)})$ by $(J^{(r)}, \varphi^{(r)})$ after calculating $\exp(\varepsilon L_{X_r})$. Using the expansion of the exponential close to $\varepsilon = 0$:

$$H^{(r)} = \left(1 + \varepsilon L_{X_r} + \frac{1}{2}\varepsilon^2 l_{X_r}^{(2)} + \dots\right)\left(h_0^{(r-1)}\left(J^{(r)}\right) + \varepsilon h_1^{(r-1)}\left(J^{(r)}, \varphi^{(r)}\right)\right).$$

If we equate the expression with (3.46) we find order by order:

$$\begin{aligned} &\varepsilon^0 : h_0^{(r)}\left(J^{(r)}\right) = h_0\left(J^{(r-1)}\right) \\ &\varepsilon^1 : L_{X_r} h_0\left(J^{(r)}\right) + h_1\left(J^{(r-1)}, \varphi^{(r+1)}\right) = h_1^{(r)}\left(J^{(r)}\right) \end{aligned}$$

We notice that the generic form of the preceding equations is:

$$h_j^{(r)}\left(J^{(r)}\right) = F_j^{(r)}\left(J^{(r)}, \varphi^{(r)}\right) + L_{X_r} h_0\left(J^{(r)}\right) \tag{3.47}$$

where $F_j^{(r)} = F_j^{(r)}(J^{(r)}, \varphi^{(r)})$ are known functions depending on the solutions of previous orders and $h_0(J^{(r)})$ is the unperturbed part of the Hamiltonian written in terms of the new variables $J^{(r)}$. By the definition of the Lie-operator (3.47) also gives:

$$\begin{aligned} h_j^{(r)}\left(J^{(r)}\right) &= F_j^{(r)}\left(J^{(r)}, \varphi^{(r)}\right) + \left\{h_0\left(J^{(r)}\right), X_r\left(J^{(r)}, \varphi^{(r)}\right)\right\} \\ &= F_j^{(r)}\left(J^{(r)}, \varphi^{(r)}\right) + \frac{\partial h_0}{\partial J^{(r)}} \frac{\partial X_r}{\partial \varphi^{(r)}} \\ &= F_j^{(r)}\left(J^{(r)}, \varphi^{(r)}\right) + \omega\left(J^{(r)}\right) \frac{\partial X_r}{\partial \varphi^{(r)}} \end{aligned} \tag{3.48}$$

We follow the argumentation of the previous section taking into account the effect of resonant terms with Fourier order $\leq K$. Due to the analyticity of the functions involved we have for any $F_j^{(r)}$:

$$F_j^{(r)}\left(J^{(r)}, \varphi^{(r)}\right) = \sum_{k \in \mathbb{Z}^n} \hat{f}_k\left(J^{(r)}\right) e^{i \cdot \left(k_1 \varphi_1^{(r)} + \dots + k_n \varphi_n^{(r)}\right)} \tag{3.49}$$

Let us denote by $\bar{F}_j^{(r)}$ the mean part of $F_j^{(r)}$ defined by:

$$\bar{F}_j\left(J^{(r)}\right) = \frac{1}{(2\pi)^n} \int_{\mathbb{T}^n} F_j^{(r)}\left(J^{(r)}, \varphi^{(r)}\right) d\varphi^{(r)}$$

and by $\tilde{F}_j^{(r)}$ the zero average part defined by:

$$\tilde{F}_j\left(J^{(r)}, \varphi^{(r)}\right) = F_j^{(r)}\left(J^{(r)}, \varphi^{(r)}\right) - \bar{F}_j\left(J^{(r)}\right)$$

Moreover, we may split $\tilde{F}_j^{(r)}$ with respect to some positive Fourier order $K \in \mathbb{Z}$ into:

$$\tilde{F}_j^{(r),\leq K} = \tilde{F}_j^{(r),\leq K}\left(J^{(r)}, \varphi^{(r)}\right) = \sum_{k\in\mathbb{Z}^n, |k|\leq K} \hat{f}_k\left(J^{(r)}\right) e^{ik\cdot\varphi^{(r)}}$$

and

$$\tilde{F}_j^{(r),>K} = \tilde{F}_j^{(r),>K}\left(J^{(r)}, \varphi^{(r)}\right) = \sum_{k\in\mathbb{Z}^n, |k|>K} \hat{f}_k\left(J^{(r)}\right) e^{ik\cdot\varphi^{(r)}}$$

where we wrote $k\cdot\varphi^{(r)} = k_1\varphi_1^{(r)} + \cdots + k_n\varphi_n^{(r)}$ for short. Finally we split $\tilde{F}_j^{(r),\leq K}$ into its resonant and nonresonant contributions. Let

$$F_j^{(r),\leq K,\mathrm{nr}} = F_j^{(r),\leq K,\mathrm{nr}}\left(J^{(r)}, \varphi^{(r)}\right) = \sum_{k\in\mathbb{Z}^n, |k|\leq K, \omega\cdot k\neq 0} \hat{f}_k\left(J^{(r)}\right) e^{ik\cdot\varphi^{(r)}}$$

and

$$F_j^{(r),\leq K,\mathrm{re}} = F_j^{(r),\leq K,\mathrm{re}}\left(J^{(r)}, \varphi^{(r)}\right) = \sum_{k\in\mathbb{Z}^n, |k|\leq K, \omega\cdot k = 0} \hat{f}_k\left(J^{(r)}\right) e^{ik\cdot\varphi^{(r)}}$$

where $\omega\cdot k = \omega_1 k_1 + \cdots + \omega_n k_n$ and $\omega_i = \partial h_0/\partial J^{(r)}$ are the unperturbed frequencies of the system. In this setting the following equality holds true:

$$F_j^{(r)} = \bar{F}_j^{(r)} + F_j^{(r),\leq K,\mathrm{nr}} + F_j^{(r),\leq K,\mathrm{re}} + \tilde{F}_j^{(r),>K}$$

and

$$\tilde{F}_j = F_j^{(r),\leq K,\mathrm{nr}} + F_j^{(r),\leq K,\mathrm{re}}$$

To solve (3.48) we first make the following observations:

1. the functions $\bar{F}_j^{(r)}$ do not contribute to $\partial H^{(r)}/\partial\varphi^{(r)}$ and thus are kept as normal form terms. Formally we identify them (see (3.46)) with:

   $$h_j^{(r)}\left(J^{(r)}\right) = \bar{F}_j^{(r)}\left(J^{(r)}\right)$$

   for $j \leq r$.
2. the Fourier harmonics $F_j^{(r),\leq K,\mathrm{nr}}$ can be removed from (3.48) by a proper choice of $X_r$. Writing $X_r$ in Fourier component form:

   $$X_j\left(J^{(r)}, \varphi^{(r)}\right) = \sum_{k\in\mathbb{Z}^n, |k|\leq K, \omega\cdot k\neq 0} \chi_{j,k}\left(J^{(r)}\right) e^{ik\cdot\varphi^{(r)}}$$

   we find in terms of the Fourier coefficients:

   $$\chi_{j,k} = -i\frac{\hat{f}_k}{\omega\cdot k} \quad \text{for} \quad |k| \leq K\,, \quad k\cdot\omega \neq 0$$

3. the functions $F_j^{(r),\leq K,\mathrm{re}}$ cannot be removed from (3.48) since $\omega \cdot k = 0$ would induce a zero divisor in the definition of $X_j$. Thus they must be retained as part of the normal form. This will modify (3.46) in the following way:

$$\begin{aligned} H^{(r),\mathrm{re}} &= H^{(r),\mathrm{re}}\left(J^{(r)}, \varphi^{(r)}\right) \\ &= h_0\left(J^{(r)}\right) + \cdots + \varepsilon^r h_r^{(r)}\left(J^{(r)}, \varphi^{(r)}\right) + \varepsilon^{r+1} h^{(r+1)}\left(J^{(r)}, \varphi^{(r)}\right) \end{aligned} \tag{3.50}$$

where $\varepsilon^j h_j^{(r)}(J^{(r)}, \varphi^{(r)})$ is defined by:

$$h_j^{(r)}\left(J^{(r)}, \varphi^{(r)}\right) = F_j^{(r),\leq K,\mathrm{re}}\left(J^{(r)}, \varphi^{(r)}\right)$$

with $j \leq r$ and thus only consists of fully resonant terms. We may also write:

$$H^{(r),\mathrm{re}}\left(J^{(r)}, \varphi^{(r)}\right) = Z^{(r)}\left(J^{(r)}; \varepsilon\right) + Y^{(r)}\left(J^{(r)}, \varphi^{(r)}; \varepsilon\right) + R^{(r+1)}\left(J^{(r)}, \varphi^{(r)}; \varepsilon\right)$$

for short, where

$$Y^{(r)}\left(J^{(r)}, \varphi^{(r)}; \varepsilon\right) = \sum_{j=1}^{r} \varepsilon^j F_j^{(r),\leq K,\mathrm{re}}\left(J^{(r)}, \varphi^{(r)}\right)$$

If $Y^{(r)} = Y^{(r)}(J^{(r)}, \varphi^{(r)}; \varepsilon) = 0$ we recover the nonresonant normal form (3.45). For nonvanishing $Y^{(r)}$ the normal form equations become:

$$\begin{aligned} \dot{J}^{(r)} &= -\frac{\partial H^{(r)}}{\partial \varphi^{(r)}} = -\frac{\partial Y^{(r)}}{\partial \varphi^{(r)}} + O(\varepsilon)^{r+1} \\ \dot{\varphi}^{(r)} &= \frac{\partial H^{(r)}}{\partial J^{(r)}} = \Omega^{(r)}\left(J^{(r)}\right) + \frac{\partial Y^{(r)}}{\partial J^{(r)}} + O(\varepsilon)^{r+1} \end{aligned} \tag{3.51}$$

The normalized system contains just Fourier modes which are in resonance. Neglecting orders $O(\varepsilon)^{r+1}$ it defines a simplified resonant model to the original system defined by (3.44).

### 3.4.2
### Mapping method

It is possible to adapt the concepts of perturbation theory to the mapping case. We follow the notation and algorithm outlined in [79, 80]. See also [81] for normal forms in mappings based on generating functions. Consider a map in $\mathbb{R}^{2n}$, with $n$ the dimension, which is defined by:

$$\begin{aligned} p' &= f(p, q) \\ q' &= g(p, q) \end{aligned} \tag{3.52}$$

where $(p, q) \in \mathbb{R}^n \times \mathbb{R}^n$ and $(p', q')$ denote the updated values for known $(p, q)$ after one iteration of the mapping equations. The functions $f, g$ are assumed to be real analytic. In addition, we require that there exists a fixed point at the origin, such that:

$$f(0,0) = g(0,0) = 0$$

and that the mapping is symplectic:

$$R \cdot J \cdot R^{-1} = J$$

where $R$ denotes the Jacobian matrix of (3.52) and the symplectic matrix $J$ is defined as:

$$J = \begin{pmatrix} 0 & I_n \\ -I_n & 0 \end{pmatrix}$$

and $I_n$ is the $n \times n$ identity matrix. It is natural to introduce complex coordinates

$$z_k = q_k + i p_k\,, \quad \tilde{z}_k = q_k - i p_k\,, \quad \text{with} \quad k = 1, \ldots, n$$

where $z_k$ and $\tilde{z}_k$ are complex conjugated to each other. We also write $z = (z_1, \ldots, z_n)$ and $\tilde{z} = (\tilde{z}_1, \ldots, \tilde{z}_n)$ for short. In addition, we define the complex conjugated functions $F_k = f_k + i g_k$ and $\tilde{F}_k = f_k - i g_k$ with $k = 1, \ldots, n$. Since the original coordinates $(p, q)$ are real we are interested in the subspace $z = \tilde{z}$, where $F = \tilde{F}$ and the map (3.52) can be written as:

$$z' = F(z, \tilde{z}) \tag{3.53}$$

where $F = (F_1, \ldots, F_n)$. For the subsequent discussion we have to introduce the following notations. Let $l = (l_1, \ldots, l_n)$, $m = (m_1, \ldots, m_n)$ and $l, m \in \mathbb{N}^n$. We denote by $z^l = z_1^{l_1} \cdot \cdots \cdot z_n^{l_n}$ and similar for $\tilde{z}^m$. Moreover, we define by

$$|l + m| = \sum_{j=1}^{n} l_j + m_j$$

the order of the monomial proportional to $z^l \tilde{z}^m$. In this setting we are able to define

$$\begin{aligned} [F]_N &= (F_1, \ldots, F_k) \quad \text{where} \quad z^l z^m \quad \text{and} \quad |l + m| = N \in \mathbb{N} \\ [F]_{\leq N} &= (F_1, \ldots, F_k) \quad \text{where} \quad z^l z^m \quad \text{and} \quad |l + m| \leq N \in \mathbb{N} \\ [F]_{> N} &= (F_1, \ldots, F_k) \quad \text{where} \quad z^l z^m \quad \text{and} \quad |l + m| > N \in \mathbb{N} \end{aligned}$$

Assume that we expand $F_k$, with $k = 1, \ldots, n$, into Taylor series around $z = \tilde{z} = 0$:

$$F_k(z, \tilde{z}) = \sum_{l,m \in \mathbb{N}^n} a_{l,m}^k z_1^{l_1} z_2^{l_2} \ldots z_n^{l_n} \tilde{z}_1^{m_1} \tilde{z}_2^{m_2} \ldots \tilde{z}_n^{m_n} \tag{3.54}$$

where $a_j^k \in \mathbb{C}$ are complex coefficients. Since $F = (F_1, \ldots, F_n)$ we also have:

$$[F]_N = \sum_{|l+m|=N} a_{l,m} z_1^{l_1} z_2^{l_2} \ldots z_n^{l_n} \tilde{z}_1^{m_1} \tilde{z}_2^{m_2} \ldots \tilde{z}_n^{m_n}$$

and similarly for $[F]_{\leq N}$ and $[F]_{>N}$, that is, we have:

$$F(z,\tilde{z}) = \sum_{j=1}^{\infty}[F]_j \tag{3.55}$$

We seek to find a nonlinear transformation $(z,\bar{z}) \to (\zeta,\bar{\zeta})$ defined by:

$$z = \Phi(\zeta,\bar{\zeta}) = \zeta + \sum_{j\geq 2}[\Phi]_j(\zeta,\bar{\zeta})$$

such that the transformed map

$$U = U(\zeta,\bar{\zeta}) = \sum_{j\geq 2}[U]_j(\zeta,\bar{\zeta})$$

becomes a twist map

$$\zeta' = U(\zeta,\zeta') = e^{i\Omega(\zeta\cdot\bar{\zeta})}\zeta \tag{3.56}$$

with a new frequency defined by:

$$\Omega = \Omega(\zeta\cdot\bar{\zeta}) = \sum_{j\geq 2}[\Omega]_j(\zeta\cdot\bar{\zeta})$$

Thus the frequency $\Omega$ just depends on products of the form $\zeta\cdot\bar{\zeta}$. From

$$F(z,\bar{z}) = F(\Phi(\zeta,\bar{\zeta}),\bar{\Phi}(\zeta,\bar{\zeta}))$$

we obtain the functional equation:

$$F(\Phi(\zeta,\bar{\zeta}),\bar{\Phi}(\zeta,\bar{\zeta})) = \Phi(e^{i\Omega(\zeta\cdot\bar{\zeta})}, e^{-i\Omega(\zeta\cdot\bar{\zeta})}) \tag{3.57}$$

which defines $U$ and $\Phi$. If we substitute in the above $F$, $\Phi$, $U$ by $[F]_N$, $[\Phi]_N$ and $[U]_N$ we get instead at order $r$ the equation

$$\left[F\circ[\Phi]_{\leq N}\right]_r - \left[[\Phi]_{\leq N}\circ[U]_{\leq N}\right]_r = 0 \tag{3.58}$$

which can be solved for $2 \leq r \leq N$ in a recursive way. As the initial solution we get:

$$[\Phi]_1 = \zeta$$

and

$$[U]_1 = e^{i\omega}\zeta$$

The relation between $[U]_r$ and $[\Omega]_r$ is simply given by the formula:

$$[U]_r = \left[e^{i[\Omega]_{\leq N}}\zeta\right]_r$$

The remainder of the mapping

$$R_N = R_N(\zeta, \bar{\zeta}) = \sum_{j>N} [R]_j(\zeta, \bar{\zeta})$$

is given by:

$$R_N = \left[F \circ [\Phi]_{\leq N}\right]_{>N} - \left[[\Phi]_{\leq N} \circ [U]_{\leq N}\right]_{>N}$$

To solve (3.58) we introduce the linear operator:

$$\Delta[\Phi]_r = [\Phi]_r \left(e^{i\omega}\zeta, e^{-i\omega}\bar{\zeta}\right) - e^{i\omega}[\Phi]_r$$

and the projector $\Pi_1$

$$\Pi_1[T]_r = [U]_r$$

where $[T]_r$ is an arbitrary polynomial in $(\zeta, \bar{\zeta})$. The two operators have the following special properties:

$$\begin{aligned} \Pi_1[U]_r &= [U]_r \\ \Pi_1\Delta[\Phi]_r &= 0 \end{aligned}$$

We write

$$\begin{aligned} [S]_r &= [U]_r + \Delta[\Phi]_r \\ &= \{F \circ [\Phi]_{\leq N}\}_r - \{[\Phi]_{\leq N} \circ [U]_{\leq N}\}_r + [U]_r + \Delta[\Phi]_r \end{aligned}$$

and from the requirement $[S]_r = 0$ we obtain

$$\begin{aligned} [U]_r &= \Pi_1[S]_r \\ \Delta(1 - \Pi_1)[\Phi]_r &= (1 - \Pi_1)[S]_r \end{aligned}$$

where $(1 - \Pi_1)$ denotes the inverse of $\Pi_1$. From the inverse of $\Delta$ we get

$$(1 - \Pi_1)[\Phi]_r = \Delta^{-1}(1 - \Pi_1)[S]_r$$

which defines $[\Phi]_r$ at normalization order $r$. The inverse of a monomial term in $[S]_r$, say

$$s_{l,m}\zeta_1^{k_1} \cdot \cdots \cdot \zeta_n^{k_n} \bar{\zeta}_1^{k_{n+1}} \cdot \cdots \cdot \bar{\zeta}_n^{k_{2n}}$$

is given by:

$$\frac{s_{l,m}}{e^{i[\omega_1(k_1-k_2)+\cdots+(k_{2n-1}-k_{2n})]}} \zeta_1^{k_1} \cdot \cdots \cdot \zeta_n^{k_n} \bar{\zeta}_1^{k_{n+1}} \cdot \cdots \cdot \bar{\zeta}_n^{k_{2n}} \tag{3.59}$$

where $k_1 + \cdots + k_{2n} = r$ and $s_{l,m} \in \mathbb{C}^n$. The normal form components $\Pi_1[\Phi]_r$ of $[\Phi]_r$ are still undetermined. They are used to ensure that the transformation $\Phi$

and the normalized mapping $U$ are symplectic up to the order $N$. For this reason we add $\Pi_1[\Phi]_r$ proportional to unknown coefficients, say $\alpha$, to $(1-\Pi_1)[\Phi]_r$ to get:

$$[\Phi]_r = (1-\Pi_1)[\Phi]_r + \Pi_1[\Phi]_r$$

and determine the $\alpha$ from the requirement

$$\begin{pmatrix} \{[\Phi]_r, [\Phi]_r\} & \{[\Phi]_r, [\bar{\Phi}]_r\} \\ \{[\bar{\Phi}]_r, [\Phi]_r\} & \{[\bar{\Phi}]_r, [\bar{\Phi}]_r\} \end{pmatrix} = \begin{pmatrix} 0 & I_n \\ -I_n & 0 \end{pmatrix}$$

where $\{f, g\}$ for polynomials $f, g$ in $(\zeta, \bar{\zeta})$ is defined, analogously to the continuous case, as:

$$\{f, g\} = \sum_{j=1}^{n} \frac{\partial f}{\partial \zeta_j} \frac{\partial g}{\partial \bar{\zeta}_j} - \frac{\partial f}{\partial \bar{\zeta}_j} \frac{\partial g}{\partial \zeta_j}$$

Once we solved for the inverse for $s_{l,m}$ we also obtain $[\Phi]_r$ and $[U]_r$ and the normal form (3.56) can be easily obtained. The normal form is not defined for terms which would introduce a zero divisor in (3.59). They play the role of resonant terms in a similar way as in the continuous case. For a further reading on the topic see for example [79, 80].

# 4
# The Stability Problem

In this chapter we are interested in the different kinds of stability found in dynamical systems useful for the applications in celestial mechanics. We first introduce the concept of linear stability including basic definitions, the notion of Lyapunov stability, and the definition of the stable and unstable manifolds of hyperbolic equilibria. In the next step we state the stability problem we have in mind when dealing with Hamiltonian systems that can be be composed of an integrable (by definition stable) part and a perturbation. This will lead to a slightly different concept of stability which we call *persistence of a system against additional perturbations* and requires further understanding of the way resonances interact in phase space and will lead us directly to two famous stability theorems found in the last century: the KAM and Nekhoroshev theorems. The aim of the present chapter is not to provide the reader with all details and the proofs, but rather introduce to the definitions and concepts behind them. The part on linear stability is based on the introduction of [82], the part on resonant interaction is motivated on [83, Chapter 4]. In the section on the KAM theory we followed the way of presentation given in [84, 85], the section on the Nekhoroshev stability is based on [86, 87] and [88].

## 4.1
## Review on Different Concepts of Stability

The usual concept of stability is the one according to Lyapunov [89]. We consider an $n$-dimensional autonomous system of ordinary equations of motion of the form:

$$\dot{x} = f(x) \tag{4.1}$$

where $x = (x_1, \ldots, x_n)$ are the coordinates and $f = (f_1, \ldots, f_n)$ is a vector function of the vector $x$, which defines the derivatives of the system. A given solution, say $x_s = x_s(t)$, of (4.1) is said to be Lyapunov stable if for any $\varepsilon > 0$ there exists $\delta = \delta(\varepsilon)$ such that if

$$\|x(t_0) - x_s(t_0)\| < \delta$$

*Celestial Dynamics*, First Edition. R. Dvorak and C. Lhotka.

at given time $t = t_0$ then

$$\|x(t_0) - x_s(t_0)\| < \varepsilon$$

for any $t \geq t_0$. Here $\|\cdot\|$ denotes the standard Euclidean norm. It is said to be asymptotically stable, if for

$$\|x(t_0) - x_s(t_0)\| < \delta$$

one has

$$\lim_{t\to\infty} \|x(t_0) - x_s(t_0)\| = 0$$

Another concept is the one of linear stability. Let us assume that there exists an equilibrium solution of (4.1), say $x_* = (x_{1*}, \ldots, x_{n*})$, such that

$$f(x_*) = 0$$

and thus $\dot{x} = 0$ for $x = x_*$. The Jacobian of $f$, in short $J_f = Df$ is defined by:

$$J_f(x_*) = Df|_{x_*} = \begin{pmatrix} \frac{\partial f_1}{\partial x_1} & \cdots & \frac{\partial f_1}{\partial x_n} \\ \cdots & \cdots & \cdots \\ \frac{\partial f_n}{\partial x_1} & \cdots & \frac{\partial f_n}{\partial x_n} \end{pmatrix}_{x_*}$$

and the linearized system of (4.1) is then given by:

$$\dot{y} = J_f(x_*)y$$

where $y = (y_1, \ldots, y_n)$ is related to $x$ by $y = x - x_*$. We investigate the determinant of

$$\det\left(J_f(x_*) - \lambda\, I_n\right) = 0$$

which is called the characteristic equation. Here $I_n$ is the identity matrix and $\lambda = (\lambda_1, \ldots, \lambda_n)$ are the eigenvalues of $J_f(x_*)$ and the corresponding eigenvectors $(\hat{y}_1, \ldots, \hat{y}_n)$ are defined by the equations:

$$J_f(x_*)\hat{y}_k = \lambda_k \hat{y}_k \quad \text{with} \quad k = 1, \ldots, n$$

Let us denote by $\lambda_c \subseteq \lambda$, with $\lambda_c = \{\lambda : \mathrm{Re}(\lambda) = 0\}$, the eigenvalues with zero real part, $\lambda_+ \subseteq \lambda$, with $\lambda_+ = \{\lambda : \mathrm{Re}(\lambda) > 0\}$, those with positive real part, and by $\lambda_- \subseteq \lambda$, with $\lambda_- = \{\lambda : \mathrm{Re}(\lambda) < 0\}$, those with negative real part. The equilibrium solution $x_*$ is said to be *hyperbolic* if $\lambda_c = 0$; it is called an *attractor* if $\lambda = \lambda_-$, a *repeller* if $\lambda = \lambda_+$ and a *saddle* if $\lambda_+, \lambda_-$ are both strictly positive. Since points in the vicinity of an attractor are attracted, it may be called stable, while a repeller is unstable by definition.

In two dimensions with $n = 2$ we have $\lambda = (\lambda_1, \lambda_2)$ and we are able to distinguish between the following cases:

(A) for $\lambda_1, \lambda_2$ real negative the equilibrium $x_*$ is said to be a *stable node*
(B) for $\lambda_1, \lambda_2$ real positive numbers the equilibrium $x_*$ is said to be an *unstable node*.
(C) for $\lambda_1, \lambda_2$ real numbers with opposite sign the equilibrium $x_*$ is said to be a *saddle* or a hyperbolic equilibrium point.
(D) for $\lambda_1, \lambda_2$ complex numbers with negative real parts the equilibrium $x_*$ is said to be a *stable focus* or a stable spiral point.
(E) for $\lambda_1, \lambda_2$ complex numbers with positive real parts the equilibrium $x_*$ is said to be an *unstable focus* or an unstable spiral point.
(F) for $\lambda_1, \lambda_2$ complex with zero real parts the equilibrium $x_*$ is called a *center* or an elliptic equilibrium point.

In higher dimensions the classification becomes more complicated since it allows for mixed-type classifications of the preceding definitions.

A further concept of stability important in the stability problem is the one of stable and unstable manifolds. Let $x_*$ be a saddle point and let us denote by $\Phi(t, x)$ the flow given at time $t$ with initial condition $x_0 = (x_{10}, \ldots, x_{n0})$. The *stable manifold* $\mathfrak{W}^s = \mathfrak{W}^s(x_*)$ is the set of initial conditions of (4.1) which reach $x_*$ for $t \to \infty$ while the *unstable manifold* $\mathfrak{W}^u = \mathfrak{W}^u(x_*)$ is the set of points which end up at $x_*$ for $t \to -\infty$. Mathematically speaking the sets may be defined:

$$\mathfrak{W}^s(x_*) = \left\{x \in \mathbb{R}^n : \lim_{t\to\infty} \Phi(t, x) = x_*\right\}$$

$$\mathfrak{W}^u(x_*) = \left\{x \in \mathbb{R}^n : \lim_{t\to-\infty} \Phi(t, x) = x_*\right\}$$

The intersection of $\mathfrak{W}^s$ and $\mathfrak{W}^u$ is called a *homoclinic point* if the sets belong to the same equilibrium point $x_*$, the intersection is called a *heteroclinic point* if $\mathfrak{W}^s$, $\mathfrak{W}^u$ belong to different equilibria. A formal criterion for chaos (and therefore instability) is the existence of heteroclinic points which also implies the nonintegrability of the system.

In the following discussion we are interested in another concept of stability which is important for dynamical systems which admit a Hamiltonian function of the form:

$$H = H(p, q) = h_0(p) + h_\varepsilon(p, q)$$

where $p = (p_1, \ldots, p_n)$, $q = (q_1, \ldots, q_n)$ are conjugated variables and $\varepsilon$ parametrizes the strength of the perturbation $h_\varepsilon(p, q)$ of the integrable approximation $h_0(p)$. Since for $\varepsilon = 0$ the $p = p(0)$ are conserved we aim to find simple bounds on $|p(t) - p(0)| \leq \alpha$ valid within a stability time $t \leq T$. Moreover we assume that the inequalities are valid within a threshold of the small parameter, say $\varepsilon \leq \varepsilon_0$. In other words we ask for the existence of a parameter domain in which the con-

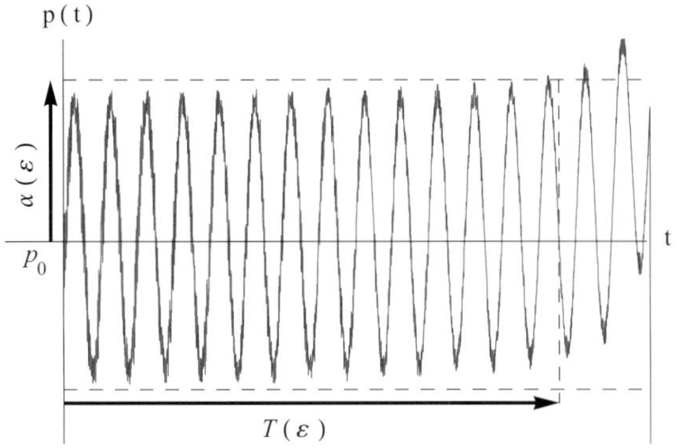


**Figure 4.1** Schematic view of the stability problem for nearly conserved quantities $p$. See text for the definitions of the symbols.

served quantities are still nearly conserved. A schematic view of the main stability problem addressed in this chapter is given in Figure 4.1. In a more formal way the stability statement is of the form:

$$\varepsilon < \varepsilon_0 : \|p(t) - p(0)\| < \alpha(\varepsilon) \quad \text{for any} \quad |t| \leq T(\varepsilon) \tag{4.2}$$

where $\varepsilon_0 \in \mathbb{R}$ is the limiting parameter, $\alpha(\varepsilon)$ is the maximal deviation of $p$ from the initial condition and $T(\varepsilon)$ is called the stability time. The functions $\alpha$, $T$ are themselves function of $\varepsilon$ with the special property that for $\varepsilon \to 0$ the deviation $\alpha(\varepsilon) \to 0$ as $T(\varepsilon) \to \infty$ and $\alpha(\varepsilon) \to \infty$ as $T(\varepsilon) \to 0$ for $\varepsilon \to \varepsilon_0$. We define the concept of stability in between: an initial condition $p_0 \in A$ is called *perpetually stable* in the parameter regime $\varepsilon < \varepsilon_0$ if $\alpha > 0$ is finite for $T \to \infty$. An initial condition $p_0 \in A$ is called *unstable* if $\alpha \to \infty$ for finite times $T$. In between the concept of *eternal stability* and *instability* we have *marginal* or *weak stability* if there is a simple relationship between the deformation $\alpha$ and the stability time $T$ and *long term* or *strong stability* if $\alpha$ remains small for long times $T$. The latter concept is important for physical systems where the life time of the underlying dynamical system is shorter than the finite stability time $T$, in which case we call the initial condition *practically stable*. An important case of the latter is the concept of *exponential stability*: in this case the stability time $T$ becomes exponentially long for small $\alpha$.

## 4.2 Integrable Systems

Formally the solution of a system of differential equations of the form given by:

$$\frac{dx_i}{dt} = f_i(x) , \quad \text{with} \quad i = 1, \ldots, n \quad \text{and} \quad x = (x_1, \ldots, x_n)$$

is given in its implicit form [83]:

$$\int_{x(0)}^{x(t)} \frac{dx_i}{f_i(x)} = \int_0^t dt \tag{4.3}$$

If the integrals on the left-hand side of the preceding equality can be solved in terms of analytic functions and moreover, the resulting expressions can be inverted, the system is said to be integrable (see [90, Chapter 4]). As a standard example, one takes the 1-dimensional problem $f(x) = x$ such that the integral becomes $\ln(x(t)) - \ln(x(0)) = t$, which gives the solution $x(t) = x(0)e^t$. In general it is difficult to decide if a system of ordinary differential equations is integrable or not since the latter would require proving that there is no method to either perform the integration or to invert the resulting relations.

For Hamiltonian systems the concept of integrability can be stated in another, usually much simpler, way. Given a Hamiltonian system of the form $H(y, x)$ with

$$\dot{x}_i = \frac{\partial H}{\partial y_i}, \quad \dot{y}_i = -\frac{\partial H}{\partial x_i}, \quad \text{with} \quad i = 1, \dots, n$$

and $x = (x_1, \dots, x_n)$, $y = (y_1, \dots, y_n)$ the system is said to be integrable if it admits $n$ independent constants of motion $\Phi_1(y, x), \dots, \Phi_n(y, x)$ such that $\{\Phi_i, \Phi_j\} = 0$ for $i \neq j$. The finding of integrals of motion is usually much simpler than the finding of the explicit form of the solutions. Nevertheless, again there is no general method to prove or disprove the presence of additional integrals of motion if only fewer $m < n$ are known. A famous example is that of the search for the global third integral (see [91, 92] and [93–96]) of a three degrees of freedom Hamiltonian describing the motion of a star in a simplified galactic potential [97], which finally turned out not to exist. The results were shown by numerical simulations which admit chaotic trajectories (the presence of chaos is a good indicator to exclude the complete integrability of a dynamical system).

In the following discussion on the concept of integrability we will mainly deal with conservative dynamical systems (which can be derived from a potential) of the following kind:

1. The autonomous one dimensional case ($n = 1$) for which the Hamiltonian function itself is the integral of motion to the system which can usually be identified with the total energy of the system. In general, however, $H$ can be any function depending on the conjugated variables $(y, x)$.
   Standard examples of the present case are the Hamiltonian of the harmonic oscillator:

$$H(y, x) = \frac{y^2}{2} + \omega^2 \frac{x^2}{2} \tag{4.4}$$

   where $\omega$ defines the angular frequency of oscillation, and the Hamiltonian of the harmonic repulser

$$H(y, x) = \frac{y^2}{2} - \omega^2 \frac{x^2}{2} \tag{4.5}$$

both being time-independent and of dimension one. In both cases $H(y, x)$ is a first integral of motion, thus, every orbit is a subset of the level sets defined by the constant $E = H(y(0), x(0))$. In the case of (4.4) and for $E = 0$ the orbit degenerates to a point located at $y = x = 0$, which is also the only equilibrium of the system. For $E > 0$ the level curves are ellipses centered around the origin. They correspond to libration (oscillation) around the origin with a libration amplitude proportional to the semimajor axis of the ellipse. In case of the Hamiltonian (4.5) and $E = 0$ the level curves are two straight lines intersecting at the origin. The origin itself is an unstable equilibrium. The straight line starting in the first quadrant and the line of the third quadrant represent the orbits which are asymptotic to the origin for $t \to -\infty$ and going to $+\infty$ for $t \to \infty$. The straight line of the second quadrant and the one of the fourth quadrant represent the orbits starting from $+\infty$ at $t = -\infty$ and going to the origin for $t \to +\infty$. Usually the former is called the unstable the latter the stable manifold of the system. For $E < 0$ the orbits are along hyperbolae coming from $+\infty$ (at $t = -\infty$) getting close to without reaching the origin and being pushed back to $+\infty$ for $t \to +\infty$ while for $E > 0$ the orbits go through the origin. Typical phase portraits of the harmonic oscillator and repulser (for $\omega = 1$) are shown in Figure 4.2a and Figure 4.2b, respectively.

2. Hamiltonian systems of the form $H(p_1, \ldots, p_n)$ where $n$ is the dimension and $p = (p_1, \ldots, p_n)$, $q = (q_1, \ldots, q_n)$ are the generalized momenta and coordinates, respectively. The canonical equations reduce to the simple form:

$$\dot{p}_i = -\frac{\partial H}{\partial q_i} = 0\,, \quad \dot{q}_i = \frac{\partial H}{\partial p_i} = \omega_i(p_1, \ldots, p_n)$$

and therefore $p_i(t) = p_i(0)$ and $q_i(t) = \omega_i(p_1(0), \ldots, p_n(0))t$ with $i = 1, \ldots, n$. All momenta $p_i$ are therefore integrals of motion and the angles $q_i$ evolve linearly with time, $\omega_i$ being related to the frequency of the motion. Mathematically

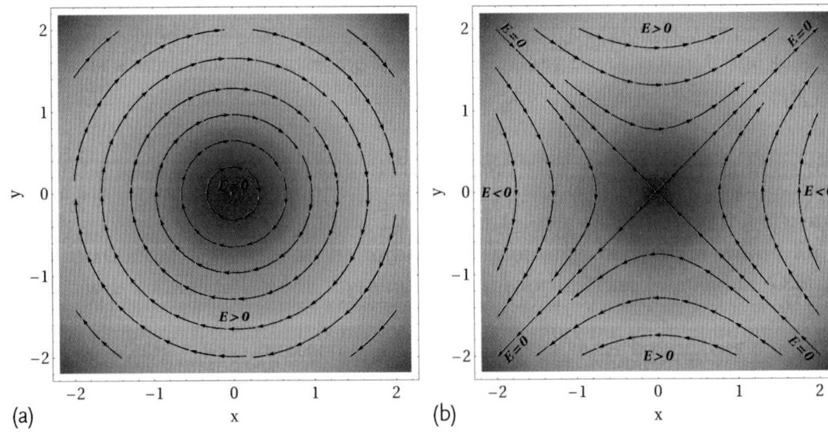


**Figure 4.2** Phase portrait of the harmonic oscillator (a) and the harmonic repulser (b), respectively. See text.

speaking, motion takes place on a torus $\mathbb{T}^n$, the time evolution on the different phase planes $(p_i, q_i)$ being independent from each other. If the motion just takes place on invariant level sets in $\mathbb{T}^n$ then the phase space is said to be foliated in invariant tori. Note that $\mathbb{T}$ is topologically equivalent to the motion on the circle while motion on $\mathbb{T}^2$ can be seen as motion on the surface of a doughnut (see Figure 4.3).

Remark 4.1

The above systems already include the concept of a constructive tool to show the integrability of a Hamiltonian system $H(y, x)$: if it is possible to find a change of coordinates from $(y_i, x_i) \to (p_i, q_i)$, with $i = 1, \dots, n$, such that in the new variables the Hamiltonian takes the form $H(p, q) = H(p)$ then the system in its original variables is also called integrable. The original system is said to be integrable, since $p_i(y, x)$ with $i = 1, \dots, n$ are integrals of motion. Important cases of this kind are found in celestial mechanics, that is, the two body problem in terms of Delaunay or the free rotator in terms of the so-called Serret-Andoyer variables.

The main parameters in systems of the form $H = H(p)$ are the frequencies of motion $(\omega_1, \dots, \omega_n)$ which themselves depend on the initial conditions $p(0) = (p_1(0), \dots, p_n(0))$. Assume that it holds a relationship of the form.

$$k \cdot \omega = \sum_{j=1}^{n} k_j \omega_j = 0 \tag{4.6}$$

where $k = (k_1, \dots, k_n) \in \mathbb{Z}^n$ and $\omega = (\omega_1, \dots, \omega_n) \in \mathbb{R}^n$.

If the condition (4.6) has only a unique integer solution $k = (0, \dots, 0)$ then the motion will densely cover the torus. This is true if the frequencies are completely nonresonant, that is, if no meaningful integer ratio between the frequencies can be identified. The motion is called quasi-periodic[1)], and after a suitable time the complete torus will be densely covered by the motion. On the contrary, if it is possible to find an integer vector $k = (k_1, \dots, k_n) \in \mathbb{Z}^n$ such that (4.6) is true, then for a nonzero integer vector $k$, the motion is called completely resonant. In that case the motion will fall back to its initial state after a finite time $T$, called the period

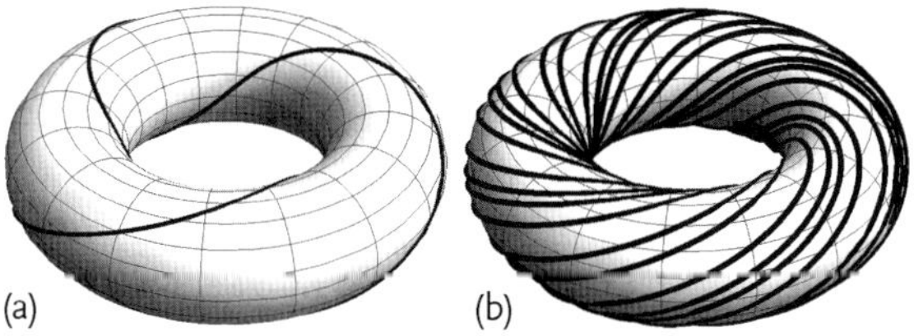

**Figure 4.3** The case of (3 : 2) resonant motion (a) and the case of quasi-periodic motion (b) for $n = 2$.

1) In former times also called conditionally periodic, see e.g. [98].

of motion of the system. To be more precise it is possible to express $n - 1$ angles in terms of a new angle such that the Hamiltonian is periodic in the new angle with period $T$. The intermediate case where (4.6) is fulfilled for $(k_1, \ldots, k_m)$ with $0 < m < n$ means that the motion is fully resonant on some subsets, and quasi-periodic on the remaining ones. Again it is possible to introduce $n - m$ angles to express the resonant $m$ angles instead of using the full set of $n$ angles. For this reason the projection of the motion on the $(n - m)$-dimensional torus is dense, while the projection of the remaining $m$-dimensional torus is resonant. The system is usually called resonant of multiplicity $m$. One may define the quantity

$$\min_{1 \leq j \leq m} |k^j| \equiv \left|k_1^j\right| + \cdots + \left|k_m^j\right|$$

which is called the order of the resonance. We show for $n = 2$ the case of a periodic $3 : 2$ resonant motion in Figure 4.3a and the case of quasi-periodic motion in Figure 4.3b.

The main stability theorems strongly rely on the concept of further resonance or nonresonance conditions. For this reason it is important to understand how the motion close to resonance behaves. We therefore investigate the simplest formulation of the resonance phenomena – the pendulum equation, which can be seen as another important case of integrable systems. The Hamiltonian of the standard mathematical pendulum takes the form:

$$H = \frac{p^2}{2} - A\cos(q) \tag{4.7}$$

where $(p, q)$ is the conjugated pair of momentum and coordinate, while $A$, the amplitude, is a free parameter of the system. From the canonical equations of motion:

$$\dot{q} = \frac{\partial H}{\partial p} = p$$

$$\dot{p} = -\frac{\partial H}{\partial q} = -A\sin(q)$$

and by the identification of the momentum with the velocity in the system $p = \dot{q}$ we find

$$\ddot{q} = \dot{p} = -A\sin(q)$$

The relation to the standard form of the equation of motion of the mathematical pendulum is given by identifying the angle $\theta$ with the generalized coordinate $q$ as well as by the relation $A = gl^{-1}$, where $g$ is the acceleration due to gravity and $l$ is the length of the pendulum. In this setting the above equation reads:

$$\ddot{\theta} + \frac{g}{l}\sin(\theta) = 0 \tag{4.8}$$

which is the Newtonian formulation of the problem. For small angles $|\theta| \ll 1$ and initial conditions $\theta_0 = \theta(0)$, $\dot{\theta}_0 = 0$ the solution takes the form:

$$\theta(t) = \theta_0 \cos\left(\sqrt{\frac{g}{l}}t\right), \quad |\theta_0| \ll 1$$

which recalls the motion of the simple harmonic oscillator with the period of oscillation:

$$T_0 = 2\pi\sqrt{\frac{l}{g}}, \qquad |\theta_0| \ll 1$$

which is, at least in the small angle approximation, independent of the initial amplitude $\theta_0$. For arbitrary amplitudes it is possible to integrate and invert (4.8) as done in (4.3) to find:

$$\frac{dt}{d\theta} = \frac{1}{\sqrt{2}}\sqrt{\frac{l}{g}}\frac{1}{\sqrt{\cos(\theta) - \cos(\theta_0)}}$$

which gives:

$$T = 4\sqrt{\frac{l}{g}}\frac{1}{\sqrt{2}}\int_0^{\theta_0}\frac{1}{\sqrt{\cos(\theta) - \cos(\theta_0)}}d\theta$$

The implicit solution can be written in terms of elliptic functions of the first kind. The value $T$ can be computed using the series representation:

$$T = 2\pi\sqrt{\frac{l}{g}}\sum_{n=0}^{\infty}\left[\frac{(2n)!}{(2^n n!)^2}\right]^2 \sin^{2n}\left(\frac{\theta_0}{2}\right)$$

As one can see the period also depends on the initial amplitude $\theta_0$. Note that dynamical systems where the period $T$ does not depend on the initial conditions are sometimes called isochronous systems while systems, where the period depends on the initial conditions are called nonisochronous. In terms of Hamiltonian dynamics, the associated frequency $\omega$ does not depend on $p$ in the former, while it does in the latter.

A typical phase portrait of the pendulum for $A = 1$ is shown in Figure 4.4. For $E = -1$ the orbit is the stable (elliptic) equilibrium located at $p = q = 0$. The surrounding phase space structure around recalls the harmonic oscillator. For $-1 < E < 1$ the orbit lies in the librational regime of motion. For $E = 1$ the orbit starts at the unstable equilibrium $q = \pm\pi$ with stable and unstable manifolds emanating from it. Close to the saddle point the motion looks like the motion of the harmonic repulser, with the difference that the unstable and stable manifolds coincide further away from it. Thus for $t \to \pm\infty$ the orbits starting close to the manifolds are asymptotic to the unstable equilibrium point. The curves which represent the manifolds are called the separatrices, since they separate the rotational and the librational regimes of motion, which is found for energies $E > 1$. We call the maximum distance in between the separatrix width. Above the separatrix the angle $q$ circulates from 0 to $2\pi$ with positive time derivative and the frequency of motion decreases monotonically with the distance from the resonance. Inside the librational regime of motion the angle $q$ oscillates around 0 with constant period.

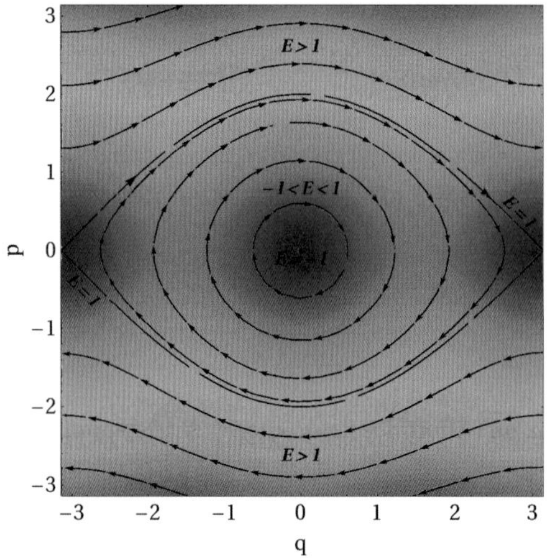


**Figure 4.4** Phase portrait of the mathematical pendulum – see text.

In the rotational regime of motion below the separatrix the $q$ circulates again but from $2\pi$ to 0 with negative time derivative. Again the period increases with the distance from the separatrix. To this end, the period on the separatrix tends to infinity $T \to \infty$, which means that it will take forever for an orbit starting on the unstable equilibrium to circulate back to itself. Moreover, the motion on the separatrix close to $q = 0$ is fast while it becomes exponential slowly close to $q = \pm\pi$.

Remark 4.2

In celestial mechanics it is often possible to approximate a real dynamical system close to a resonance by a Hamiltonian of the form [83]:

$$H_{\text{res}} = \alpha I + \beta \frac{I^2}{2} + \gamma \cos(\phi + \delta)$$

where $(I, \phi)$ are the action-angle variables of the system and the parameters $\alpha, \beta, \gamma, \delta$ are constants or slowly changing variables with time. Neglecting the influence of $\alpha, \delta$ we clearly see the correspondence to the pendulum Hamiltonian. The separatrix width, the distance of the unstable and stable manifolds along the line of intersection through the stable equilibrium, is given by $4(\gamma\beta^{-1})^{1/2}$; the frequency of libration is $(\beta\gamma)^{1/2}$.

## 4.3 Nearly Integrable Systems

In this section we investigate nearly (also called quasi) integrable Hamiltonian systems. Let us denote by $(p, q) \in \mathbb{R}^{2n}$ the generalized momenta and coordinates, respectively, and by $H(p, q; \varepsilon)$ the Hamiltonian function where $\varepsilon \in \mathbb{R}$ is a small parameter. Moreover, we assume that for $\varepsilon = 0$ the function $H(p, q; 0) = h_0(p)$ depends on the momenta only and that $H(p, q; \varepsilon)$ is analytic in $\varepsilon$. We thus can expand the analytic function around $\varepsilon = 0$ into Taylor series:

$$H = h_0(p) + \varepsilon h_1(p, q) + \frac{1}{2}\varepsilon^2 h_2(p, q) + \dots \tag{4.9}$$

where $h_k$ with $k = 0, 1, \dots$ collects terms of order $\varepsilon^k$. For convenience we also define $h_\varepsilon \equiv h_\varepsilon(p, q) = H(p, q; \varepsilon) - h_0(p)$ to write (4.9) in the shorter form:

$$H(p, q) = h_0(p) + h_\varepsilon(p, q) \tag{4.10}$$

Let us define by $D_r \subset \mathbb{R}^{2n}$ the domain of analyticity of $H \equiv H(p, q)$ where $D_r = \{(p, q) \in \mathbb{R}^{2n} : |p| < r, |q| < r\}$ and $r$ is the analyticity radius. In this setting we define the norm of a function $f(p, q)$ by:

$$\| f(p, q)\|_{r_0} = \sup_{|p|<r_0, |q|<r_0} |f(p, q)|$$

where $r_0 < r$. From the definition of $h_\varepsilon$ we have $\|h_\varepsilon(p, q)\|_{r_0} \leq C \cdot \varepsilon$, where $C$ is a constant. We furthermore assume, that $\varepsilon \ll \|h_0(p)\|_{r_0}$. For $\varepsilon = 0$ the canonical equations read:

$$\begin{aligned} \dot{q}_i &= \frac{\partial h_0(p)}{\partial p_i} \equiv \omega(p) \\ \dot{p}_i &= -\frac{\partial h_0(p)}{\partial q_i} = 0 \end{aligned} \tag{4.11}$$

which can be solved by direct integration:

$$\begin{aligned} q_i(t) &= \omega(p_0)t + q_{i0} \\ p_i(t) &= p_{i0} \end{aligned} \tag{4.12}$$

with $i = 1, \dots, n$. Here $p_0 = (p_{10}, \dots, p_{n0})$, $q_0 = (q_{10}, \dots, q_{n0})$, in short $(p_0, q_0) = (p(0), q(0))$ are the initial conditions at given time $t = 0$. We call $p_{i0}$ an integral of motion and $\omega(p_0)$ the frequency of the solution. However, for $\varepsilon \neq 0$ the system of equations (4.11) is of the form:

$$\begin{aligned} \dot{q}_i &= \omega(p) + \frac{\partial h_\varepsilon(p, q)}{\partial p_i} \\ \dot{p}_i &= -\frac{\partial h_\varepsilon(p, q)}{\partial q_i}, \quad \text{with} \quad i = 1, \dots, n \end{aligned} \tag{4.13}$$

In general there are no simple methods to construct the general solution of the preceding system but it is possible to bound the deviation of $p$ for given $p(0)$ up to a stability time $T$ and thus provide a stability theorem of the form (4.2). Different techniques exist in the literature to tackle the problem mainly based on some kind of normal-form theory.

## 4.4 Resonance Dynamics

There is a strong relationship between the concept of stability and the interaction of resonances in dynamical systems. It is therefore crucial to understand how different resonances interact in phase space to understand theorems of the form (4.2). The basic model that describes a resonance is that of the simple pendulum. In fact, many dynamical models can be reduced to a similar form like in (4.7). Thus motion close to resonance lies either inside the separatrix, in the librational regime of motion, or outside in the rotational regime. For the further discussion we will always denote motion as being resonant if it lies inside the separatrix and nonresonant otherwise. From the discussion of the standard integrable systems we already know that the separatrix width is proportional to the square root of the coefficient of the resonant term. In this section we would like to relate these results to arbitrary Hamiltonian functions of the form (4.10).

Since we can assume that (4.10) is analytic in the domain of interest we may expand the lowest-order term $h_1(p, q)$ into the Fourier series of the form:

$$h_1(p, q) = \sum_{k \in \mathbb{Z}^n} \hat{h}_{1,k}(p) e^{ik \cdot q} \tag{4.14}$$

where $\hat{h}_{1,k} = \hat{h}_{1,k}(p)$ are analytic functions in $p = (p_1, \ldots, p_n)$ and $q = (q_1, \ldots, q_n)$ as well as $k = (k_1, \ldots, k_n)$. For fixed $p = p_0$ the unperturbed frequency $\omega_0 = \omega_0(p_0)$ becomes:

$$\omega_0 = \frac{\partial h_0}{\partial p}(p_0)$$

which is a fixed vector $\omega_0 = (\omega_{10}, \ldots, \omega_{n0})$. Let us identify with $k_* = (k_{1*}, \ldots, k_{n*})$ the integer combination for which the resonant condition (4.6) is fulfilled. In the generic case all harmonics far from resonance can be transformed into higher orders, but the resonant one remains. The existence of such a transformation essentially tells us that their influence on the dynamics is present at higher orders in $\varepsilon$, while the resonant modes are of order $\varepsilon$. To a first-order approximation we may neglect the other Fourier harmonics and get as an approximation of (4.10), close to the initial conditions $p_0$, the much simpler Hamiltonian:

$$h_0(P) + \varepsilon \left[ c(p_0) e^{-iQ} + \bar{c}(p_0) e^{iQ} \right]$$

where we introduced $Q = k_* \cdot q$, which is called the resonant angle and $P = p - p_0$, which is the small deviation from the exact resonance position. Here the constant $c(p_0)$ (as well as its complex conjugate $\bar{c}(p_0)$) is related to the coefficient in the Fourier series expansion by $c(p_0) = \hat{h}_{1,k_*}(p_0)$, which we evaluate at the exact resonance condition, located at $p_0$. If we moreover expand $h_0(P)$ around $P = 0$ (which is the location of the center of the resonance) we get up to second order the approximate expression:

$$a(p_0)P + b(p_0)\frac{P^2}{2} + \varepsilon[c(p_0)e^{-iQ} + \bar{c}(p_0)e^{iQ}]$$

In addition, we may also assume (without loss of generality) that the dominant term of the Fourier series expansion takes the form $c(p_0)\cos(Q + Q_0)$ to get the Hamiltonian:

$$a(p_0)P + b(p_0)\frac{P^2}{2} + \varepsilon c(p_0)\cos(Q + Q_0) \tag{4.15}$$

which is very similar to the standard pendulum Hamiltonian as given in (4.7)[2]. A schematic view of the phase space structure is given in Figure 4.5. The approximation (4.15) close to the initial condition $p_0$ therefore defines a pendulum-like motion with a stable equilibrium located at $P = 0$ and $Q - Q_0 = 0$ and an unstable equilibrium located at $P = 0$, $Q - Q_0 = \pm\pi$. The angle $Q$ circulates above the separatrix with positive derivative, while it does with negative derivative below the separatrix. The resonance width (along the line through the stable equilibrium) is given by $4\sqrt{\varepsilon c/b} \propto \sqrt{\varepsilon}$ the frequency of libration close to the resonance is given by $\sqrt{\varepsilon b c}$ which is again proportional to $\sqrt{\varepsilon}$ at the exact resonance and decreases while reaching the separatrix. We may locally define a stable and unstable manifold as well. With (4.15) we are therefore able to locally approximate the motion around any initial condition $p_0$ of a Hamiltonian system of the form (4.10). The question remains, in which domain centered around $p_0$ and up to which time $T$ the approximated motion follows the real solution, which can be seen as a perturbed problem of the form (4.15) by other harmonics.

Remark 4.3

We can already note from (4.15), that the resonance width from the exact resonance location depends on the actual value of $p_0$ as well as on the size of the perturbation $\varepsilon$, while the location of the resonance (in terms of $p$) is determined by the integrable approximation, only. In other words, $h_0$ already defines which resonances are close to each other (it gives and ordering inbetween different sets of integer vectors $k$ by means of (4.6)), while $h_\varepsilon$ defines the possible amplitudes of libration close to it. While $\varepsilon$ is a parameter to the system, $c(p_0)$ is due to the form of $h_\varepsilon$ and just depending on the actual choice of $p_0$.

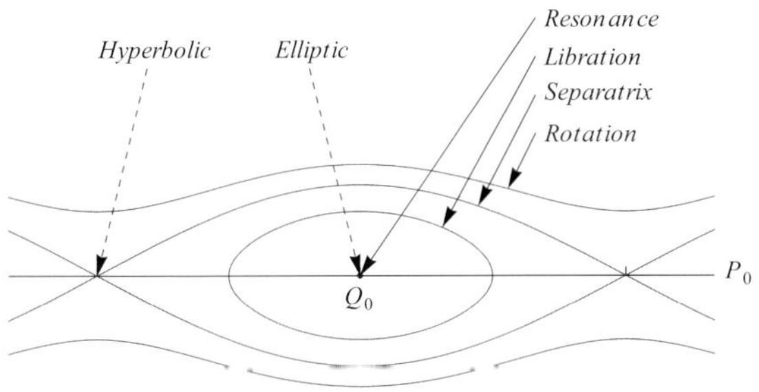


**Figure 4.5** Schematic view of the phase portrait of the pendulum-approximation close to a resonance located at $P_0$.

2) The form $\cos(Q + Q_0)$ can always be obtained since $e^{i\varphi} = \cos(\varphi) + i\sin(\varphi)$ and from the identity $\cos(\varphi + \varphi_0) = \cos(\varphi)\cos(\varphi_0) - \sin(\varphi)\sin(\varphi_0)$.

To see how long the approximation (4.15) is valid in comparison to the real system (4.10), we need to understand how different approximations of the form (4.15) interact with each other. For this reason let us repeat the series of approximations outlined previously but for a different initial condition, say $p_0'$. From (4.6) we will obtain another integer vector $k_*'$ that will define an exact resonance location in the action space $p$ and thus another approximation of the form (4.15) but for different local variables $(P', Q')$ and with different constants $a', b', c'$ being themselves functions of $\varepsilon$. To understand the interaction of these two resonances on the overall dynamics, let us investigate in more detail the following cases:

1. the new pendulum approximation is comparable in size to the former one (in terms of $a, a', b, b', c, c'$, that is, the separatrix width), but is located far away from the former initial condition $p_0$ such that the two resonances are separated by many rotational-invariant curves (see Figure 4.6a). If all the higher-order terms that we neglected are small enough, the motion will be trapped (at least for some time) either close to $p_0$ or close to $p_0'$. The two resonances are said to act more or less independently as long as the distance $|p_0 - p_0'|$ is larger as the width of the dominant resonance.
2. the second initial condition $p_0'$ lies close to $p_0$ but the approximation (4.15) shows that $c'/b' \ll c/b$ such that the resonant motions are still separated by some rotational tori and the time scales of libration are quite different (see Figure 4.6b). This always may happen. Depending on the actual form of $h_\varepsilon$ the Fourier expansion (4.14) for $p_0'$ very close to $p_0$ may always give an additional integer vector $k_*'$ such that (4.6) is nearly fulfilled if one lets $k_j$ with $j = 1, \ldots, n$ just become large enough. In that case the convergence of the sum given in (4.14) defines the ratios $c/c'$, $b/b'$ and that is, the width of nearby resonances.
3. no rotational curves separate the two local resonant models based on the initial conditions $p_0$ and $p_0'$ (the two separatrices touch each other, see Figure 4.7a. Thus the motion starting close to the resonance located at $p_0$ may be trapped af-

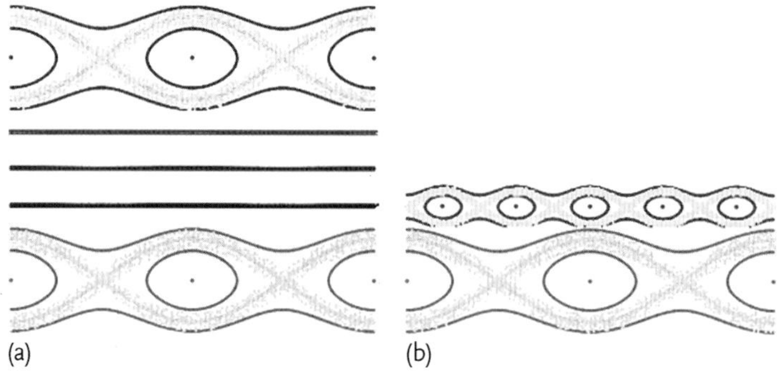

**Figure 4.6** Schematic view of resonance interaction between two nearby resonances: (a) The resonances are well separated by rotational invariant tori. (b) The resonances are close but the Fourier orders of resonances are different (see text case 1., 2.).

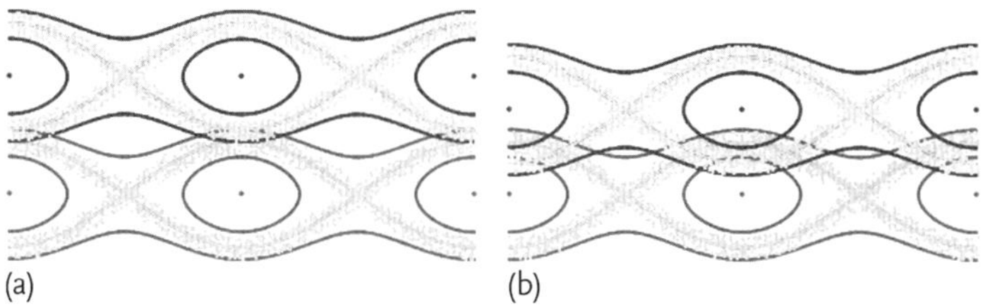

Figure 4.7 Schematic view of resonance interaction between two close-by resonances. (a) The two separatrices of the pendulums approximations touch and open diffusion channels between them. (b) The separatrices overlap, the pendulum approximations are no more valid (see text case 3., 4.).

ter some while by the resonance located at $p_0'$. The trapping may either happen due to the fact that the motion comes too close to one separatrix of the nearby resonance is then pushed into the other due to the higher-order terms (which we neglected in (4.15)) or due to the fact that the constants $a, b, c$ or $a', b', c'$ vary with time (due to the nonlinear effects) in such a way that the separatrices overlap for some time and the motion follows the librational curves of the other resonance.

4. last but not least we assume that the two initial conditions $p_0, p_0'$ are close enough such that the two respective separatrices overlap from the beginning (see Figure 4.7b). If this is the case then the approximation in terms of (4.15) is no longer valid and one has to include additional degrees of freedom to investigate the motion, depending on the specific problem.
5. To this end we show a hybrid case, in between case 1. and case 3., where the separatrix overlapping of different resonances connects two main resonances of similar order (see Figure 4.8).

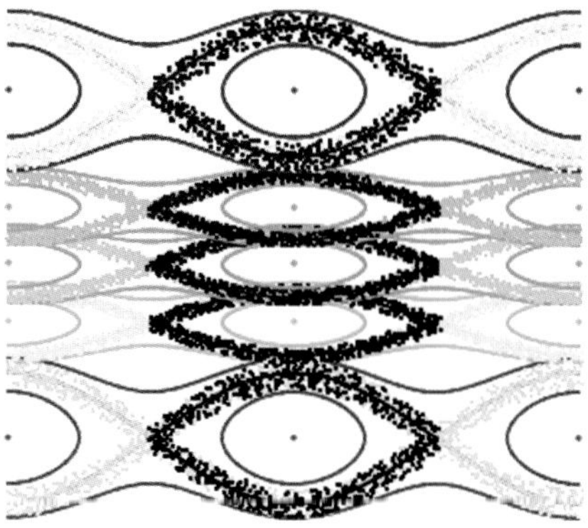

Figure 4.8 Schematic view of resonance interactions. The picture shows the case where higher-order resonances connect the two main resonances which may lead to instability of the motion on long time scales.

Remark 4.4

All cases may give rise to chaotic motion. The approximation (4.15) is perturbed by higher-order terms, consisting of higher-order harmonics also. The effect of the perturbation destroys the exact location of the separatrices – local chaos is present. In case 1. the higher-order harmonics may connect the two main resonances, opening a diffusion channel, which is a consequence of the case 2. The effect becomes stronger in situations where resonances of equal order start interacting as it is true in case 3., 4. Usually it is not possible to predict the time or the amount of events when the change from one resonance to another happens. One can assume the jumps to be of the kind of a random walk [99]. This sort of unpredictability can be seen as the origin of generic instability in dynamical systems.

The approximations used are of course an oversimplification of the original problem. This is true, since the Hamiltonian (4.15) is indeed integrable, while in general the original system (4.10) is not. As it was already pointed out, the nonintegrability is somehow hidden in the higher-order Fourier harmonics as well as the various Taylor series expansions which were done to low order only.

The idea to consider integrable single resonance models to analyze the different regimes of phase space can be used to define a simple but effective criterion to decide if there is chaos or not. The idea dates back to [100–103] to estimate the threshold value $\varepsilon_{\text{th}}$ at which global chaos is present. The method is therefore called Chirikov-criterion and was used extensively, for example [104]. However, the numerical simulation shows that chaos appears much before the critical values predicted by the above schematic view of the of phase space geometry. This is due to the fact that to arrive at (4.15), one has to neglect all the higher-order harmonics as well as the deformations and mutual interactions in between nearby resonances. The former can be improved by the use of normal forms and perturbation theory and taking into account the higher-order harmonics, the latter needs a much deeper understanding of the processes involved when resonances are coupled. Very sophisticated theories exist, their content is beyond the scope of this book. See [99] or [105] for further information.

We conclude this section by the investigation of a simple double resonance model with the following structure:

$$\begin{aligned} H(p_1, p_2, p_3, q_1, q_2, q_3) &= \frac{p_1{}^2}{2} + \frac{p_2{}^2}{2} + p_3 \\ &+ \varepsilon \left[ L\cos(l_1 q_1 + l_2 q_2 + l_3 q_3) + M\cos(m_1 q_1 + m_2 q_2 + m_3 q_3) \right] \end{aligned} \tag{4.16}$$

where $L, M \in \mathbb{R}$ and $l_i, m_i \in \mathbb{Z}$ with $i = 1, \ldots, 3$. The condition for the exact resonance location in the space $(p_1, p_2, p_3)$ turns out to be of the simple form:

$$p_1 k_1 + p_2 k_2 + k_3 = 0$$

where $k_1, k_2, k_3 \in \mathbb{Z}$. We look for two independent solutions of the preceding equation of the same Fourier order. We set $L = M = 1$ and identify $(l_1, l_2, l_3) =$

$(-1, 0, -1)$ as well as $(m_1, m_2, m_3) = (-1, 0, 1)$ both being located at $p_1 = -1$ and $p_2 = 1$, respectively. With this we also have $p_2 =$ const and thus investigate (4.16), where we choose the initial conditions to cover the regions of interest, in the plane $(q_1, p_1)$ only. The phase portrait which reminds of 1. is shown in Figure 4.9, obtained for $\varepsilon = 0.01$. We provide magnifications of the regimes of motion close two the two main resonances in Figure 4.9b,c, respectively. If we increase the perturbation parameter to $\varepsilon = 0.11$ motion close to the separatrices already admit strong but still locally confined chaotic motions (see Figure 4.10 and compare with case 3. and Figure 4.7). To this end increasing the parameter to $\varepsilon = 0.2$, then large chaotic layers are visible, as shown in Figure 4.11. The local pendulum approximations (4.15) are no longer valid, and chaotic layers surround the former resonance location.

The mathematical theory behind the interacting of resonances, is the one of heteroclinic intersections, which are the intersections of the unstable manifold of one resonance with the stable manifold of the other resonance (different from homoclinic intersections, which are the crossings of unstable and stable manifolds of one single resonance). A general solution to the problem of heteroclinic intersections does not yet exist, since it is difficult and nearly impossible to tackle the problem with standard perturbation theory (see for example Melnikov method [18]). Usually, one has to investigate the topic numerically, case by case. The experiment is similar to the one of the single resonance problem. One computes the stable and unstable manifolds numerically which turn out to be lobes of complex structure. At

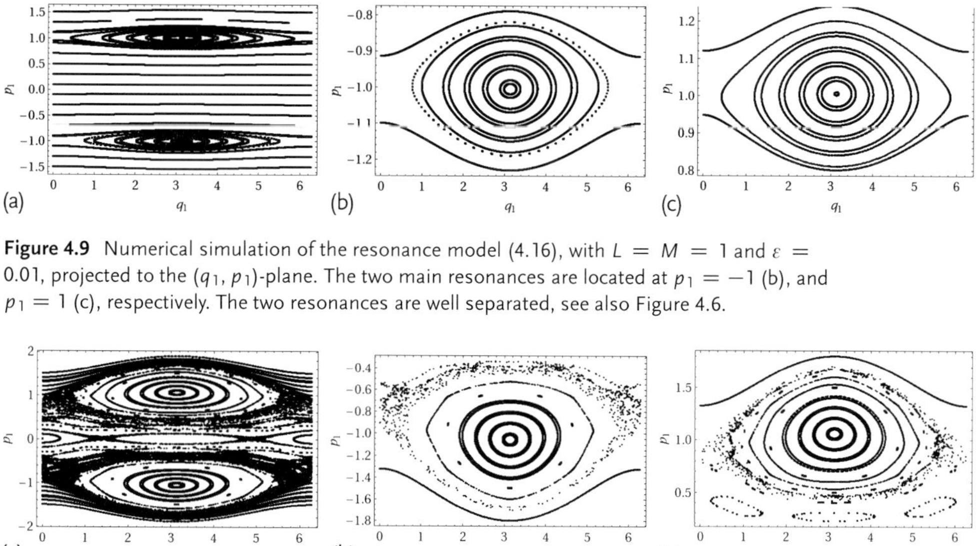


**Figure 4.9** Numerical simulation of the resonance model (4.16), with $L = M = 1$ and $\varepsilon = 0.01$, projected to the $(q_1, p_1)$-plane. The two main resonances are located at $p_1 = -1$ (b), and $p_1 = 1$ (c), respectively. The two resonances are well separated, see also Figure 4.6.

**Figure 4.10** Numerical simulation of the resonance model (4.16), with $L = M = 1$ and $\varepsilon = 0.11$, projected to the $(q_1, p_1)$-plane. The two main resonances are located at $p_1 = -1$ (b), and $p_1 = 1$ (c), respectively. The two resonances interact, regimes of locally confined chaos is present close to the separatrices; see also Figure 4.7.

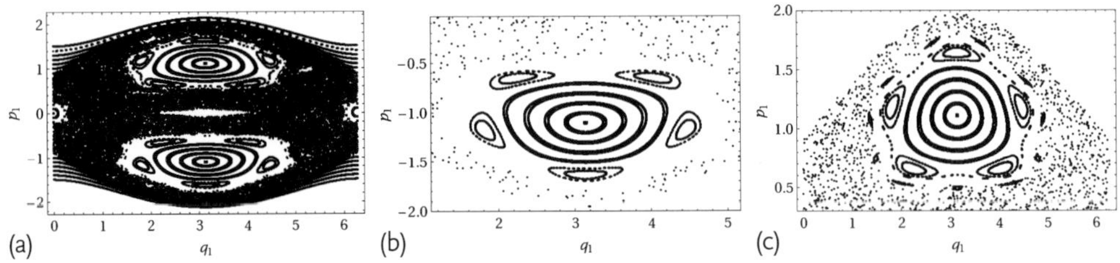


**Figure 4.11** Numerical simulation of the resonance model (4.16), with $L = M = 1$ and $\varepsilon = 0.2$, projected to the $(q_1, p_1)$-plane. The two main resonances are located at $p_1 = -1$ (b), and $p_1 = 1$ (c), respectively. The two separatrices of the two pendulum approximations overlap, chaos is not confined anymore to one of the resonances.

the threshold value of the perturbation $\varepsilon$ the two resonances are connected through transversal intersections of their lobes, which are called the heteroclinic points. In fact, as pointed out above, chaos is present already at this stage. The question remains, if chaotic trajectories are confined to lie in a local domain of the phase space or if large scale chaos is present already at this stage.

An answer already exists, when the number of degrees of freedom is small enough: in two dimensional systems the presence of invariant rotational tori between different resonant initial conditions already gives a full stability criterion of the form (4.2), that is, for $T = \infty$, even if the motion is chaotic. If one could prove the existence of a rotational invariant torus, separating the phase space region, the maximum distance an orbit starting close to $p_0$ could reach is given by the distance to the invariant torus. However, in higher dimensional systems, the rotational tori do not isolate the different resonances anymore. Although they may still serve as a barrier between the different kind of motion, diffusion channels may exist and allow to diffuse from one region of phase space to another on finite time scales in an arbitrary way.

## 4.5 KAM Theorem

The KAM theorems (after Kolmogorov [106], Arnold [107] and Moser [108]) are stability theorems valid for a Cantor set of initial conditions – the so-called Kolmogorov set. Resonances related to the small divisor problem play a central role in the proofs. Classical formulations of the theorem are concerned with nearly integrable systems of the form (4.10) which are real analytic in all its arguments. Let us assume that $(p, q)$ are action angle like variables of dimension $n$ which for $\varepsilon = 0$ admit the solution of the form (4.12). Every solution being a straight line can be identified with the motion on a torus, since the angle variables $q$ are defined mod $2\pi$. The functions $\omega = (\omega_1, \ldots, \omega_n)$ define the constant frequencies (or winding numbers) for given initial condition $p_0$. Solution of the form (4.12) are sometimes called Kronecker-tori. Being the perturbation $\varepsilon = 0$ the phase space is foliated into an $n$-parameter family of invariant tori, on which the Hamiltonian flow is constant.

Usually, solutions of the form (4.12) are connected to solutions in terms of other variables, which are stemming from the physical derivation of the problem. In most examples, found in nature, the coordinate transformation between the set of original variables and the corresponding action-angle variables is periodic in $q$. It is thus natural to expand this transformation in terms of a Fourier series into the generic form:

$$\sum_{k \in \mathbb{Z}^n} a_k(p_0) e^{k \cdot q_0 + t[k \cdot \omega(p_0)]}, \quad a_k \in \mathbb{R}^{2n}$$

where $\cdot$ is the dot product. As a consequence the solution (4.12) in terms of some "original" variables is quasi-periodic or periodic in time $t$. However, some fundamental properties of dynamical systems admitting quasi-periodic solutions are usually hidden when working in the original set of variables. Let us, for the moment, assume the transformation from original to action-angle variables is well defined and investigate the solution of the form (4.12). Different situations exist:

1. All $\omega_i$ are independent of the actions $p$ – the system is called isochronous, which means that changing the initial conditions does not change the frequencies involved. Thus, for an isochronous system $\partial \omega_i / \partial p = 0$, for all $i = 1, \ldots, n$. A typical example is the one of the coupled harmonic oscillator.
2. All $\omega_i$ depend on the actual value of the actions – the system is called non-isochronous (or anisochronous); we find the opposite situation as in 1., the first partial derivatives $\partial \omega_i / \partial p \neq 0$ for all $i = 1, \ldots, n$. This is the case, for example, for the coupled pendulum.
3. The system is isochronous with respect to some dimension, while it is anisochronous with respect to other dimensions, thus $\partial \omega_j / \partial p = 0$, while $\partial \omega_k / \partial p \neq 0$ for $j \neq k$ and $j, k = 1, \ldots, n$. For example one can easily construct a Hamiltonian that defines the coupling between a simple harmonic oscillator with a pendulum dynamics, the integrable parts being isochronous and anisochronous, respectively.

Note that the preceding concepts are defined with respect to the integrable approximation of the problem ($\varepsilon = 0$), while the "real" perturbed frequencies may change an isochronous system to be anisochronous (but not vice versa). Typical examples of celestial mechanics show that some of the actions are only introduced with the perturbation, for example the two body problem, which leads us to the following case:

4. Some of the $\omega_j$ are zero, which means that in addition to the fact $\partial \omega_j / \partial p = 0$, the integrable approximation $\partial h_0 / \partial p_j = 0$ for some $j \in (1, \ldots, n)$. The system is called degenerate in the actions $p_j$, which on the one hand shows that in the integrable limit the $p_j$ reduce to parameters of the system, while they are

much more difficult to handle if one wants to implement standard techniques of perturbation theory.

For the following discussion we will therefore assume that the integrable part of the Hamiltonian function is nondegenerate, that is, we require that

$$\det\left(\frac{\partial^2 h_0(p)}{\partial p^2}\right) \neq 0$$

With this we guarantee that the frequency of the motion $\omega(p)$ varies with the momenta $p$ as $p$ varies by a small amount, that is,

$$\omega(p_0) \neq \omega(p_0 \pm \Delta p)$$

where $\Delta p$ is a (small) variation of the initial condition.
In the next step we investigate different types of relations of the form (4.6):

5. the frequencies $\omega_i$ are nonresonant, or mathematically speaking rationally independent:

$$\omega \cdot k \neq 0\,, \quad k \in \mathbb{Z}^n$$

Thus, the motion densely fills the torus, labeled by the action $p_0$, and the corresponding flow is ergodic. If the system is isochronous, then the foregoing relation implies that the nonresonant regime of motion depends only on the fixed $\omega$. From number-theoretical arguments one can show that the system being nonresonant implies that $\omega_i$ with $i = 1, \ldots, n$ must be irrational numbers, thus the term rationally independent. On the contrary, if the relation is fulfilled for an anisochronous system and given $p_0$, changing $p_0$ does not imply that the new frequencies are still nonresonant. Since the irrational numbers are embedded in the set of real numbers it will therefore always be possible to find $p_0 \pm \Delta p$ close to $p_0$ such that above relation is no longer fulfilled.

6. the opposite case of 5. – the frequencies are rationally dependent, it exists (at least) one relation of the form:

$$\omega \cdot k = 0\,, \quad k \in \mathbb{Z}^n\,, \quad k \neq 0$$

It is possible to generalize the concepts to finite Fourier orders:

7. the frequencies are nonresonant up to some finite Fourier order:

$$\omega \cdot k \neq 0\,, \quad k \in \mathbb{Z}^n \quad \text{with} \quad |k| < K$$

where the Fourier order $K$ is defined as the absolute sum $K = \sum_{i=1}^{n} |k_i|$. This case will become important in the next section, where we will deal with the

Nekhoroshev theorem. Again the choice of $K$ depends on the choice of $\omega$ in the isochronous case and as well as on $p_0$ in the anisochronous case. Thus, fixing $\omega$ allows to define the nonresonant regime up to Fourier order $K$ and vice versa fixing $K$ defines the regime of the frequency space being nonresonant up to order $K$.

The presence of small divisors is strongly related to the presence of resonances in the original Hamiltonian system. Small divisors harm the convergence of the various series approximations, which are needed to prove the main stability theorems. What concerns the KAM theorem we will even need a stronger nonresonance condition as given in 7.:

8. The frequency of the motion is diophantine which is a stronger nonresonance condition of the form:

$$|\omega(p_0)\cdot k| \geq \gamma(|k|)^{-\mu}$$

for any integer vector $k \in \mathbb{Z}^n \backslash \{0\}$ and for some positive constants $\gamma > 0$ and $\mu > n-1$, where $|k| = |k_1| + \cdots + |k_n|$. From the preceding relation it is clear, that a set of numbers $\omega$ which is diophantine is nonresonant up to any Fourier order.

9. Another condition for diophantine frequencies is that the frequencies involved are Bruno numbers. If we denote by $r_n/s_n$ the convergence of an irrational number $\alpha$ a number is said to be a Bruno number if the condition:

$$\sum_{n=0}^{\infty} \frac{\log s_{n+1}}{s_n} < +\infty$$

holds. This condition is the optimal diophantine condition to linearize an analytic power series of the form

$$f(z) = e^{2\pi i \alpha} z + O(z^2)$$

If $\alpha$ satisfies the Bruno condition it is possible to find another power series $g(z) = z + O(z^2)$ such that:

$$g^{-1} \circ f \circ g(z) = e^{2\pi i \alpha} z$$

As we can see, the number-theoretical properties of the frequencies involved in our dynamical system are strongly related to the concept of perturbation theory. Let us exploit a little bit the notion of a Diophantine condition. We restrict ourselves to the case $n = 2$ and investigate the ratio $\rho = \omega_1/\omega_2$. For this reason we consider the error of the approximation defined by:

$$|s\rho - r| < C < \frac{1}{2}$$

where the bound $C < 1/2$ follows from a basic result of number theory. Let us furthermore define the profile[3] of a real number $\alpha$ as the function defined on $\mathbb{Z}$ which maps to any $s \in \mathbb{Z}$ the minimal error of approximation $|s\alpha - r|$ by scanning over $r \in \mathbb{Z}$. A plot of the function for the transcendental numbers $\pi$ and $e$ is shown in Figure 4.12; the same plots for the irrational numbers $\phi$ and $\sqrt{2}$ are shown in Figure 4.13. The respective plots[4] show the first $10^3$ values for $s$ versus the error on a double logarithmic scale. The dashed lines correspond to the functions $s^{-1}$ and $(s\sqrt{5})^{-1}$; they follow from different theorems of number theory, the theorem of Dirichlet and the theorem of Hurwitz for the approximation of irrational numbers, respectively. In addition we show the best approximating values found by the continuous fractions. As we can see, the approximation of irrational numbers by integers follows simple laws that can be used to prove the convergence of the various series expansions found in perturbation theory, while the approximation of real or rational numbers does not.

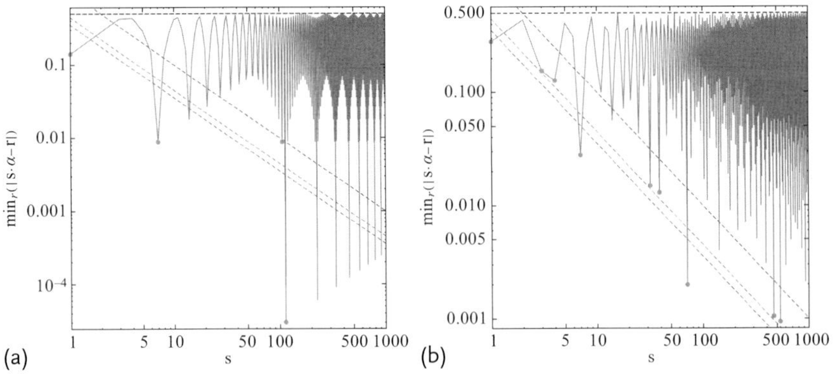


**Figure 4.12** Profile of transcendental numbers ((a) $\pi$ and (b) $e$).

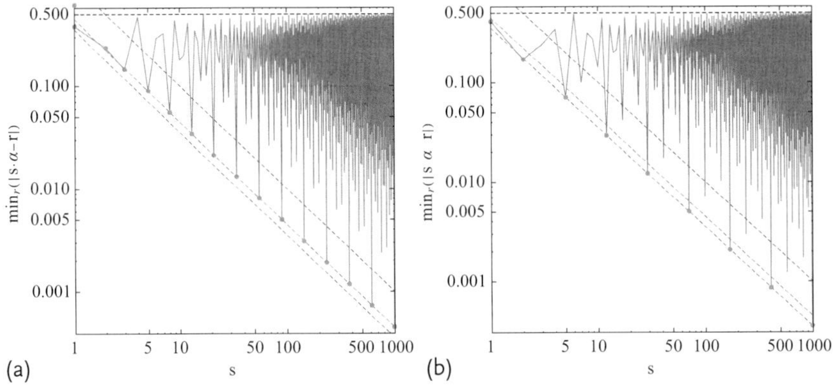


**Figure 4.13** Profile of irrational numbers ((a) golden ratio $\phi$ and (b) $\sqrt{2}$).

3) The term profile was introduced in 2012 by F. Vrabec (personal communication).
4) We do not show the profile for rational numbers, since they are trivial.

We are now able to state the theorem in its simplest form. KAM theorem ensures the persistence of an invariant surface for the perturbed system with the frequency equal to the unperturbed frequency $\omega(p_0)$. The proof of the theorem is rather technical but constructive in that sense that it also includes an explicit algorithm to obtain the invariant surface. The statement of the theorem (after Kolmogorov) is as follows:

**Theorem (Kolmogorov)**

Given a Hamilton system of the form (4.10) being nondegenerate and fixing the initial condition $p_0$ such that the unperturbed frequency $\omega(p_0)$ is diophantine, for the perturbation $\varepsilon$ sufficiently small it is possible to find an invariant torus on which the motion is quasi-periodic with the same constant vector frequency $\omega(p_0)$ as in the integrable approximation.

The experience shows that with increasing values of the perturbing parameter the invariant surface with frequency $\omega(p_0)$ becomes more and more distorted and displaced, until the parameter $\varepsilon$ reaches a critical value $\varepsilon_0$ at which the torus breaks up. The KAM theorem provides both an estimate on the distortion, which is about proportional to $\sqrt{\varepsilon}$, as well as a lower bound on the breakdown threshold $\varepsilon_0$. It has been shown with computer-assisted implementations of the KAM-algorithm that the theory can provide rigorous lower bounds on the system parameters which are consistent with the physical values [82].

The interested reader is referred to [98, 109–115] as well as [90, 116, 117] for original formulations of the stability theorem. See [118] and references within for a detailed list of publications on the topic of KAM theorem and celestial mechanics in modern times.

## 4.6 Nekhoroshev Theorem

The Nekhoroshev theorem [119, 120] is the basis on which stability theorems of the form (4.2) can be derived, that is, for higher dimensional systems, where the KAM-theorem no longer separates the phase space. The Nekhoroshev theorem does not provide a tool to prove eternal stability ($T \to \infty$) but it allows, as we will see, to bound the motion on exponentially long time scales. The proof ensures the stability for times long enough, since usually the life time of a system is much shorter than exponential $T$. The theorem in its original form is formulated for very generic dynamical systems: systems that are close to be integrable, depending on a small parameter, defined in terms of suitable (but arbitrary) action-angle variables, analytic and steep. Because of this the theorem may be applied to many kinds of conservative systems, which we see in nature: it was applied on systems on the quantum level or to bound the motion of various objects in our Solar system.

Let us point out, from the very beginning, the difference between the concept of exponential stability and the Nekhoroshev theorem. As we will see, later on, the

Nekhoroshev theorem requires a special geometry of the phase space and thus restricts the class of Hamiltonian functions for which the theorem can be applied. The main idea of covering the phase space by different "blocks" allows to find a stability theorem of the form (4.2) for an open set of initial conditions. To obtain this stability result one needs to understand the geometry of the phase space, that is, the one of the action space (see below). It is exactly here, where we require special properties of the Hamiltonian flow, like steepness or convexity. From that point of view, the Nekhoroshev theorem also requires that the frequencies depend on the actions – as we have seen – the integrable approximation must be anisochronous. In contrast, exponential stability estimates can be obtained also for isochronous dynamical systems that do not require a special geometry of the phase space structure. If it is possible to obtain a normal form of the original Hamiltonian which admits an exponential small remainder, a stability theorem of the form (4.2) can still be obtained. However, in the latter case, the main ingredient due to Nekhoroshev, the covering of the phase space, is missing. What is usually called Nekhorosohev theorem is therefore an ensemble of theorems for anisochronous systems. If the stability on exponentially long times can be shown also for isochronous systems, we will refer from now on to exponential estimates or isochronous versions of Nekhoroshev theorem instead.

The following concepts are needed in this discussion:

1. Let us assume that the function $f = f(p, q)$ is analytic and that $(p, q)$ are action-angle variables of dimension $n$. We may thus find $f$ in terms of a Fourier series expansion:

$$f = \sum_{k \in \mathbb{Z}^n} a_k(p) e^{ik \cdot q}$$

where $a_k$ are the amplitudes and $k \cdot q$ label the different Fourier modes. It is important to note that, due to the analyticity, the size of a single Fourier component $a_k$ decreases exponentially with $|k|$. Thus we are always able to split $f$ into two parts, say $f = f^{\leq K} + f^{>K}$, where:

$$f^{\leq K} = \sum_{k \in \mathbb{Z}^n, |k| \leq K} a_k(p) e^{ik \cdot q}$$

consists only of a finite set of harmonics and

$$f^{>K} = \sum_{k \in \mathbb{Z}^n, |k| > K} a_k(p) e^{ik \cdot q}$$

which is again an infinite sum but exponential small by the right choice of $K$, due to the analyticity. To obtain exponential stability it is therefore safe to neglect the part $f^{>K}$, which will only contribute to the dynamics after an exponentially long time. In case of (4.10) it means that we expand $h_\varepsilon$ into Fourier series but only keep $h_\varepsilon^{\leq K}$. It means that we essentially approximate (4.13) by the equations of motion:

$$\dot{q} = \omega(p) + \frac{\partial h_\varepsilon^{\leq K}(p,q)}{\partial p} + O(e^{1/\varepsilon})$$

$$\dot{p} = -\frac{\partial h_\varepsilon^{\leq K}(p,q)}{\partial q} + O(e^{1/\varepsilon})$$

where $O(e^{1/\varepsilon})$ is due to the exponential decay of the Fourier coefficients. As a result we have only to deal with a finite set of Fourier harmonics, that is, a limited set of possible resonances of the form (4.6) and thus with a smaller set of small divisors in the perturbation theory. Usually, $h_\varepsilon^{>K}$ is called the "ultraviolet cutoff". It can be bounded, by means of standard proofs of Fourier analysis, by $|h_\varepsilon^{>K}| < Ce^{-cK}$, where $c, C > 0$ are constants which are depending on the actual form of $f$.

2. In the next step we properly identify the domains in terms of $p$ where (4.6) is fulfilled, as well as the domain, in terms of $p$, where (4.6) cannot be fulfilled (up to the threshold $K$). We call the former the resonant, the latter the nonresonant domain modulo $K$. For this reason we need the notion of an $r$-dimensional sublattice $\Lambda \in \mathbb{Z}^n$ that is the set of integer vectors $(k_1, \ldots, k_r)$ with $|k_1| + \cdots + |k_r| < K$ and $r = 1, \ldots, n$. On its basis we define the resonant manifold $\mathbb{M}_\Lambda$ as:

$$\mathbb{M}_\Lambda = \{p \in D : k \cdot \omega(p) = 0 \quad \text{for} \quad k \in \Lambda\}$$

The parameter $r$ relates to the multiplicity of the resonance, and $\mathbb{M}_\Lambda$ defines the set of all possible resonances in the domain $D$ for a given threshold value $K$. As we can see from (4.15) the amplitude is, at least, of the order $O(\sqrt{\varepsilon})$ and any initial condition $p_0$ which is $\varepsilon$-close to the exact resonance location is trapped by it. For this reason we define the resonant regime $R_\Lambda$ based on the resonant manifold as a subset of $D$ in such a way to obtain:

$$R_\Lambda = \{p \in D : |\omega(p) \cdot k| < \delta_s \quad \text{for} \quad s = 1, \ldots, r\}$$

Moreover, we require $\delta_1 < \delta_2 < \cdots < \delta_n$, which will be of importance later on. To this end, we define the resonant zones $Z_\Lambda$ recursively in the following way. Denoting by $Z_\Lambda = R_\Lambda$ for $\Lambda = \mathbb{Z}^n$ and for $\dim \Lambda = n-1, \ldots, 1$ we get:

$$Z_\Lambda = R_\Lambda \setminus \bigcup_{\Lambda' : \dim \Lambda' = r+1} R_{\Lambda'}$$

and moreover

$$Z_0 = D \setminus \bigcup_{\Lambda' : \dim \Lambda' = 1} R_{\Lambda'}$$

which defines the nonresonant domain. A schematic view of the resonant manifolds and resonant zones for the case $d = 1$ is given in Figure 4.14, where the resonant zones and domains coincide. The resonant manifolds are shown in black

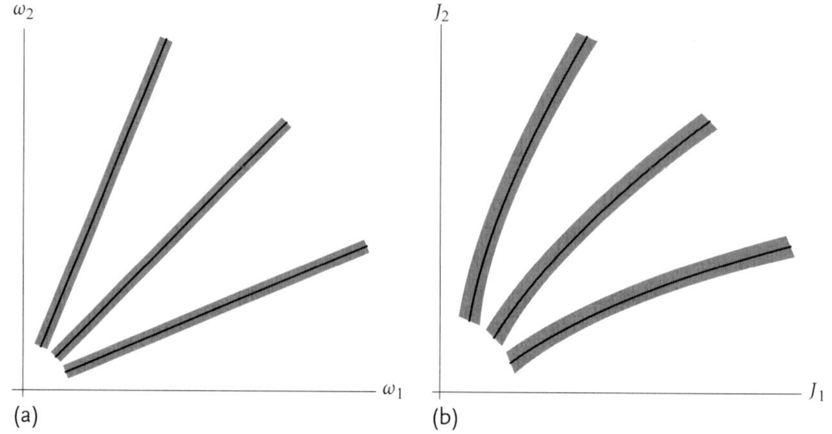


**Figure 4.14** Resonant zones (shown in gray) and resonant manifold (black) in the one dimensional case. See text for their definitions.

(which are lines), the resonant zones and domains coincide and are indicated by the dashed region. On the left the different domains are given in the frequency space $(\omega_1, \omega_2)$ which directly translates into the actions space $(p_1, p_2)$ by the relation $\omega = \omega(p)'$. A schematic view of the different resonant domains for the case $d = 2$ is shown in Figure 4.15. Here the resonant zones and the resonant domains do not coincide. The resonant regime is clearly visible. With this we are able to define the so-called geometry of resonances.

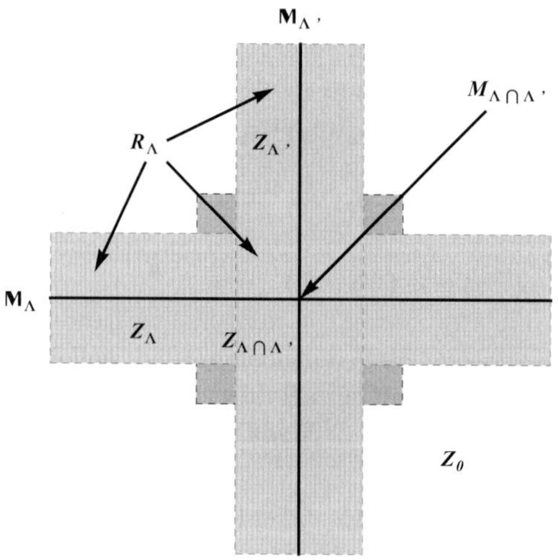


**Figure 4.15** Resonant zones (shown in gray) and resonant manifold (black), the two-dimensional case – see text for further definitions.

In the nonresonant zone we are able to find a canonical transformation to obtain ($o$ for order of normalization):

$$H^{(o)}(p,q) = h_0(p) + \varepsilon g^{(o)}(p) + \varepsilon^{o+1} f^{(o+1)}(p,q)$$

Thus we are able to construct a solution of the form (4.12) which is valid up to $\varepsilon^{o+1}$. On the contrary, in the resonant zone we are able to find a transformation such that:

$$H^{(o)}(p,q) = h_0(p) + \varepsilon g_\Lambda^{(o)}(p,q) + \varepsilon^{o+1} f^{(o+1)}(p,q)$$

where $g_\Lambda$ has only Fourier components in $\Lambda$:

$$g_\Lambda(p,q) = \sum_{k\in\Lambda} \hat{g}_k(p) e^{ik\cdot q}$$

In both cases, if $\varepsilon^{o+1} f^{(o+1)}$ is exponentially small, the corresponding action will stay close to their initial value for exponential times.

3. Let us assume that we are able to construct a normal form with an exponentially small remainder in each resonant domain $Z_\Lambda$:

   $$H_\Lambda(p,q) = h_0(p) + \varepsilon g_\Lambda(p,q) + O(e^{-1/\varepsilon})$$

   The corresponding canonical equations for the action-like variables are of the form:

   $$\dot{p} = \varepsilon \sum_{k\in\Lambda} ik\hat{g}_k(p) e^{ik\cdot q} + O(e^{-1/\varepsilon})$$

   so that $\dot{p}$ is almost parallel to the resonant lattice $\Lambda$.

Thus, the distance $\|p(t) - \Pi_\Lambda(p(0))\|$ stays small for an exponential time, while motion is fast on the complement motion. We call the manifold on which the fast motion takes place the plane of fast drift, which we label for future reference by $\Pi_\Lambda$.

4. The original version of the theorem relies on the concept of steepness, which is a restriction on the geometry in phase space of the integrable approximation. The steepest among analytic functions are convex functions, followed by quasi-convex functions or functions satisfying the three-jet nondegeneracy, which can be formalized in the following way: assume a function $f = f(y_1, \ldots, y_n)$ and by $'$ the derivative of it with respect to all the $y_j$ with $j = 1, \ldots, n$; the properties of steepness in the special cases, mentioned above reduce to (see [88]):

$$f'' u \cdot u = 0 \Rightarrow u = 0\,, \qquad \text{means } f \text{ is convex}$$

$$f' \cdot u = 0 \quad \text{or} \quad f'' u \cdot u = 0 \Rightarrow u = 0\,, \qquad \text{means } f \text{ is quasi–convex}$$

$$f' u = 0 \quad \text{or} \quad f'' u \cdot u = 0$$

$$\text{or} \quad \sum_{i,j,k} \frac{\partial^3 f}{\partial y_i \partial y_j \partial y_k} = 0 \Rightarrow u = 0\,, \qquad \text{means } f \text{ is three-jet}$$

A typical example of a function $f$ which is convex is $f(y_1, y_2) = (y_1^2 + y_2^2)/2$. Quasi-convexity is important for systems with time extended phase space, since they are of the generic form $f(y_1, y_2, y_3) = (y_1^2 + y_2^2)/2 + y_3$, where $y_3$ plays the role of the action conjugated to time. To this end a function $f(y_1, y_2, y_3) = (y_1^2 + y_2^2)/2 + y_3^3/6$ is neither convex nor quasi-convex but satisfies the three-jet condition. A typical example for a function $f$ which is neither convex, quasi-convex nor steep is given by $f(y_1, y_2) = (y_1^2 - y_2^2)/2$.

For the following discussion we will only require quasi-convexity. The main purpose is the fact that for quasi-convex systems the plane of fast drift lies transversally to the resonant manifold $\mathbb{M}_\Lambda$ (see Figure 4.16). If $\Pi_\Lambda$ would be tangent, for some vectors $\xi = \sum_i c_i k_i \in \Pi_\Lambda$ it would be orthogonal to $f'' k_s$ which are then orthogonal to $\mathbb{M}_\Lambda$. Thus $f''\xi \cdot \xi = 0$ and $\omega \cdot \xi = 0$ which is in conflict with the condition, that $f$ is quasi-convex. In addition, since we assume $h_0$ to be convex or quasi-convex, the unperturbed part of the Hamiltonian (4.10) has an extremum in $p_*$ which is the intersection of the resonant manifold with the plane of fast drift. To formalize the idea, for a vector $\xi$ parallel to $\Lambda$ one may expand $h_0$ into a Taylor series up to order two to get:

$$h_0(p_* + \xi) = h_0(p_*) + \omega(p_*) \cdot \xi + \frac{1}{2} h_0''(p_*)\xi \cdot \xi + O(\|\xi\|)^3$$

We find the extremum, since $\omega \cdot \xi = 0$ but not $1/2h_0''(p_*)$, which has a definite sign. To this end the energy conservation of $H(p, q)$ ensures that the motion "living" on the plane of fast drift oscillates at most by orders of $\varepsilon^{1/2}$, since the energy oscillates not more than orders of $\varepsilon$. A typical example for a convex structure on the plane of fast drift is shown in Figure 4.17a. Here the resonant manifold $\mathbb{M}_\Lambda$ lies normal to the plane of fast drift $\Pi_\Lambda$, the intersection defines to location of $p_*$. Due to convexity and the conservation of the energy motion starting close to $p_*$ stays close to $\mathbb{M}_\Lambda \cap \Pi_\Lambda$. The opposite situation is shown in Figure 4.17b, where the manifold $\mathbb{M}_\Lambda$ lies tangent to $\Pi_\Lambda$ and the hyperbolic structure opens channels of fast motion leaving $p_*$ on linear time scales.

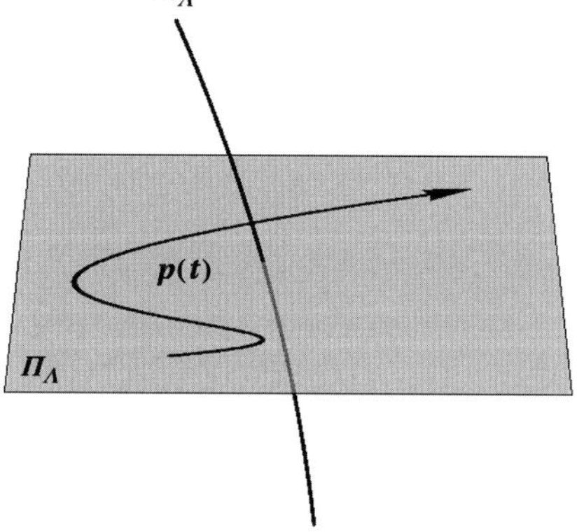


**Figure 4.16** The plane of fast drift lies transversally to the resonant manifold – see text.

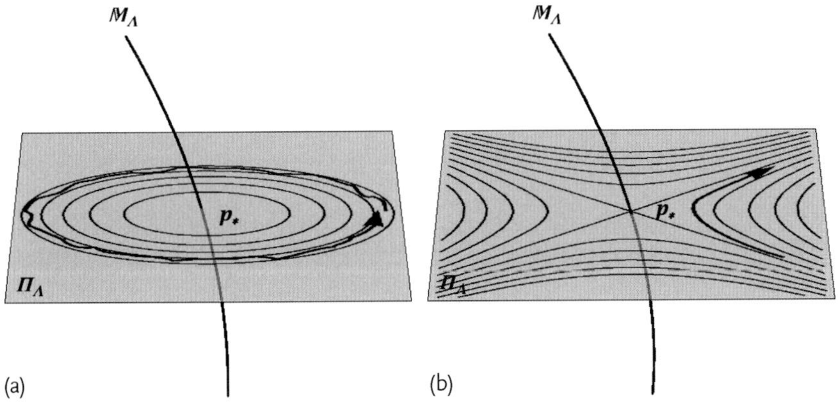


**Figure 4.17** Typical motion on the plane of fast drift. (a) convexity and energy conservation locally bounds the motion. (b) the lacking of convexity opens fast diffusion channels.

5. All our investigations are based on the assumption, that it is possible to replace the dynamics close to resonance by a simple pendulum model (see (4.15)). However, we still need an argument, that this is possible at least for time scales proportional to $e^{1/c}$, if we want to prove the confinement of the actions for an exponentially long time. For this reason one has to show, that one can exclude the presence of other resonances of the same multiplicity close by. It can be done by showing that within the same $\delta_r$ this is true, provided that $\varepsilon$ is small enough and $\delta_1 < \delta_2 < \cdots < \delta_n$. With this we can ensure that resonant regions of larger multiplicity are larger than resonant regions with smaller multiplicity. In that case, the motion of multiplicity $r$ will not enter resonant regions of multiplicity $r-1$, which again would allow the motion trapped close to one resonance to enter the resonant region of another resonance of same multiplicity.

In conclusion, if we are able to cover the action space by resonant zones in such a way to be able to neglect the other harmonics of the perturbation, besides the main resonant one, then we are able to locally confine the motion to the plane of fast drift. Due to convexity and energy conservation the motion in the plane of fast drift is bounded too. If the covering of the action space is such to avoid the interaction of different resonances, that is, to keep resonances of same multiplicity sufficiently far away from each other, then we are able to prove the conservation of the actions on exponentially long times on the basis of classical perturbation theory.

Different formulations of the more recent Nekhoroshev-like theorems exist, their form mainly depending on the assumptions made on the dynamical system of interest, as well as the method used to obtain the proof. In its simplest formulation the theorem can be stated in the following way.

**Theorem (Nekhoroshev)**

Let us consider a Hamiltonian of the form (4.10) analytic in a complex neighborhood of $D_r$ as specified above. If $h_0$ is convex, then there exist positive constants $\varepsilon_0, a, b, T_0, R_0$, such that for any $\varepsilon \leq \varepsilon_0$, and for any motion $(p(t), q(t))$ with $(p(0), q(0)) \in A \subset D_r$, $A$ an open set, it holds:

$$|p(t) - p(0)| \leq R_0 \varepsilon^a$$

for any time $t$ satisfying:

$$|t| \leq T_0 \exp\left(\frac{\varepsilon_0}{\varepsilon}\right)^b$$

The values of the constants $a, b$ depend on the convexity properties of $h_0$, the parameters $R_0, T_0, \varepsilon_0$ on the actual form of the perturbation.

Remark 4.5

The stability theorem does not exclude chaotic motions, that is, the variable $p(t)$ can possibly change in a chaotic way. Moreover the statement is true not only for one initial condition but rather it is for an open domain of initial conditions lying in $A$. The confinement of the actions strongly depends on the actual value of $R_0$ and the value of $a$, while the stability time depends on the inverse of the size of the perturbation as well as the parameter $T_0$ and the exponent $b$. Usually the quantities $a, b$ are fixed (by the form of the unperturbed Hamiltonian), while the parameters $R_0, T_0, \varepsilon_0$ change with the strength of the perturbation.

As one can see, the theorem is valid for all $\varepsilon \leq \varepsilon_0$ but does not necessarily give a long-term stability for arbitrary $\varepsilon$. This becomes clear in the limit, when the perturbation parameter tends to the critical one, $\varepsilon \to \varepsilon_0$ such that the stability time essentially becomes proportional to $T_0$ and the exponential results in $e^b$. Usually $R_0$ and $T_0$ are related such that for increasing $R_0$ we get better estimates in terms of $T_0$ while decreasing $R_0$ will also mean a decrease in $T_0$. It depends on the application to choose the parameters such to optimize the results.

Proofs on the initial formulation of the theorem are given in [121]. Extensions of this theorem were found in the case of isochronous systems with elliptic equilibria by [122] or isochronous symplectic mappings with elliptic fixed points [80]. The theorem was further generalized by [123, 124] in nonisochronous formulations to systems with elliptic equilibria. Other versions of the theorem, together with their proofs, are based on specific classes of dynamical systems, that is, done by [125] (for time dependent potentials), [126, 127]. From the mid 90s the theorem was extended to more general dynamical systems or connected to KAM-theorem in [128, 129]. The list is far from being complete see also [130] and references within for further information.

## 4.7 The Froeschlé–Guzzo–Lega Hamiltonian

For the purpose of demonstration we take the simple example of a 3-degree of freedom Hamiltonian of the form (see [131] and also [131]):

$$H(p, q) = \frac{p_1^2}{2} + \frac{p_2^2}{2} + p_3 + \frac{\varepsilon}{\cos(q_1) + \cos(q_2) + \cos(q_3) + 4} \quad (4.17)$$

which is usually called the Froeschlé–Guzzo–Lega Hamiltonian. The unperturbed part (for $\varepsilon = 0$) is quasi-convex, the perturbation is analytic and can be expanded in terms of multidimensional Fourier series into the form:

$$H(p, q) = \frac{p_1^2}{2} + \frac{p_2^2}{2} + p_3 + \varepsilon \sum_{k_1} \sum_{k_2} \sum_{k_3} c_{k_1,k_2,k_3} e^{i(k_1 q_1 + k_2 q_2 + k_3 q_3)} \quad (4.18)$$

where $c_{k_1,k_2,k_3} \in \mathbb{C}$ are complex numbers not depending on $p_i (i = 1, 2, 3)$. It therefore resembles the simplest form of a perturbation which consists of all possible Fourier harmonics from the beginning.

The structure of the resonance (or Arnold) web is defined by the relations:

$$p_1 k_1 + p_2 k_2 + k_3 = 0 \quad (4.19)$$

for $k_i \in \mathbb{Z}$ and $p_i \in \mathbb{R}$. Without knowing the solution of the problem we can already confine the motion by purely geometrical arguments as indicated previously:

The integrable part of the Hamiltonian (4.17) defines a parabola along the $p_3$-axis with its maximum for $\varepsilon = 0$ located at the value of the energy $E = H(p, q)$. For $\varepsilon \neq 0$ we find a maximum deviation from the energy surface of order $O(\varepsilon)$. By the conservation of the energy the motion of every initial condition is therefore bounded within the strip $E \pm \varepsilon$ along the surface of the parabola. The resonant planes, defined by (4.19), of dimension two intersect the parabola transversally and moreover intersect themselves, depending on the actual choice of $k_i$ with $i = 1, 2, 3$. We plot the web of resonances up to order three on the surface of the Hamiltonian in Figure 4.18a. We immediately see that the web of resonances is symmetric with respect to $p_1 = 0$ and $p_2 = 0$. Moreover, we are able to omit the information along the $p_3$-axis, since all resonant planes lie parallel to it. The resulting Arnold web given in the first positive quadrant is shown in Figure 4.18b. Each black line defines resonant values for (4.19) for $|k| < 8$, while gray lines define the resonant values in the space $(p_1, p_2)$ for $K = 9$. With this we neglect in (4.18) all resonances of order greater than $K$ which we thus define to be nonresonant in the following discussion (in real applications $K$ is much larger).

We sketch the only possible different regimes of motion of Figure 4.18 in Figure 4.19. For fixed $K \in \mathbb{Z}$ and $|k| < K$ motion starting in the nonresonant domain (I) is not affected by the presence of resonances of order less equal than $K$. The initial condition is nonresonant up to order $K$, we formally define it in terms of the frequency vector $\omega$ as:

$$(\mathrm{I}) = \left\{ p : \omega(p) \cdot k > \varepsilon^{1/2} \quad \text{for all } k \text{ with } |k| < K \right\}$$

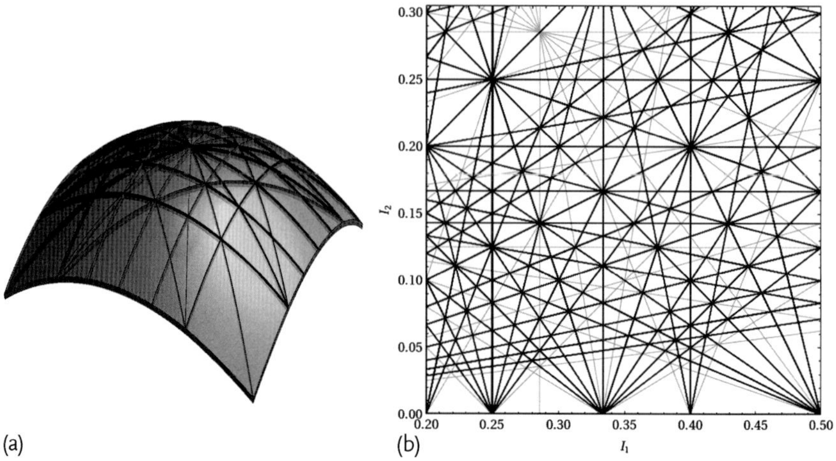


**Figure 4.18** The web of resonances on the energy level of the Froeschlé–Lega–Guzzo Hamiltonian (a), projection to the $(p_1, p_2)$-plane (b).

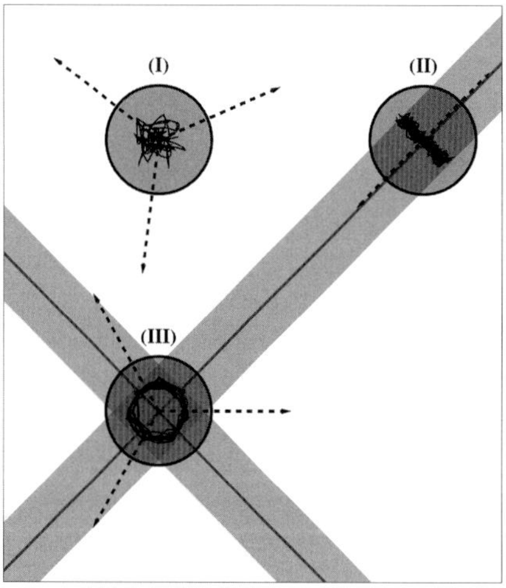


**Figure 4.19** Schematic view of the resonant web for the Froeschlé–Lega–Guzzo Hamiltonian. Nonresonant motion (I); single resonant motion (II); double resonant motion (III). Motion along the planes of fast drift is indicated by black curves. The dotted arrows locate the possible direction of diffusion.

In this domain the motion stays close to its initial condition for given stability time $T$ until it leaves the domain (indicated by the dotted arrows in the light gray ball around $p(0)$ in Figure 4.19). The domain as well the stability time can be constructed by a nonresonant normal form and its remainder.

In presence of a single resonance of order less equal than $K$ the domain becomes:

$$(\text{II}) = \{p : \omega(p) \cdot k_* = 0 \quad \text{for one unique } |k_*| \leq |k| < K\}$$

while for all other $k \neq k_*$ the domain is again said to be nonresonant up to order $K$. The fast dynamics is trapped within the resonance along the plane of fast drift motion (see Figure 4.19). By the requirement of steepness the fast plane drift lies transversal to the resonance. Along this line the actions vary with time, but if they would enter the nonresonant domain ($I$), then they would freeze again and follow the same stability theorem as in the nonresonant case. The important effect of the remainder in the single resonant domain therefore mainly acts along the resonant line (indicated by dashed arrows). It must again be bounded within a stability domain for given stability time $T$ based on a resonant normal form construction.

We are left to understand the motion starting in the dark shaded region (III) of Figure 4.19. We call it the double resonant domain, where two independent resonance conditions are fulfilled, that is

$$(\text{III}) = \{p : \omega(p) \cdot k_* = 0, \omega(p) \cdot k_\times = 0 \quad \text{for unique } |k_*|, |k_\times| \leq |k| < K\}$$

In contrast to the single resonance domains, it is not possible to locate a unique plane of fast drift motion. Fast motion takes place along arbitrary directions spanned by the intersection of the two resonant domains. In principle it is thus possible to enter both, the nonresonant and resonant domains after a short while. But if motion enters the single-resonant domain, then the fast drift motion would again be restricted to lie transversally to one of the resonant lines and, moreover, the motion would again enter, transversal toil, the nonresonant domains, where all the frequencies become fixed. The same is true for motion starting in the double resonant regimes which would exit the initial domain directly to enter the nonresonant domain. We are thus safe if the remainder of the "double resonant" normal form bounds the motion indicated along the dashed arrows for given stability time $T$.

The preceding argumentation can be generalized to define other domains, where more than 2 resonances intersect in a straightforward way. Each domain isolates the motion within a resonant-region of multiplicity $m$, where $m$ angles are reso-

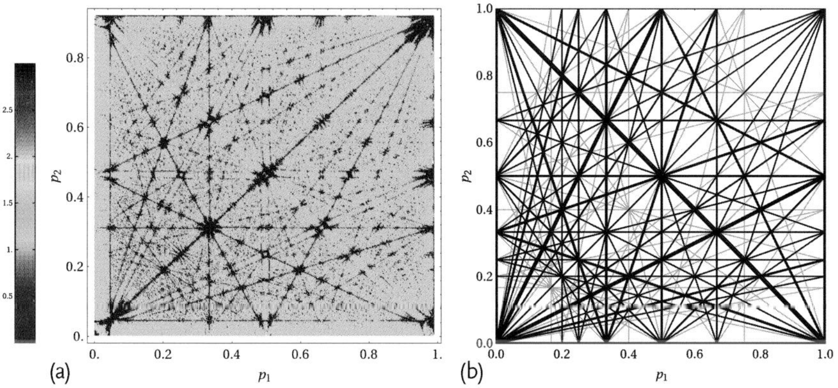


**Figure 4.20** The case $\varepsilon = 0.005$. (a) MEGNO-chaos indicator (blue: quasi-periodic, red: chaotic). (b) resonant web obtained from analytic theory.

nant and the $n - m$ remaining one are not. Any of these domains is far from all completely resonant manifolds, except a finite number. We thus can identify, for any set of integer vectors $k$ with $|k| < K$ a lattice $\Lambda$, spanned by $k$, the resonant regime $R_\Lambda$ and the resonant zone $Z_\Lambda$. The covering allows to approximate the different regimes by simple Hamiltonians which are well understood. Moreover, we also guarantee, that the local approximations are valid for exponentially long time.

We investigate the resonance-web structure for different values of the perturbation parameter as shown in Figure 4.20 ($\varepsilon = 0.005$) and Figure 4.21 ($\varepsilon = 0.01$). The left plots of the respective figures were produced by integrating the canonical equations of motion of (4.17) together with the variational equations for initial conditions $(p_1, p_2) \in (0, 1)$. The times series were analyzed using a chaos indicator, the MEGNO method [47] (see also Chapter 3). The color blue relates to quasi-periodic orbits, yellow labels regular and red labels chaotic regimes of motion. We clearly see that chaos is strongest close to lines which clearly remind of the ones found by the graphical method as shown in Figure 4.18 and based on condition (4.19). The lines indicate the location of the separatrices of the simple pendulum approximations valid close to the initial condition. The separatrix width can be estimated by simple arguments (from the size of the coefficient in the Fourier series expansion). From the comparison of Figure 4.20 with Figure 4.21 we also clearly see that increasing $\varepsilon$ the width of the resonant lines (which correspond to the separatrices of the pendulum approximations in (4.15)) increases and we have bigger chaotic domains of motion. A magnification of Figure 4.21 is shown in Figure 4.22. We still see quasi-periodic orbits (indicated by blue dots) as predicted from KAM-theory and the resonant domains can still be separated as required from Nekhoroshev theory. However, close to the crossings of resonances of same multiplicity chaos increases and allows for diffusion from one resonance to another (compare with Figures 4.9–4.11).

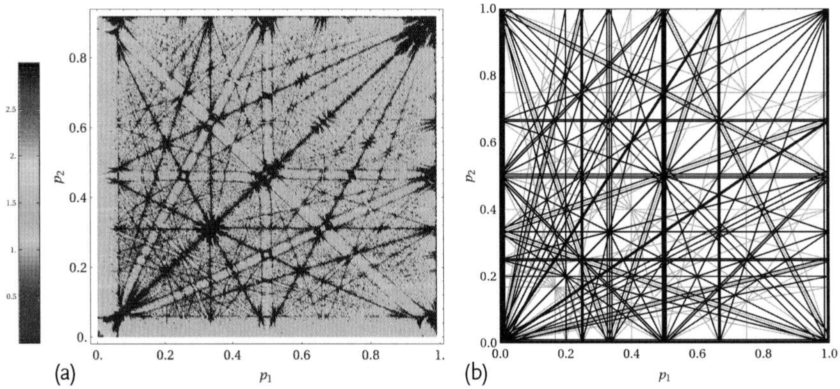


**Figure 4.21** The case $\varepsilon = 0.01$. (a) MEGNO-chaos indicator (blue: quasi-periodic, red: chaotic). (b) resonant web obtained from analytic theory.

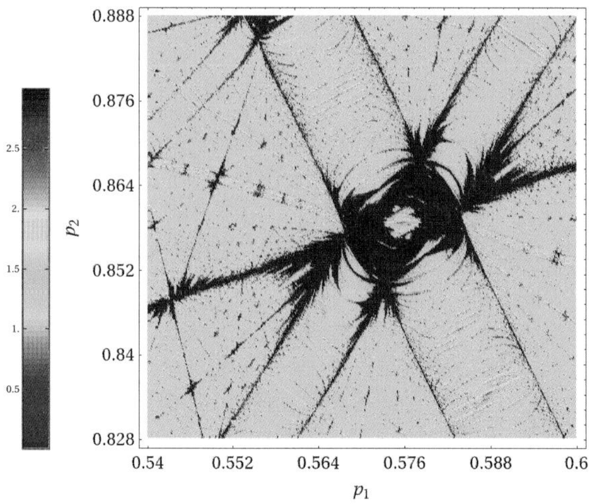


**Figure 4.22** Magnification of Figure 4.21. We clearly see the nonresonant, resonant and double resonant regimes of motion.

Numerical simulations of (4.17) were partly made on the local computing resources (Cluster URBM-SYSDYN) at the University of Namur (FUNDP, Belgium). The numerical graphs were produced by C. Hubaux using a symplectic integration scheme. For further information see [131, 132].

# 5
# The Two-Body Problem

Due to the fact that the planet Mars has a relatively large eccentric orbit the observations of this planet by the Danish astronomer Tycho Brahe (1546–1601) allowed Johannes Kepler (1571–1632) to find out, that the motion of this planet is not a circle. This fortunate circumstance led Johannes Kepler to formulate his first law, that the motion of the planets is elliptic with the Sun in one focus and his second law that the radius vector sweeps equal areas in equal times. Both were published in 1609 in his book 'Astronomia Nova' which he dedicated to the emperor Rudolf II in Prague.

Using the Lagrange equations of motion for a three dimensional space according to 3 degrees of freedom we can write

$$\frac{d}{dt}\frac{\partial L}{\partial \dot{x}_j} - \frac{\partial L}{\partial x_j} = 0\,, \quad j = 1, 2, 3 \tag{5.1}$$

with the Lagrangian (compare Chapter 2)

$$L = T(\dot{\boldsymbol{r}}) - V(\boldsymbol{r}) \tag{5.2}$$

and where $\boldsymbol{r} = (x_1, x_2, x_3)^{-\mathrm{T}}$ connects the central body $m_1$ with $m_2$ (see Figure 5.2), where the two masses are the Sun and a planet) and the velocity can be written as $\dot{\boldsymbol{r}} = (\dot{x}_1, \dot{x}_2, \dot{x}_3)^{-\mathrm{T}}$. In case of a pure gravitational Kepler motion $\mu$ is the reduced mass which can be written as

$$\frac{1}{\mu} = \frac{1}{m_1} + \frac{1}{m_2}$$

the potential function $U$[1] is independent of the velocity and the kinetic energy is simply

$$T = \tfrac{1}{2}\mu\dot{\boldsymbol{r}}^2$$

Because of the fact that the potential energy $V$ is dependent only on the distance $r$ spherical symmetry is provided and one of the angles is cyclic[2]. In addition it is

1) $U = -V$.
2) The rotation about its axes is without influence on the solution.

practical to describe the problem – now a 1-body problem with a central force function and only 2 degrees of freedom – with polar coordinates in the plane. It makes it now simpler to describe due to the fact that the total angular momentum vector is conserved (see Chapter 8).

$$g = \boldsymbol{r} \times \dot{\boldsymbol{r}} \tag{5.3}$$

In the new coordinates $\theta$ and $r = |\boldsymbol{r}|$ the Lagrangian setting $m = m_2$ reads

$$L = T - V = \tfrac{1}{2}m(\dot{r}^2 + r^2\dot{\theta}) - V(r) \tag{5.4}$$

and because of (5.4) the respective equations of motion are

$$\frac{d}{dt}(m\dot{r}) - mr\dot{\theta}^2 + \frac{\partial V}{\partial r} = 0 \tag{5.5}$$

$$\frac{d}{dt}(mr^2\dot{\theta}) = 0 \tag{5.6}$$

From the last equation we find that $mr^2\dot{\theta}$, the angular momentum $|\mathbf{g}| = c$ is constant; the second Kepler law (KII) follows immediately

$$r^2\dot{\theta} = c \tag{5.7}$$

Besides the constant angular momentum the energy integral $T - U = H$ exists in this problem which is a conservative system with a Hamiltonian independent of the time $t$:

$$\tfrac{1}{2}m\left(\dot{r}^2 + r^2\dot{\theta}\right) + V(r) = h \tag{5.8}$$

Thus in principle the problem is solved, but it is NOT possible to derive $r$ and $\theta$ in a simple form, which is the topic of this section.

## 5.1 From Newton to Kepler

The former considerations are valid for any central force function; now we discuss the case that $V \sim r^{-1}$, from which one derives the equation of motions (see [133]):

$$\frac{d^2\boldsymbol{r}}{dt^2} = -\frac{\partial V}{\partial \boldsymbol{r}} \tag{5.9}$$

Due to Newton (1642–1726)

> "The acceleration of an object as produced by a net force is directly proportional to the magnitude of the net force ($F$), in the same direction as the net force, and inversely proportional to the mass of the object"

one can write $a = F/m$, where

$$F = -k^2(m_1 + m_2)\frac{d^2 r}{dt^2} = -\kappa^2 a \quad (5.10)$$

with $k$ the gravitational constant[3)]

$$\ddot{r} = -\kappa^2 \frac{r}{r^3} \quad (5.11)$$

Now we use (5.11) and perform a vectorial multiplication with $r$ to get

$$r \times \ddot{r} = -\frac{\kappa^2}{r^3}(r \times r) = 0 \quad (5.12)$$

from which follows by integration that the angular momentum vector defined above ($g = r \times \dot{r}$) is a constant and is perpendicular to the plane of motion ($|g| = c$) Another constant of motion can be found from the equation of motion (5.11)

$$\ddot{r} \times g = -\frac{\kappa^2}{r^3}(r \times g) = -\frac{\kappa^2}{r^3}[r \times (r \times \dot{r})] \quad (5.13)$$

After some minor algebraic operation we find

$$\ddot{r} \times g - \kappa^2 \frac{d}{dt}\left(\frac{r}{r}\right) \quad (5.14)$$

and because of

$$\frac{d}{dt}(\dot{r} \times g) = (\ddot{r} \times g) \quad (5.15)$$

one finally obtains by integrating

$$\dot{r} \times g = \frac{\kappa^2}{r} r + f \quad (5.16)$$

The constant vector $f$ is the so-called *Laplace–Runge–Lenz vector*, which is perpendicular to g; $|f| = d =$ const. This is the vector connecting the Sun with the perihelion position of the planet (compare Figure 5.1). Note that $gf = 0$. A scalar multiplication with $r$ leads to

$$r \cdot (\dot{r} \times g) = \frac{\kappa^2}{r}(r \cdot r) + (r \cdot f) \quad (5.17)$$

Because of

$$r \cdot (\dot{r} \times g) = g \cdot (\dot{r} \times r) = g \cdot g = c^2 \quad (5.18)$$

and $r \cdot f = rd\cos(\phi - \phi_0)$; consequently

$$c^2 = r\left[\kappa^2 + d \cdot \cos(\phi - \phi_0)\right] \quad (5.19)$$

3) In astronomy, more precise in celestial mechanics, quite often $k^2$ is expressed in AU, Solar masses and days, from what one derives $k = 0.017\,202\,098\,95$; in the following we will use $\kappa^2 = k^2(m_1 + m_2)$; $m_1$ and $m_2$ are the two masses involved.

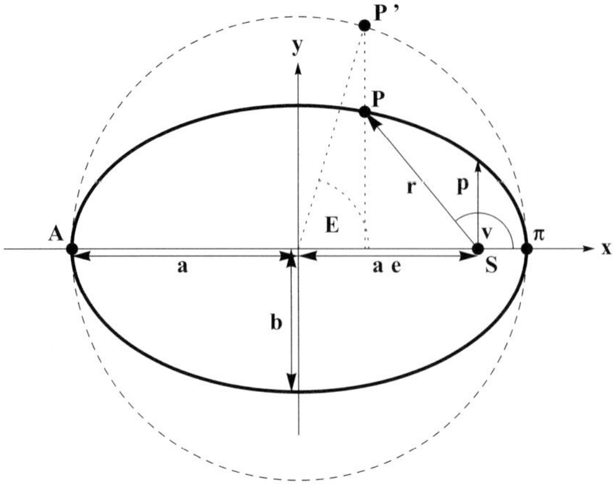


**Figure 5.1** Configuration of the two-body problem; $\pi$ is the perihel; $A$ is the aphel; $\overline{A\pi}$ is the line of apsides.

from which we can find an equation for $r$

$$r = \frac{c^2/\kappa^2}{1 + (d/\kappa^2)\cos(\phi - \phi_0)} \tag{5.20}$$

We thus have found the first and second law of Kepler (KI and KII); both were derived from Newton's law of gravitation and are valid for all problems with an inverse power law for the potential.

## 5.2 Unperturbed Kepler Motion

In the beginning of the chapter we said that Kepler found his first law empirically, whereas we derived it directly from Newton's law of gravitation. KI reads in the 'classical' form

$$\boxed{r = \frac{p}{1 + e\cos\nu}}$$

where $r$ is the distance between the planet and the central star. Furthermore

- $e$ is the eccentricity of the ellipse
- $\nu$ the true anomaly, the angle counted from the perihelion position
- $a$ is the large semimajor axis
- $b$ the semiminor axis
- $p$ the parameter of the ellipse[4] (Figure 5.1).

4) Called semi latus rectum; note that comparing KI with (5.20) $p = c^2/\kappa^2$ and the eccentricity $e = d/\kappa^2$.

The geometrical properties of the ellipse can be derived as follows:

$$a = \frac{p}{1-e^2}, \qquad b = \frac{p}{\sqrt{1-e^2}}$$

$$p = \frac{b^2}{a}, \qquad e = \sqrt{1-\frac{b^2}{a^2}}$$

$$\frac{b}{a} = \sqrt{1-e^2} \tag{5.21}$$

To derive KII we can use simple geometrical considerations: from Figure 5.2 one can see how the radius vector wipes over the area of the ellipse which can be computed by

$$\Delta A = \frac{1}{2} r(r+\Delta r)\sin\Delta v \sim \frac{1}{2} r^2 \Delta v$$

or – introducing $N$, the number of small triangles shown in this figure

$$\Delta A = \frac{a b \pi}{N}$$

from what follows for $N \to \infty$ and $n = 2\pi/P$, where $P$ is the orbital period

$$r^2 \Delta v \sim c\Delta t$$

respectively

$$\boxed{r^2 \frac{dv}{dt} = c}$$

which is the second law of Kepler (KII).

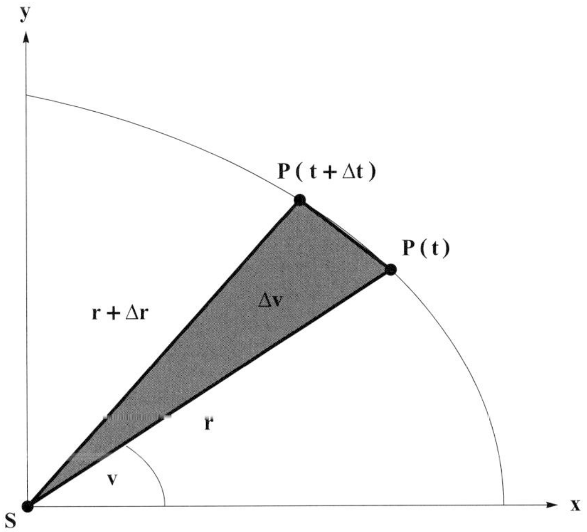


**Figure 5.2** Motion of a planet along the ellipse

In addition we can now derive the following relations for the constant angular momentum

$$c = \frac{2\pi ab}{P} = nab = \frac{nap}{\sqrt{1-e^2}} \tag{5.22}$$

A direct way to find the relation between the semimajor axis $a$ and the mean motion $n$ uses (5.20) and KI to express the constant $c$ as

$$c^2 = \kappa^2 p = \kappa^2 a(1-e^2) \tag{5.23}$$

and from (5.22)

$$c^2 = n^2 a^2 b^2 = n^2 a^4 (1-e^2) \tag{5.24}$$

From the two last equations it follows immediately the third Kepler law (KIII) in the following form

$$\boxed{n^2 a^3 = \frac{4\pi^2}{P^2} a^3 = \kappa^2}$$

which we can be express in words: 'the square of the orbital period $P$ is proportional to the cube of its semimajor axis $a$'.

## 5.3 Classification of Orbits: Ellipses, Hyperbolae and Parabolae

A very useful relation for the velocity (which we will denote as v) of a planet (asteroid, comet) can be found which serves to determine whether an orbit is elliptical, hyperbolic or parabolic. Differentiation of KI leads to

$$\dot{r} = \frac{pe \sin v}{(1 + e \cos v)^2} \dot{v}$$

and using KII we replace $\dot{v}$

$$\dot{r} = \frac{c}{p} e \sin v \tag{5.25}$$

from which we derive

$$\dot{r}^2 = \frac{c^2}{p^2} e^2 (1 - \cos^2 v) \tag{5.26}$$

Furthermore from KII it follows

$$r\dot{v} = \frac{c}{r} = \frac{c}{p}(1 + e \cos v) \tag{5.27}$$

and by squaring this expression

$$r^2\dot{v}^2 = \frac{c^2}{p^2}(1 + 2e\cos v + e^2\cos^2 v) \tag{5.28}$$

The velocity reads

$$\mathrm{v}^2 = \dot{r}^2 + r^2\dot{v}^2 \tag{5.29}$$

where we can replace the first and second term and get for the velocity

$$\mathrm{v}^2 = \frac{c^2}{p^2}(1 + 2e\cos v + e^2) \tag{5.30}$$

which can be expressed also as

$$\mathrm{v}^2 = \frac{c^2}{p}\left(2\frac{1 + e\cos v}{p} - \frac{1 - e^2}{p}\right) = \frac{c^2}{p}\left(\frac{2}{r} - \frac{1 - e^2}{p}\right) \tag{5.31}$$

We substitute $p$ by $c^2/\kappa^2$ and take for $p = a(1 - e^2)$[5]; thus we find the important *velocity relation*

$$\boxed{\mathrm{v}^2 = \kappa^2\left(\frac{2}{r} - \frac{1}{a}\right)}$$

This relation can be used to classify an orbit whether it is bounded (Ellipse) or unbounded (Parabola, Hyperbola); also it can be expressed using the constant energy $h$

$$h = \frac{\mathrm{v}^2}{2} - \kappa^2\frac{1}{r} \tag{5.32}$$

From theory it is well known that the energy of a conservative dynamical systems tells about the qualitative behavior of an orbit:

- with $h < 0$ the orbit is bounded (elliptic)
- with $h > 0$ the orbit is unbounded (hyperbolic)
- with $h = 0$ the orbit is parabolic.

Using now the velocity relation we can classify just from the velocity and the position of a celestial body (e.g. a comet) whether it is coming from infinity (in fact from the Oort cloud[6]) or has an elliptic orbit because of Jupiter who is capturing comets:

$$\mathrm{v}^2 > \frac{\kappa^2}{r}, \quad \mathrm{v}^2 < \frac{\kappa^2}{r}, \quad \mathrm{v}^2 = \frac{\kappa^2}{r} \tag{5.33}$$

5) For a hyperbolic motion $a$ can be regarded as negative.
6) This is believed to be the reservoir consisting of some billions of comets far out at approximately 50 000 astronomical units (AU) from the sun. From time to time a perturbation brings a comet to a hyperbolic orbit leading it into the inner part of the Solar system.

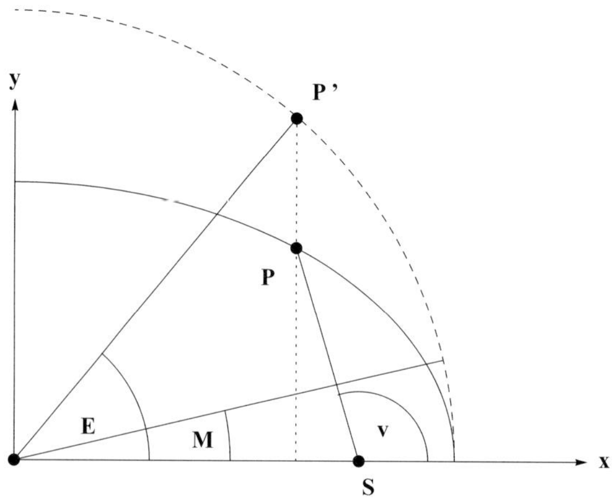


**Figure 5.3** Relation between true, mean and eccentric anomaly. Note that the "fictitious angle" $M = nt$ cannot be found in an exact way by geometrical construction.

Thus, just verifying the velocity in a special distance to the Sun, a comet can be classified according to its velocity as being in a hyperbolic, elliptic or parabolic orbit. For circular orbits the velocity is constant $\mathrm{v} = \kappa/\sqrt{a}$.

A different representation for $c$ can be derived, in the planar problem, when we introduce rectangular coordinates

$$x = r\cos v\,, \quad y = r\sin v \tag{5.34}$$

$$r = \sqrt{x^2 + y^2}\,, \quad \tan v = \frac{y}{x} \tag{5.35}$$

from which we can find via implicit differentiation

$$\frac{\dot{v}}{\cos^2 v} = \frac{\dot{v}\,r^2}{x^2} = \frac{\dot{y}x - y\dot{x}}{x^2} \tag{5.36}$$

and finally for $c$

$$c = \dot{y}x - y\dot{x} \tag{5.37}$$

Now we replace the velocity using the velocity relation, which simply can be derived from the energy-relation knowing that the potential reads $V(r) = \kappa^2 r^{-1}$.

## 5.4 Kepler Equation

To derive the famous Kepler equation we have to define the mean anomaly $M = n \cdot t$ and the eccentric anomaly $E$ (see Figure 5.3)

$$r = a(1 - e\cos E) \tag{5.38}$$

where we set $E = 0°$ if the planet is in perihelion and $E = 180°$ when it is in the aphelion position. From KI we find

$$p = r + er\cos\nu$$

and consequently

$$er\cos\nu = a(1-e^2) - a(1 - e\cos E)$$

which leads to

$$r\cos\nu = a(\cos E - e) \tag{5.39}$$

$y = r\sin\nu$ can now be obtained by using the former results

$$r^2 - r^2\cos^2\nu = a^2\left[(1 + e^2\cos^2 E) - (\cos^2 E + e^2)\right]$$

and consequently

$$r^2(1-\cos^2) = a^2(1-e^2)(1-\cos^2 E)$$

where from we find

$$r\sin\nu = a\sqrt{1-e^2}\sin E \tag{5.40}$$

From the definition of $E$ and (5.38) we derive

$$\cos\nu = \frac{\cos E - e}{1 - e\cos E}$$

which now gives us the possibility to express $\nu$ as a function of $E$ and the eccentricity $e$ by computing

$$\begin{aligned} 1 - \cos\nu &= \frac{(1+e)(1-\cos E)}{1-\cos E} \\ 1 + \cos\nu &= \frac{(1-e)(1-\cos E)}{1+\cos E}\ . \end{aligned} \tag{5.41}$$

Making use of a well known trigonometric formula[7] we find the relation between the two angles $\nu$ and $E$

$$\tan\frac{\nu}{2} = \sqrt{\frac{1+e}{1-e}}\tan\frac{E}{2} \tag{5.42}$$

To find the relation between $E$ and $M$ we use KI in a form $dr$

$$dr = \frac{r^2 e}{p}\sin\nu\, d\nu \tag{5.43}$$

7) $\tan(\alpha/2) = (1-\cos\alpha)/(1+\cos\alpha)$.

On the other hand from KII we can replace $dv$ by $dt$

$$r^2 dv = c dt = \frac{nap}{\sqrt{1-e^2}} dt$$

from which we derive

$$dr = \frac{na}{\sqrt{1-e^2}} e \sin v \, dt \tag{5.44}$$

where we replace $\sin v$ (5.40) and get

$$dr = \frac{nae}{\sqrt{1-e^2}} \frac{a}{r} \sin E \sqrt{1-e^2} dt \tag{5.45}$$

From the definition of $E$ (5.38) we find

$$dr = ae \sin E dE \tag{5.46}$$

Using (5.45) and (5.46) we find

$$ae \sin E dE = \frac{na^2 e}{r} \sin E dt$$

and consequently

$$dE = \frac{na}{r} dt$$

respectively by using the definition (5.38)

$$(1 - e \cos E) dE = n dt \, . \tag{5.47}$$

Integration of this equation finally gives

$$E - e \sin E = n(t - t_0) \tag{5.48}$$

and defining the fictitious angle $M = nt$, the mean anomaly, for $t_0 = 0$, the position in the perihelion

$$\boxed{E - e \sin E = M}$$

which is the famous transcendental *Kepler equation* (KE). It cannot be solved in closed form but there exist different ways to solve it numerically; even today one finds new algorithms to solve it efficiently.

From the KIII and KE we derive directly $t = t(E)$

$$t = \sqrt{\frac{a^3}{\kappa^2}} \cdot (E - e \sin E) \tag{5.49}$$

and from (5.47) and (5.38) we get immediately

$$\frac{dt}{dE} = \sqrt{\frac{a}{\kappa^2}} r \tag{5.50}$$

## 5.5 Complex Description

We can write the equation of motion of the two-body problem (5.11) differently when we write for the position vector $\boldsymbol{r}$ a complex number $\boldsymbol{x} \in \mathbb{C}$ after [134]

$$x = \begin{pmatrix} x_1 \\ x_2 \end{pmatrix}$$

which reads in complex notation

$$\boldsymbol{x} = x_1 + i x_2$$

and the equation of motion can be written as

$$\ddot{\boldsymbol{x}} + \kappa^2 \frac{\boldsymbol{x}}{r^3} = 0 \tag{5.51}$$

With $r = |\boldsymbol{x}|$ the energy can be expressed as

$$\frac{1}{2}|\dot{\boldsymbol{x}}|^2 - \frac{\kappa^2}{r} = -h \tag{5.52}$$

We now introduce the Levi-Civita regularization, which can be used for motion in the plane. The principal operation is $\boldsymbol{x} = \boldsymbol{u}^2$, $\boldsymbol{u} \in \mathbb{C}$ which reads

$$x_1 + i x_2 = (u_1 + i u_2)^2 \tag{5.53}$$

which means, that the distances of the physical plane are squared, whereas the angles are halved (see Figure 5.4). The Levi-Civita regularization leads to a simple map of $\mathrm{x} \to \mathrm{u}$

$$\begin{aligned} x_1 &= u_1^2 - u_2^2 \\ x_2 &= 2u_1 u_2 \end{aligned} \tag{5.54}$$

In detail the transformation from the old variable $x$ in the physical plane to the new variable $\boldsymbol{u}$ may be written $\boldsymbol{x} = L(\boldsymbol{u})\boldsymbol{u}$, where $L(\boldsymbol{u})$, the $L$-matrix is the matrix of transformation

$$L(\boldsymbol{u}) = \begin{pmatrix} u_1 & -u_2 \\ u_2 & u_1 \end{pmatrix}$$

When we now introduce a new fictitious time[8)] by

$$\frac{dt}{ds} = r \tag{5.55}$$

8) This transformation of the time was first introduced by Sundman; note that differentiation with respect to this time will be denoted by $'$.

which means that the velocity at collision is multiplied by the distance $r$ and thus vanishes.

Furthermore the velocity in the new coordinates reads

$$x' = L(u)'u + L(u)u' = 2L(u)u' \tag{5.56}$$

and for the acceleration after another differentiation with respect to the new time

$$x'' = 2L(u'') + 2L(u')u' \tag{5.57}$$

Up to now this was just a formal introduction of the parameter space without reference to the two-body problem. In the equation of motion (5.51)[9] we accomplish the Sundman transformation of the time introduced in (5.55)

$$rx'' - r'x' + \kappa^2 x = 0 \tag{5.58}$$

Then we implement the Levi-Civita transformation

$$r = (u, u) = |u|^2 , \quad r' = (u, u)' = 2(u, u') \tag{5.59}$$

and express also the energy-relation in the new variables

$$h = \frac{\kappa^2 - 2|u'|^2}{|u|^2} \tag{5.60}$$

The velocity after Levi-Civita can be expressed as $x' = 2L(u)u'$, where the orthogonal $L$-matrix can now be used to express the Keplerian motion in the form of a harmonic oscillator

$$u'' + \frac{\kappa^2/2 - (u', u')}{(u, u)} u = 0 \tag{5.61}$$

or explicitly written in two coordinates with the frequency $\omega = h/2$

$$u_1'' + \frac{h}{2} u_1 = 0 , \quad u_2'' + \frac{h}{2} u_2 = 0 \tag{5.62}$$

Inserting the results from Eqs. (5.39) and (5.40) the elliptic motion in the physical plane can be described by

$$x = a\left[(\cos E - e) + i\sqrt{1 + e^2} \sin E\right] \tag{5.63}$$

and in the complex parameter plane

$$u = \sqrt{a(1 - e)} \cos \frac{E}{2} + i\sqrt{a(1 + e)} \sin \frac{E}{2} \tag{5.64}$$

9) Note that the position vector $r$ has to be replaced by the complex number $x$.

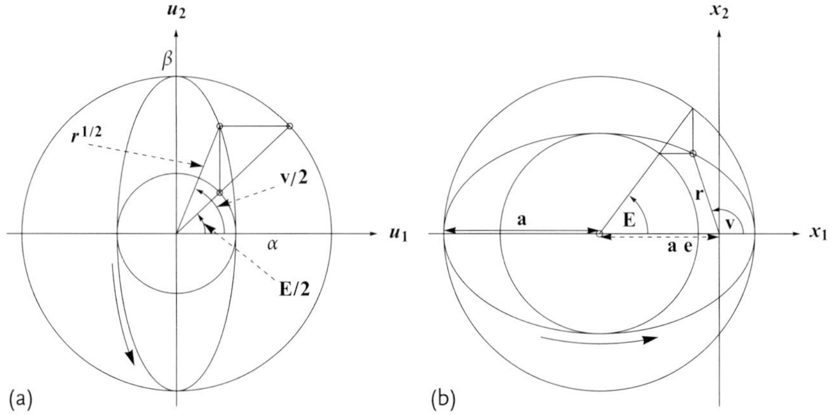


**Figure 5.4** The two-body problem in the parameter plane (a) and in the physical plane (b); after [134].

### 5.5.1 The KS-Transformation

For the motion in space a more general regularization method was introduced by Kustaanheimo and Stiefel (called KS-transformation) where the $L-$ matrix is a $4 \times 4$ matrix

$$L(\boldsymbol{u}) = \begin{pmatrix} u_1 & -u_2 & -u_3 & u_4 \\ u_2 & u_1 & -u_4 & -u_3 \\ u_3 & u_4 & u_1 & u_2 \\ u_4 & -u_3 & u_2 & -u_1 \end{pmatrix}$$

One may ask why the extension to a 4-dimensional space is necessary and what is the fourth coordinate $x_4$. Using again the transformation $x = L(\boldsymbol{u}).\boldsymbol{u}$ leads to

$$\begin{aligned} x_1 &= u_1^2 - u_2^2 - u_3^2 + u_4^2 \\ x_2 &= 2(u_1 u_2 - u_3 u_4) \\ x_3 &= 2(u_1 u_3 + u_2 u_4) \\ x_4 &= 0 \end{aligned} \tag{5.65}$$

In this form the regularized equations of motion can also be expressed as perturbed harmonic oscillator when additional forces are acting. This is quite useful when they are – like in Solar system dynamics – due to the perturbations of other planets (we refer to the details in the book [134]).

## 5.6
## Motion in Space and the Keplerian Elements

Until now, we did not take into account, that the motion of a planet takes place in the three dimensional space, with three degrees of freedom. It was not necessary, because one can reduce it to a motion on a plane, via an adequate change of variables. In three dimensions we need 6 *orbital elements*, which are the following three action-like variables:

1. semimajor axis $\boldsymbol{a}$
2. eccentricity of the orbit $\boldsymbol{e}$
3. inclination $\boldsymbol{i}$ (of the orbital plane, with respect to an inertial system)

and the three angle-like variables

1. the longitude of the ascending node $\boldsymbol{\Omega}$
2. the argument of the perihelion $\boldsymbol{\omega}$
3. the true anomaly $\boldsymbol{\nu}$.

These elements of the two-body problem are plotted in Figures 5.5 and 5.1. We already discussed in detail the three Kepler laws, which govern the motion of a planet. What we are interested in is now the position of a planet on the celestial sphere. To compute from the given orbital elements for a given time $t$ the right ascension $\alpha$ and declination $\delta$ of a planet (asteroid) we need to compute step by step:

1. The eccentric anomaly $E$ from the mean anomaly $M$ of the celestial body for the instant of time we want to know via the Kepler equation

$$E - e \sin E = n(t - T) = M(t_0) + n(t - t_0) \tag{5.66}$$

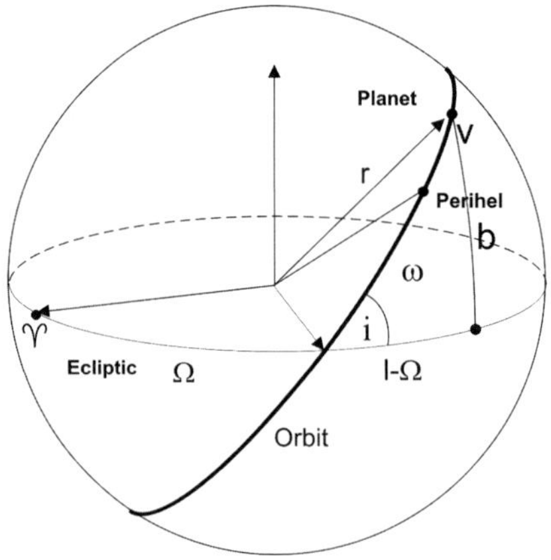


**Figure 5.5** The orbital elements.

2. The distance $r$, from the Sun is

$$r = a(1 - e\cos E) \tag{5.67}$$

3. The true anomaly $v$ giving the position in the orbit

$$\tan\frac{v}{2} = \sqrt{\frac{1+e}{1-e}}\tan\frac{E}{2} \tag{5.68}$$

4. The so-called argument of the latitude $u = \omega + v$
5. The heliocentric polar coordinates $r, l, b$ (see Figure 5.5)

$$\begin{aligned} \cos b \cos l &= \cos u \cos \Omega - \sin u \sin \Omega \cos i \\ \cos b \sin l &= \cos u \sin \Omega - \sin u \cos \Omega \cos i \\ \sin b &= \sin u \sin i \end{aligned} \tag{5.69}$$

6. The rectangular heliocentric coordinates $x$, $y$ and $z$

$$\begin{aligned} x &= r(\cos b \cos l) \\ y &= r(\cos b \sin l) \\ z &= r \sin b \end{aligned} \tag{5.70}$$

7. Which we need to transform into the equatorial heliocentric coordinates, taking into account the variable obliquity of the ecliptic $\varepsilon \sim 23^\circ$[10)]

$$\begin{aligned} \bar{x} &= x \\ \bar{y} &= y\cos\varepsilon - z\sin\varepsilon \\ \bar{z} &= z\cos\varepsilon - y\sin\varepsilon \end{aligned} \tag{5.71}$$

8. The geocentric equatorial coordinates making use of the published coordinates of the Sun $\bar{X}$, $\bar{Y}$ and $\bar{Z}$ (from the Nautical Almanac)

$$\begin{aligned} \bar{\xi} &= \bar{x} + \bar{X} \\ \bar{\eta} &= \bar{y} + \bar{Y} \\ \bar{\zeta} &= \bar{z} + \bar{Z} \end{aligned} \tag{5.72}$$

9. Via a final transformation the right ascension $\alpha$ and declination $\delta$ of the planet

$$\begin{aligned} \rho \cos\delta \cos\alpha &= \bar{\xi} \\ \rho \cos\delta \sin\alpha &= \bar{\eta} \\ \rho \sin\delta &= \bar{\zeta} \end{aligned} \tag{5.73}$$

10) The actual value for a given date has to be computed from the respective formula given in the Nautical Almanac.

The whole procedures sketched here are nowadays written in computer programs, which take into account all – in the foregoing scheme neglected – effects like aberration precession etc. In this short section we only wanted to demonstrate how one can, in principle, compute ephemerides of a body when the six Keplerian elements are known. For a very detailed and profound description we refer to the Nautical Almanac published every year by the US Naval observatory.

## 5.7
## Astronomical Determination of the Gravitational Constant

There is quite a nice way to determine the gravitational constant only from astronomical observations! We need to know the semimajor axes of the orbit of the Earth ($a_\mathrm{E} = 1\,\mathrm{AU}$) and also from the Moon's orbit ($a_\mathrm{M} = 1/389.3\,\mathrm{AU}$), as well as the periods of the motion of the Earth–Moon system around the Sun ($U_\mathrm{E} = 365.25$ days) and also from the Moon around the Earth ($U_\mathrm{Moon} = 27.32$ days). This estimation can be done without explicitly knowing the mass of then Sun ($m_\mathrm{S}$), nor the mass of the Earth ($m_\mathrm{E}$) and of the Moon ($m_\mathrm{M}$). Using KIII for the system Sun–Earth–Moon in the following form:

$$\frac{a_\mathrm{E}^3}{U_\mathrm{E}^2} = \frac{k^2}{4\pi^2}(m_\mathrm{S} + m_\mathrm{E} + m_\mathrm{M}) \tag{5.74}$$

and now for the system Earth–Moon

$$\frac{a_\mathrm{M}^3}{U_\mathrm{M}^2} = \frac{k^2}{4\pi^2}(m_\mathrm{E} + m_\mathrm{M})\,. \tag{5.75}$$

Now we can eliminate in the first equation $m_\mathrm{E} + m_\mathrm{M}$ via (5.75) and find for the system Sun–Earth–Moon

$$\frac{a_\mathrm{E}^3}{U_\mathrm{E}^2} = \frac{k^2}{4\pi^2}\left(m_\mathrm{S} + \frac{a_\mathrm{M}^3 4\pi^2}{k^2\,U_\mathrm{M}^2}\right) \tag{5.76}$$

Substituting the values given above we find for gravitational constant in days, Solar masses and AU

$$k = \frac{2\pi}{U_\mathrm{E}}\sqrt{1 - \left(\frac{U_\mathrm{E}}{U_\mathrm{M}}\right)^2 a_\mathrm{M}^3} = 0.017\,20\ldots \tag{5.77}$$

## 5.8
## Solution of the Kepler Equation

The number of different methods to solve the Kepler equation was already unclear in the beginning of the twentieth century, when in the Bulletin Astronomique of January 1900 123 papers on Kepler's equation are listed and it is said, that this list is even incomplete. The interest to find a good solution retains up to now: two last papers on it *Sequential solution to Kepler's equation* date from September 2010 [135]. Let us briefly discuss them out of the many methods.

The Newton–Raphson method is one of the tools to solve the Kepler equation:

$$f(E) = E - e \sin E - M \tag{5.78}$$

One needs to find $f(E) = 0$ for a specific mean anomaly $M = nt$ and the parameter eccentricity $e$. Using the derivative with respect to $E$, $f'(E)$, the following iteration of finding the eccentric anomaly $E$ can be used

$$E_{\text{new}} = E_{\text{old}} - \frac{E_{\text{old}} - e \sin E_{\text{old}} - M}{1 - e \cos E_{\text{old}}} \tag{5.79}$$

For small eccentricities the method works well and converges rather fast; this is not the case for large $e$.

One could use the following more efficient method which was introduced in [136]: let us assume that the real solution is $\bar{E}$ and after $n$ approximations we found a value $E_n$ with the error $\epsilon_n = E_n - \bar{E}$. We can now make a Taylor expansion for $f(\bar{E})$ of the form

$$f(E) = f(E_n) + \epsilon_n f'(E_n) + \frac{1}{2}\epsilon_n^2 f''(E_n) + \frac{1}{6}\epsilon_n^3 f'''(E_n) + \ldots \tag{5.80}$$

It is clear that the former mentioned Newton–Raphson uses the expansion (5.80) only up to the first derivative of the function $f(\bar{E})$. A much better iteration is taking into account also the second or even third derivative which leads to the computation when we define the correcting terms $\delta_{nj}$, $j = 1, 2, 3$ as follows [136]:

$$\begin{aligned}
\delta_{n1} &= -\frac{f(\bar{E})}{f'(E_n)} \\
\delta_{n2} &= -\frac{f(\bar{E})}{f'(E_n) + \frac{1}{2}\delta_{n1} f''(E_n)} \\
\delta_{n3} &= -\frac{f(\bar{E})}{f'(E_n) + \frac{1}{2}\delta_{n2} f''(E_n) + \frac{1}{6}\delta_{n2}^2 f'''(E_n)}
\end{aligned} \tag{5.81}$$

and the new value is computed by $E_{n+1} = E_n + \delta_{n3}$ which halves the necessary iteration steps. It is also clear that the convergence depends also on the initial value of $E_0$ which is more important for the Newton–Raphson method than for the extended Danby method.

# 6
# The Restricted Three-Body Problem

In this chapter we introduce the restricted three-body problem in various formulations. We derive the equations of motion in the Newtonian and Hamiltonian framework in both nonrotating and rotating reference frames. We investigate the location and stability of the equilibrium points in the rotating coordinate system and describe the topology of the configuration space by means of zero velocity curves. We provide various examples of possible configurations of the problem (different mass ratios and Jacobi constants) and also generalize the Hill's curves to the spatial problem. To this end we summarize the equations of motion of the elliptic restricted and spatial problem and provide the equations under the presence of dissipative forces found in the solar system. The Hamiltonian framework is based [137], the discussion on the stable and unstable equilibria and near-by orbits is based on [138]. We would also like to refer to [139] and references within for a further reading.

**Applications of the restricted three-body problem:** The restricted three-body problem is the backbone of many interesting applications in celestial mechanics: the motion of an asteroid, a planetesimal or a dust particle in an idealized planetary system, which consists of a central star and a planet only, the motion of a planet in a double star system, the motion of a dust particle or an artificial satellite in a planet-moon system, and the motion of stars in a simplified galactic model of two primary galaxies moving on Keplerian orbits. Typical realizations found in our Solar system are the motion of a space craft in the gravitational field formed by the Earth and the Moon, and the motion of asteroids in the Sun–Mars or Sun–Jupiter system. The first asteroids were observed in the latter system, known as Trojan (or Greek) asteroids. To this end, recent research topics, relying on the results of the study of the restricted problem, include the field of space manifold dynamics and the dynamical investigation of extra-Solar planetary systems.

**Historical note on the integrability of the restricted problem:** Many mathematicians have devoted their work to investigate the integrability of the three-body problem. L. Euler, J. Lagrange, J. Jacobi, H. Poincaré, T. Levi-Civita, G.D. Birkhoff to name a view. A special note should be given to K.F. Sundman (see [137]), who showed that the restricted problem, after regularization of the equations of motion, in principal

*Celestial Dynamics*, First Edition. R. Dvorak and C. Lhotka.

can be solved. However, the series expansions have been shown to converge slowly and the solution can be seen as of no practical use.

## 6.1 Set-Up and Formulation

Let us denote in the inertial frame of coordinates the positions of the masses $m_1$, $m_2$ by the vectors $P_1 = (\hat{x}_1, \hat{y}_1)$ and $P_2 = (\hat{x}_2, \hat{y}_2)$, respectively and let $(\hat{x}, \hat{y})$ be the position of the mass $m = m_3$ given at dimensional time $\hat{t}$. We denote the distance of the primary masses by $\alpha = m_1 l/(m_1 + m_2)$ and $\beta = m_2 l/(m_1 + m_2)$, where $l$ is the mutual distance $\overline{P_1 P_2}$. By introducing $\xi = \hat{x}/l$, $\eta = \hat{y}/l$, $t = n\hat{t}$ as well $\mu_1 = m_1/(m_1 + m_2)$, $\mu_2 = m_2/(m_1 + m_2)$ we introduce the unit of mass to be the sum of the masses of the primary bodies. Finally, our choice $l \equiv 1$, $\mu_1 + \mu_2 \equiv 1$ as well $k \equiv 1$ gives $n = 1$, $\mu_1 = 1 - \mu$ and $\mu_2 = \mu$. From

$$n^2 a^3 = k^2 (m_1 + m_2)$$

we find that one revolution period of the primaries takes $2\pi$. In this setting the Lagrangian function of the problem becomes [1)]

$$L = \frac{1}{2}\left(\dot{\xi}^2 + \dot{\eta}^2\right) + U(\xi, \eta, t)$$

where

$$T = \frac{1}{2}\left(\dot{\xi}^2 + \dot{\eta}^2\right)$$

is the kinetic energy, and $U = U(\xi, \eta, t)$ is the gravitational potential in the nonrotating coordinate frame, which is defined by:

$$U(\xi, \eta, t) = \frac{\mu_1}{r_1} + \frac{\mu_2}{r_2}$$

The potential in this form is time dependent, since the distances of the third body from the primary bodies are periodic functions of time:

$$r_1^2 = [\xi - \mu_2 \cos(t)]^2 + [\eta - \mu_2 \sin(t)]^2$$
$$r_2^2 = [\xi + \mu_1 \cos(t)]^2 + [\eta + \mu_1 \sin(t)]^2$$

Setting the generalized coordinates $q_1 = \xi$, $q_2 = \eta$ we get from $p_1 = \partial L/\partial \dot{q}_1$, $p_2 = \partial L/\partial \dot{q}_2$ the conjugated momenta $p_1 = \dot{q}_1$, $p_2 = \dot{q}_2$. From the definition of the Hamiltonian function $H = T - U$ we get:

$$H \equiv H(p_1, p_2, q_1, q_2, t) = \frac{1}{2}(p_1^2 + p_2^2) - U(q_1, q_2, t)$$

1) Note, that we use the convention that $U = -V$ to be consistent with Chapter 2 and the notation of standard text-books on the subject.

with the canonical equations ($\dot{q}_i = \partial H/\partial p_i$, $\dot{p}_i = -\partial H/\partial q_i$ with $i = 1, 2$):

$$\begin{aligned}
\dot{q}_1 &= p_1\,, \quad \dot{q}_2 = p_2 \\
\dot{p}_1 &= -\mu_1 \frac{q_1 - \mu_2 \cos(t)}{r_1^3} - \mu_2 \frac{q_1 + \mu_1 \cos(t)}{r_2^3} \\
\dot{p}_2 &= -\mu_1 \frac{q_2 - \mu_2 \sin(t)}{r_1^3} - \mu_2 \frac{q_2 + \mu_1 \sin(t)}{r_2^3}
\end{aligned}$$

The Hamiltonian in its present form is a time dependent function. To make it autonomous we introduce a system of coordinates which uniformly rotates with the mean angular motion of the primary bodies:

$$\begin{aligned}
q_1 &= Q_1 \cos(t) - Q_2 \sin(t) \\
q_2 &= Q_1 \sin(t) + Q_2 \cos(t)
\end{aligned} \tag{6.1}$$

To make the transformation symplectic we define the generating function[2)]

$$F_3(p_1, p_2, Q_1, Q_2, t) = -a_{ij} p_i Q_j$$

where the matrix[3)] $(a_{ij})$ is defined by:

$$(a_{ij}) = \begin{pmatrix} \cos(t) & -\sin(t) \\ \sin(t) & \cos(t) \end{pmatrix}$$

From $q_i = \partial F_3/\partial p_i$ and from $P_i = \partial F_3/\partial Q_i$ with $i = 1, 2$ we get the transformation of the new momenta:

$$\begin{aligned}
P_1 &= p_1 \cos(t) + p_2 \sin(t) \\
P_2 &= p_2 \cos(t) - p_1 \sin(t)
\end{aligned}$$

We substitute for $q_1$ (respectively $\xi$) and $q_2$ (or $\eta$) in $r_1$ and $r_2$ to get:

$$\begin{aligned}
r_1^2 &= (Q_1 - \mu_2)^2 + Q_2^2 \\
r_2^2 &= (Q_1 + \mu_1)^2 + Q_2^2
\end{aligned}$$

and the new Hamiltonian ($H' = H + \partial F_3/\partial t$) becomes:

$$H' \equiv H'(P_1, P_2, Q_1, Q_2) = \frac{1}{2}\left(P_1^2 + P_2^2\right) - Q_1 P_2 + Q_2 P_1 - U'(Q_1, Q_2)$$

where $U' \equiv U'(Q_1, Q_2)$ is the gravitational potential in the rotating frame of coordinates:

$$U'(Q_1, Q_2) = \frac{\mu_1}{r_1} + \frac{\mu_2}{r_2}$$

2) See section "Canonical Transformations" in Chapter 2.
3) We follow the Einstein summation convention.

Thus, in the rotating coordinate system the Hamiltonian becomes a constant of motion. We get the system of canonical equations from $\dot{Q}_i = \partial H'/\partial P_i$, $\dot{P}_i = -\partial H'/\partial Q_i$ with $i = 1, 2$:

$$\dot{Q}_1 = P_1 + Q_2\,, \quad \dot{Q}_2 = P_2 - Q_1$$
$$\dot{P}_1 = P_2 - \mu_1 \frac{Q_1 + \mu_2}{r_1^3} - \mu_2 \frac{Q_1 - \mu_1}{r_2^3}$$
$$\dot{P}_2 = -P_1 - \mu_1 \frac{Q_2}{r_1^3} - \mu_2 \frac{Q_2}{r_2^3}$$

Sometimes the terms sidereal and synodic are used to indicate the nonrotating and rotating coordinate systems, respectively. In the synodic system the Newtonian form of the equations can be derived by setting $(Q_1, Q_2) = (x, y)$ and $(P_1, P_2) = (\dot{x} - y, \dot{y} + x)$. From $(\dot{Q}_1, \dot{Q}_2)$ we get:

$$\ddot{x} - 2\dot{y} = \frac{\partial \Omega}{\partial x}$$
$$\ddot{y} + 2\dot{x} = \frac{\partial \Omega}{\partial y} \tag{6.2}$$

where we have defined the potential[4] $\Omega \equiv \Omega(x, y)$ as:

$$\Omega = \mu_1 \left( \frac{r_1^2}{2} + \frac{1}{r_1} \right) + \mu_2 \left( \frac{r_2^2}{2} + \frac{1}{r_2} \right) \tag{6.3}$$

with

$$r_1^2 = (x - \mu_2)^2 + y^2$$
$$r_2^2 = (x + \mu_1)^2 + y^2$$

The geometry of the restricted three-body problem in the synodic reference frame is shown in Figure 6.1. With successive transformation we were able to fix the position of the mass $\mu_1$ at $P_1 = (-\mu, 0)$ and the location of the mass $\mu_2$ at $P_2 = (1 - \mu, 0)$, while the position of the third mass is given at $P = (x, y)$. A

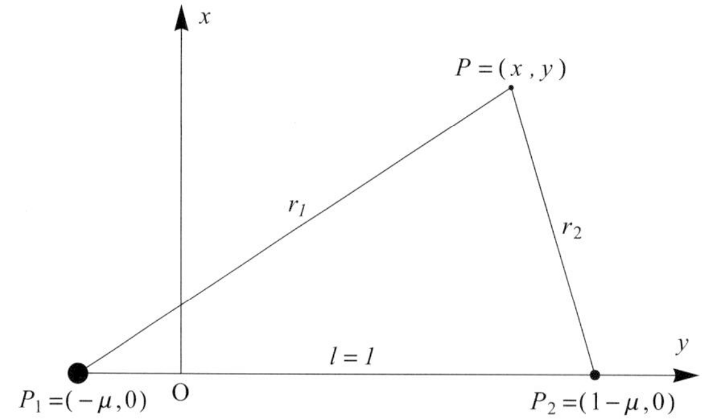


**Figure 6.1** The geometry of the restricted three-body problem in the synodic reference frame.

4) The form is true for $\mu_1 + \mu_2 = 1$ since $\mu_1 r_1^2 + \mu_2 r_2^2 = x^2 + y^2 + \mu_1 \mu_2$.

similar calculation can be done when the third body is not restricted to lie in the invariable plane of motion (the spatial problem) and for an elliptic orbit of the primary bodies (the elliptic problem). We will collect the formulae of the latter in Section 6.8 and Section 6.9.

## 6.2
## Equilibria of the System

The configuration of the circular restricted three-body problem is shown in Figure 6.2. The plot shows the geometry of the triangular and collinear equilibrium points together with the equipotential lines for $\mu = 0.05$. In this section we investigate the location and stability of these kind of equilibria based on the formulation (6.2).

An equilibrium point, say $(x_{\text{equ}}, y_{\text{equ}})$ is determined by the equations:

$$\frac{\partial \Omega}{\partial x} = \frac{\partial \Omega}{\partial r_1}\frac{\partial r_1}{\partial x} + \frac{\partial \Omega}{\partial r_2}\frac{\partial r_2}{\partial x} = 0$$
$$\frac{\partial \Omega}{\partial y} = \frac{\partial \Omega}{\partial r_1}\frac{\partial r_1}{\partial y} + \frac{\partial \Omega}{\partial r_2}\frac{\partial r_2}{\partial y} = 0$$

where the partial derivatives are given by:

$$\frac{\partial \Omega}{\partial r_1} = \mu_1\left(r_1 - \frac{1}{r_1^2}\right), \quad \frac{\partial r_1}{\partial x} = \frac{\mu_2 + x}{r_1}, \quad \frac{\partial r_1}{\partial y} = \frac{y}{r_1}$$
$$\frac{\partial \Omega}{\partial r_2} = \mu_2\left(r_2 - \frac{1}{r_2^2}\right), \quad \frac{\partial r_2}{\partial x} = \frac{x - \mu_1}{r_2}, \quad \frac{\partial r_2}{\partial y} = \frac{y}{r_2} \tag{6.4}$$

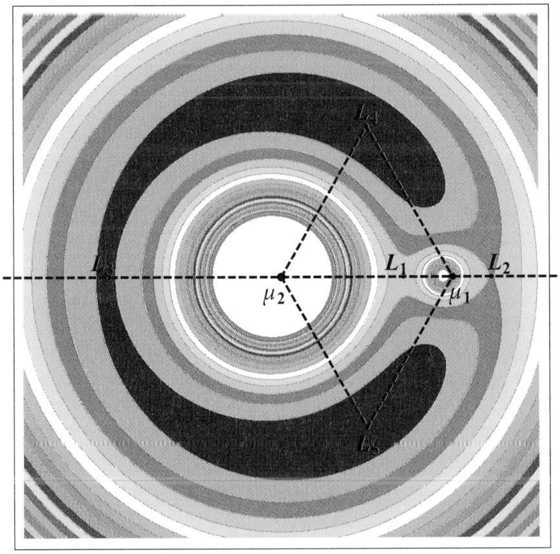


**Figure 6.2** The Lagrangian equilibrium points ($L_1, \ldots, L_5$) of the restricted three-body problem; equipotential lines are defined as the borders of equal colored regions.

From $r_1 - r_1^{-2} = 0$ and $r_2 - r_2^{-2} = 0$ we obtain the trivial solution $r_1 = r_2 = 1$. Together with our choice $l = 1$ (distance of the primaries equal to unity) the equilibrium points are located on the edges of an equilateral triangle; from $r_1^2 = (x+\mu)^2 + y^2 \equiv 1$, $r_2^2 = (x-1+\mu)^2 \equiv 1$ we find them at $x_{\text{equ}} = 1/2 - \mu$ and $y_{\text{equ}} = \pm\sqrt{3}/2$, respectively (see Figure 6.2). They are called triangular equilibria, we denote the one with positive $y_{\text{equ}}$ as $L_4$ and the one with negative $y_{\text{equ}}$ as $L_5$. Other solutions exist. They are determined by the condition

$$\left|\begin{pmatrix} \dfrac{\partial r_1}{\partial x} & \dfrac{\partial r_2}{\partial x} \\ \dfrac{\partial r_1}{\partial y} & \dfrac{\partial r_2}{\partial y} \end{pmatrix}\right| = \frac{y(\mu_1+\mu_2)}{r_1 r_2} = \frac{y}{r_1 r_2} = 0$$

which is true if $y = 0$. We can deduce from $\Omega(1-\mu) = \Omega(\mu) = \infty$ that there are at least three more minima within the intervals $-\mu_2 < x < \mu_1$, $\mu_1 < x < \infty$ and $-\infty < x < \mu_2$ (see Figure 6.3). We investigate the three cases separately.

$L_1$ For the equilibrium $L_1$ in the interval $-\mu_2 < x < \mu_1$ we find $r_1 = x + \mu_2$ and $r_2 = x - \mu_1$ and get the relation $r_2 = 1 - r_1$. From the first of (6.4) we derive

$$\frac{\partial r_1}{\partial x} = -\frac{\partial r_2}{\partial x} = 1$$

and get the condition

$$C(L_1) = \frac{\partial \Omega}{\partial r_1} - \frac{\partial \Omega}{\partial r_2} = 0$$

We introduce the ratio $\lambda = r_2/r_1$ and determine from $\lambda = (1-r_1)/r_1$ the distance $r_1 = 1/(1+\lambda)$ and $r_2 = \lambda/(1+\lambda)$; we substitute these expressions

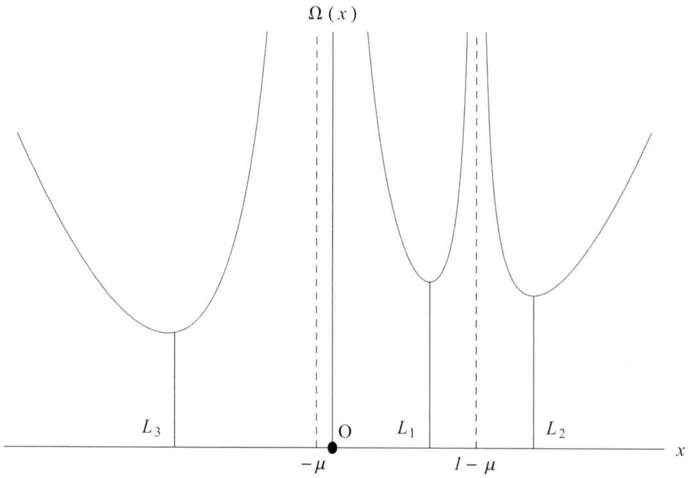


**Figure 6.3** The potential $\Omega$ along $y = 0$ for $\mu = 0.1$; location of $L_1$, $L_2$, $L_3$.

into $C(L_1)$ and finally get the equation

$$\lambda^5 + 3\lambda^4 + 3\lambda^3 = \frac{\mu_2}{\mu_1}(3\lambda^2 + 3\lambda + 1)$$

which determines $x$.

$L_2$ We repeat the calculation in the interval $\mu_1 < x$ to get the equilibrium point $L_2$. From $r_1 = x + \mu_2$ and $r_2 = r_1 - 1$ we get

$$\frac{\partial r_1}{\partial x} = \frac{\partial r_2}{\partial x} = 1$$

and the condition

$$C(L_2) = \frac{\partial \Omega}{\partial r_1} + \frac{\partial \Omega}{\partial r_2} = 0$$

We define $\lambda = (r_1 - 1)/r_1$ and get $r_1 = 1/(1 - \lambda)$, $r_2 = \lambda/(1 - \lambda)$ which we substitute into $C(L_2)$ to get

$$\lambda^3(3 - 3\lambda + \lambda^2) + (2\lambda - 1)(1 - \lambda + \lambda^2)\frac{\mu_2}{\mu_1} = 0$$

$L_3$ Finally, we determine the position for $L_3$ located in the interval $x < -\mu_2$. We have $r_1 = -(x + \mu_2)$ and $r_2 = r_1 + 1$ and get from

$$\frac{\partial r_1}{\partial x} = \frac{\partial r_2}{\partial x} = -1$$

again

$$C(L_3) = C(L_2)$$

We define $\lambda = (r_1 + 1)/r_1$ and substitute $r_1 = 1/(\lambda - 1)$ as well $r_2 = \lambda/(\lambda - 1)$ in $C(L_3)$ and get

$$-2\lambda^2 + 3\lambda^3 - 3\lambda^4 + \lambda^5 + (-1 + 3\lambda - 3\lambda^2)\frac{\mu_2}{\mu_1} = 0$$

The preceding polynomials in $\lambda$ can be solved by a numerical root finding method as well by the determination of the coefficients in a series expansion of the form $\lambda = \lambda_0 + a\lambda_0^2 + \ldots$ (case $L_1$, $L_2$) or $\lambda = 1 + a\mu_2/\mu_1 + b(\mu_2/\mu_1)^2 + \ldots$ (case $L_3$). The zeroth-order term is given by $\lambda_0 = \sqrt[3]{\mu_2/3\mu_1}$, we get

$$L_1 : \lambda = \lambda_0 + \frac{2}{3}\lambda_0^2 + \frac{2}{9}\lambda_0^3 - \frac{32}{81}\lambda_0^4 + \ldots$$

$$L_2 : \lambda = \lambda_0 + \frac{1}{3}\lambda_0^2 - \frac{1}{9}\lambda_0^3 - \frac{31}{81}\lambda_0^4 + \ldots$$

$$L_3 : \lambda = 1 + \frac{7}{12}\frac{\mu_2}{\mu_1} - \frac{35}{144}\left(\frac{\mu_2}{\mu_1}\right)^2 + \frac{3227}{20\,726}\left(\frac{\mu_2}{\mu_1}\right)^3 + \ldots$$

To determine the type of the equilibria we need to investigate the system of second order derivatives:

$$\frac{\partial^2 \Omega}{\partial x^2} = \frac{\partial^2 \Omega}{\partial r_1^2}\left(\frac{\partial r_1}{\partial x}\right)^2 + \frac{\partial \Omega}{\partial r_1}\frac{\partial^2 r_1}{\partial x^2} + \frac{\partial \Omega}{\partial r_2}\frac{\partial r_2}{\partial x^2} + \frac{\partial^2 \Omega}{\partial r_2^2}\left(\frac{\partial r_2}{\partial x}\right)^2$$

$$\frac{\partial^2 \Omega}{\partial x \partial y} = \frac{\partial^2 \Omega}{\partial r_1^2}\frac{\partial r_1}{\partial x}\frac{\partial r_1}{\partial y} + \frac{\partial \Omega}{\partial r_1}\frac{\partial^2 r_1}{\partial x \partial y} + \frac{\partial \Omega}{\partial r_2}\frac{\partial^2 r_2}{\partial x \partial y} + \frac{\partial^2 \Omega}{\partial r_2^2}\frac{\partial r_2}{\partial x}\frac{\partial r_2}{\partial y}$$

$$\frac{\partial^2 \Omega}{\partial y^2} = \frac{\partial^2 \Omega}{\partial r_1^2}\left(\frac{\partial r_1}{\partial y}\right)^2 + \frac{\partial \Omega}{\partial r_1}\frac{\partial^2 r_1}{\partial y^2} + \frac{\partial \Omega}{\partial r_2}\frac{\partial r_2}{\partial y^2} + \frac{\partial^2 \Omega}{\partial r_2^2}\left(\frac{\partial r_2}{\partial y}\right)^2$$

The partial derivatives are given by (6.4) together with:

$$\frac{\partial \Omega^2}{\partial r_1^2} = \mu_1\left(1 + \frac{2}{r_1^3}\right), \quad \frac{\partial^2 \Omega}{\partial r_1 \partial r_2} = 0, \quad \frac{\partial^2 \Omega}{\partial r_2^2} = \mu_2\left(1 + \frac{2}{r_2^3}\right)$$

$$\frac{\partial^2 r_1}{\partial x^2} = \frac{1}{r_1}\left[1 - \left(\frac{x + \mu_2}{r_1}\right)^2\right], \quad \frac{\partial^2 r_1}{\partial y^2} = \frac{1}{r_1}\left[1 - \left(\frac{y}{r_1}\right)^2\right]$$

$$\frac{\partial^2 r_2}{\partial x^2} = \frac{1}{r_2}\left[1 - \left(\frac{x - \mu_1}{r_2}\right)^2\right], \quad \frac{\partial^2 r_2}{\partial y^2} = \frac{1}{r_2}\left[1 - \left(\frac{y}{r_2}\right)^2\right]$$

$$\frac{\partial^2 r_1}{\partial x \partial y} = -\frac{(x + \mu_2)y}{r_1^3}, \quad \frac{\partial^2 r_2}{\partial x \partial y} = -\frac{(x - \mu_1)y}{r_2^3} \tag{6.5}$$

For the triangular points $L_4$, $L_5$ with $r_1 = r_2 = 1$, $x = 1/2 - \mu_2$ and $y = \pm 1/2\sqrt{3}$ we get

$$\frac{\partial \Omega}{\partial r_1} = \frac{\partial \Omega}{\partial r_2} = 0, \quad \frac{\partial^2 \Omega}{\partial r_1^2} = 3\mu_1, \quad \frac{\partial^2 \Omega}{\partial r_2^2} = 3\mu_2$$

as well as

$$\frac{\partial^2 \Omega}{\partial x^2} = \frac{3}{4}, \quad \frac{\partial^2 \Omega}{\partial x \partial y} = \pm\frac{3}{4}\sqrt{3}(1 - 2\mu), \quad \frac{\partial^2 \Omega}{\partial x^2} = \frac{9}{4}$$

Since $\mu < 1/2$ we also get

$$\frac{\partial^2 \Omega}{\partial x^2}\frac{\partial^2 \Omega}{\partial y^2} - \left(\frac{\partial^2 \Omega}{\partial x \partial y}\right)^2 = \frac{27}{16}\left[1 - (1 - 2\mu)^2\right] > 0$$

and

$$\frac{\partial^2 \Omega}{\partial x^2} > 0, \quad \frac{\partial^2 \Omega}{\partial y^2} > 0$$

As a conclusion $L_4$, $L_5$ are of center type. For the equilibria $L_1$, $L_2$, $L_3$ we find

$$\frac{\partial^2 \Omega}{\partial r_1^2} > 0, \quad \frac{\partial^2 \Omega}{\partial r_2^2} > 0$$

therefore

$$\frac{\partial^2 \Omega}{\partial x^2} > 0 \, , \quad \frac{\partial^2 \Omega}{\partial x \partial y} = 0$$

and

$$\frac{\partial^2 \Omega}{\partial y^2} < 0 \quad \text{and} \quad \frac{\partial^2 \Omega}{\partial x^2} \frac{\partial^2 \Omega}{\partial y^2} < 0$$

The collinear equilibria are therefore saddles.

## 6.3
## Motion Close to $L_4$ and $L_5$

We linearize the equations of motion to find the linear stability behavior close to the triangular equilibrium points $L_4$, $L_5$. For this reason we substitute in (6.2) $x = a + X$, $y = b + Y$ with $a = 1/2 - \mu_2$ and $b = \sqrt{3}/2$. We expand the potential around the small quantities $X$, $Y$ up to the second order:

$$\Omega\,(a+X, b+Y) = \Omega\,(a, b) + X\,\Omega_a + Y\,\Omega_b + \frac{1}{2}\left(X^2\Omega_{a,a} + 2XY\Omega_{a,b} + Y^2\Omega_{b,b}\right) + \ldots$$

where we have used the notation

$$\Omega_a = \frac{\partial \Omega}{\partial x}(a, b) \, , \quad \Omega_b = \frac{\partial \Omega}{\partial y}(a, b)$$

and

$$\Omega_{a,a} = \frac{\partial^2 \Omega}{\partial x^2}(a, b) \, , \quad \Omega_{b,b} = \frac{\partial^2 \Omega}{\partial y^2}(a, b) \, , \quad \Omega_{a,b} = \frac{\partial^2 \Omega}{\partial x \partial y}(a, b)$$

respectively. Since $\Omega_a = \Omega_b = 0$ and $\partial \Omega / \partial x = \partial \Omega / \partial X$ as well $\partial \Omega / \partial y = \partial \Omega / \partial Y$ we get the system of linearized equations:

$$\begin{aligned} \ddot{X} - 2\dot{Y} &= X\,\Omega_{a,a} + Y\Omega_{a,b} + \ldots \\ \ddot{Y} + 2\dot{X} &= X\,\Omega_{b,a} + Y\Omega_{b,b} + \ldots \end{aligned} \tag{6.6}$$

which is a linear system of ordinary differential equations. The solution can be written as $X = Ae^{\lambda t}$, $Y = Be^{\lambda t}$. If we replace $\dot{X} = \lambda A e^{\lambda t}$, $\dot{Y} = \lambda B e^{\lambda t}$ and $\ddot{X} = \lambda^2 A e^{\lambda t}$, $\ddot{Y} = \lambda^2 B e^{\lambda t}$ in (6.6) we get

$$\begin{aligned} A\left(\lambda^2 - \Omega_{a,a}\right) - B\left(2\lambda + \Omega_{a,b}\right) &= 0 \\ A\left(2\lambda - \Omega_{a,b}\right) + B\left(\lambda^2 - \Omega_{b,b}\right) &= 0 \end{aligned}$$

which can be solved, provided:

$$\begin{vmatrix} \lambda^2 - \Omega_{a,a} & -\left(2\lambda + \Omega_{a,b}\right) \\ 2\lambda - \Omega_{a,b} & \lambda^2 - \Omega_{b,b} \end{vmatrix} = 0 \tag{6.7}$$

The condition in $\lambda$

$$\lambda^4 - \lambda^2(\Omega_{a,a} + \Omega_{b,b} - 4) + \Omega_{a,a}\Omega_{b,b} - \Omega_{a,b}^2 = 0 \tag{6.8}$$

together with $\Omega_{a,a} = 3/4$, $\Omega_{a,b} = (3/4)\sqrt{3}(1-2\mu)$, $\Omega_{b,b} = 9/4$ gives

$$\lambda^4 + \lambda^2 + \frac{27}{4}\mu(1-\mu) = 0 \tag{6.9}$$

with only imaginary solutions, since

$$\lambda^2 = -\frac{1}{2} \pm \frac{1}{2}\sqrt{1 - 27\mu(1-\mu)} < 0 \tag{6.10}$$

From the above condition we conclude, that $\lambda^2$ is real only if

$$27\mu(1-\mu) \leq 1 \tag{6.11}$$

or $\mu_1/\mu_2 \simeq 1/25$. Under this assumption the roots of (6.10) become $\lambda_1 = i\nu_1$, $\lambda_2 = -i\nu_1$, $\lambda_3 = i\nu_2$, $\lambda_4 = -i\nu_2$ and the generic solution of (6.6) takes the special form

$$\begin{aligned} X(t) &= \alpha_1 \cos(\nu_1 t) + \alpha_2 \sin(\nu_1 t) + \alpha_3 \cos(\nu_2 t) + \alpha_4 \sin(\nu_2 t) \\ Y(t) &= \beta_1 \cos(\nu_1 t) + \beta_2 \sin(\nu_1 t) + \beta_3 \cos(\nu_2 t) + \beta_4 \sin(\nu_2 t) \end{aligned} \tag{6.12}$$

where $\nu_1$, $\nu_2$ define the frequencies of the periodic motion. To get the constant amplitudes $\alpha_i, \beta_i$ $(i = 1, 2, 3, 4)$ we replace (6.12) in (6.6) and compare coefficients of same Fourier components. From

$$\alpha_3 = \frac{-2\beta_4\nu_2 - \beta_3\Omega_{a,b}}{\nu_2^2 + \Omega_{a,a}}, \quad \alpha_4 = \frac{2\beta_3\nu_2 - \beta_4\Omega_{a,b}}{\nu_2^2 + \Omega_{a,a}}$$
$$\beta_1 = \frac{2\alpha_2\nu_1 - \alpha_1\Omega_{a,b}}{\nu_1^2 + \Omega_{b,b}}, \quad \beta_2 = \frac{-2\alpha_1\nu_1 - \alpha_2\Omega_{a,b}}{\nu_1^2 + \Omega_{b,b}}$$

together with the initial conditions $X_0 = X(t=0) = \alpha_1 + \alpha_3$, $\dot{X}_0 = \dot{X}(0) = \nu_1\alpha_2 + \nu_2\alpha_4$, $Y_0 = Y(0) = \beta_1 + \beta_3$, and $\dot{Y}_0 = \dot{Y}(0) = \nu_1\beta_2 + \nu_2\beta_4$ we get:

$$X_0 = -\frac{2\beta_2\nu_1 + \beta_1\Omega_{a,b}}{\nu_1^2 + \Omega_{a,a}} - \frac{2\beta_4\nu_2 + \beta_3\Omega_{a,b}}{\nu_2^2 + \Omega_{a,a}}$$
$$\dot{X}_0 = \frac{\nu_1(2\beta_1\nu_1 - \beta_2\Omega_{a,b})}{\nu_1^2 + \Omega_{a,a}} + \frac{\nu_2(2\beta_3\nu_2 - \beta_4\Omega_{a,b})}{\nu_2^2 + \Omega_{a,a}}$$

and

$$Y_0 = \frac{2\alpha_2\nu_1 - \alpha_1\Omega_{a,b}}{\nu_1^2 + \Omega_{b,b}} + \frac{2\alpha_4\nu_2 - \alpha_3\Omega_{a,b}}{\nu_2^2 + \Omega_{b,b}}$$
$$\dot{Y}_0 = -\frac{\nu_1(2\alpha_1\nu_1 + \alpha_2\Omega_{a,b})}{\nu_1^2 + \Omega_{b,b}} - \frac{\nu_2(2\alpha_3\nu_2 + \alpha_4\Omega_{a,b})}{\nu_2^2 + \Omega_{b,b}}$$

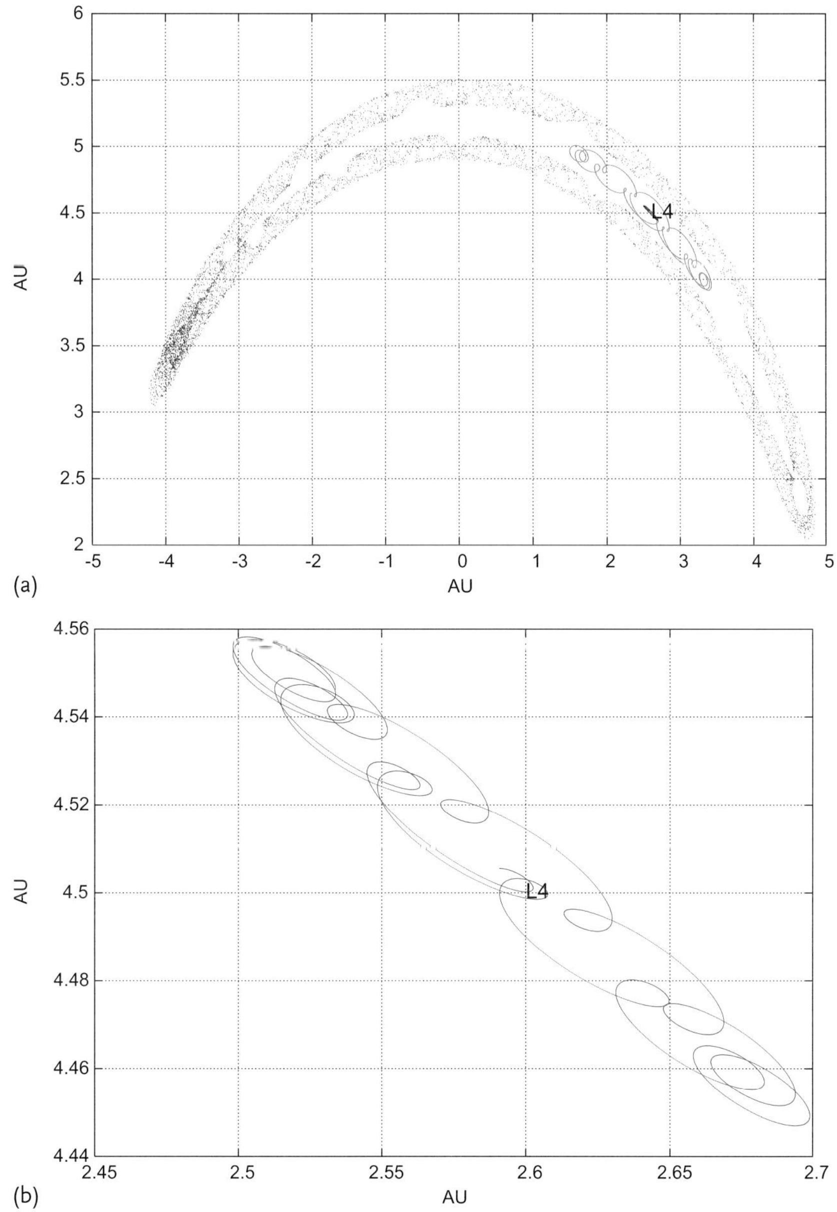


**Figure 6.4** (a): we depict three stable tadpole orbits around the Lagrange point $L_4$ for Sun–Jupiter in a rotating coordinate system. Three orbits with different initial semimajor axes (slightly different than Jupiters *a*) show their orbital behavior: the larger the difference from Jupiter's semimajor is, the larger is the amplitude of libration (from blue to red). Also the two well distinguished periods $\nu_1$, $\nu_2$ are visible. (b) zoom of the red orbit very close to $L_4$.

The system can be solved for $\alpha_1$, $\alpha_2$, $\alpha_3$, $\alpha_4$ and $\beta_1$, $\beta_2$, $\beta_3$, $\beta_4$ respectively. The solution (6.12) defines the motion of the massless body in the vicinity of $L_4$ (or $L_5$). If we start close enough to the triangular equilibria the motion is periodic in time and composed by two main periods $T_1 = 2\pi/\nu_1$ and $T_2 = 2\pi/\nu_2$ with the amplitudes given from the initial conditions. Since (6.11) does just depend on the mass ratio $\mu$ the periods are not affected by small perturbations. Moreover, since $\mu$ is small the solution for the frequencies $\nu_1$, $\nu_2$ from

$$\nu^2 = -\lambda^2 = \frac{1}{2}\left[1 \pm \sqrt{1 - 27\mu(1-\mu)}\right]$$

can be expanded around $\mu = 0$ and gives:

$$\nu_1 = 1 - \frac{27\mu}{8} - \frac{3213\mu^2}{128} - \frac{355\,023\mu^3}{1024} - \dots$$

$$\nu_2 = \sqrt{3\mu}\left(\frac{3}{2} + \frac{69\mu}{16} + \frac{13\,317\mu^2}{256} + \dots\right)$$

In the case of the Sun–Jupiter system we get with $\mu \simeq 10^{-3}$ in physical units the short period $T_1 = 11.9$ years and the long period $T_2 = 147.42$ years. Close to the triangular points of the Earth–Moon system we calculate $P_1 = 27.3$ days, $P_2 = 91.12$ days.

The orbits in the vicinity of $L_4(L_5)$ are closed only if the two periods are commensurate. In general, the superposition of the two kinds of harmonic motion in the linearized solution (6.12) fills a region around the equilibrium point, which can be bounded by an ellipse whose ratio of the major axis depends on the mass ratio of the primaries. We provide an example of this Trojan type motion in Figure 6.4 with initial conditions close to the point $L_4$.

## 6.4 Motion Close to $L_1$, $L_2$, $L_3$

In this section we adapt the method of linearization developed in the previous section to the motion around the collinear points. From a similar calculation of the previous section we find $\Omega_{a,a} > 0$, $\Omega_{a,b} = 0$ and $\Omega_{b,b} < 0$. Setting $q = -\Omega_{a,a}\Omega_{b,b}$, where $q > 0$, we can write (6.8) as

$$\lambda^4 - 2p\lambda^2 = q$$

with $\lambda^2 = p \pm \sqrt{p^2 + q}$. Moreover, setting

$$\frac{\mu_1}{r_1^3} + \frac{\mu_2}{r_2^3} = 2f$$

where $r_1$, $r_2$ are the distances of the collinear points from the masses, respectively we can write

$$\Omega_{a,a} = \mu_1\left(1+\frac{2}{r_1^3}\right) + \mu_2\left(1+\frac{2}{r_2^3}\right) = 1+4f$$
$$\Omega_{b,b} = \mu_1\left(1-\frac{1}{r_1^3}\right) + \mu_2\left(1-\frac{1}{r_2^3}\right) = 1-2f < 0$$

we find

$$p = \frac{1}{2}\left(\Omega_{a,a}+\Omega_{b,b}\right) - 2 = f-1$$
$$q = -\Omega_{a,a}\Omega_{b,b} = 8f^2-2f-1$$

and finally

$$\lambda^2 = f-1 \pm \sqrt{9f^2-4f}$$

with the solution $\lambda_1 = i\nu_1$, $\lambda_2 = -i\nu_1$, $\lambda_3 = \nu_2$, $\lambda_4 = -\nu_2$, and $\nu_1, \nu_2 \in \mathbb{R}_+$. From the generic linear solution of (6.6)

$$X = A_1 e^{i\nu_1 t} + A_2 e^{-i\nu_1 t} + A_3 e^{\nu_2 t} + A_4 e^{-\nu_2 t}$$
$$Y = B_1 e^{i\nu_1 t} + B_2 e^{-i\nu_1 t} + B_3 e^{\nu_2 t} + B_4 e^{-\nu_2 t}$$

and from $e^{\pm i\nu_1 t} = \cos(\nu_1 t) \pm i\sin(\nu_1 t)$, $e^{\pm\nu_2 t} = \cosh(\nu_2 t) \pm i\sinh(\nu_2 t)$ we get

$$X = \alpha_1\cos(\nu_1 t) + \alpha_2\sin(\nu_1 t) + \alpha_3\cosh(\nu_2 t) + \alpha_4\sinh(\nu_2 t)$$
$$Y = \beta_1\cos(\nu_1 t) + \beta_2\sin(\nu_1 t) + \beta_3\cosh(\nu_2 t) + \beta_4\sinh(\nu_2 t)$$

with

$$\alpha_1 = A_1 + A_2$$
$$\alpha_2 = i(A_1 - A_2)$$
$$\beta_1 = B_1 + B_2$$
$$\beta_2 = i(B_1 - B_2)$$

as well as

$$\alpha_3 = A_3 + A_4$$
$$\alpha_4 = A_3 - A_4$$
$$\beta_3 = B_3 + B_4$$
$$\beta_4 = B_3 - B_4$$

The coefficients, $A_1, \ldots, A_4$, $B_1, \ldots, B_4$ and $\nu_1, \nu_2$ can be obtained from the individual conditions in a similar way as we did for the linearized motion close to the triangular equilibria. As we can see, the motion around the collinear points can be seen as a composition of a periodic and a hyperbolic motion. The motion is unstable, since a small deviation from an initial periodic solution will increase the hyperbolic contribution such that the third body will drift away from the librational points.

The analysis of the motion close to $L_3$ is important for the design of transfer orbits from one primary to the other. As it is shown [140, 141] low energy orbits can be constructed with possible applications to reach the Moon. The special dynamical properties of the equilibria in the restricted three-body problem are the origin of a new branch of space science, namely the field of space manifold dynamics [142].

## 6.5
## Potential and the Zero Velocity Curves

We visualize the potential in the space $\Omega \times (x, y)$-plane in Figure 6.5 (from the direction of positive values of $\Omega$) and in Figure 6.6 (from the direction of negative $\Omega$). The different colors separate the equipotential lines, and we also indicate the locations of the five equilibria. In agreement with Figure 6.3 the potential tends to $\infty$ at the position of the two masses $\mu_1$ and $\mu_2$ – in Figure 6.5 the equipotential lines look like circles close to those critical values. In both figures we find different kinds of equipotential lines on the surface $\Omega(x, y)$. Some of them surround $\mu_1$ and $\mu_2$ only, while the others form curves of complex structure. The geometry of the potential is also useful to investigate the type of motion, which can be found in the restricted problem. It follows a natural integral of motion, the Jacobi integral.

If we multiply the first of (6.2) by $\dot{x}$ and the second by $\dot{y}$ the sum of the two becomes:

$$\ddot{x}\dot{x} + \ddot{y}\dot{y} = \frac{\partial \Omega}{\partial x}\dot{x} + \frac{\partial \Omega}{\partial y}\dot{y}$$

which can directly be integrated with respect to time. Since the right hand side of the above equation is the total derivative of the potential $d\Omega(y, x)/dt$ we immedi-

**Figure 6.5** Potential $\Omega$ in 3D for $\mu = 0.103$ from z+ view.

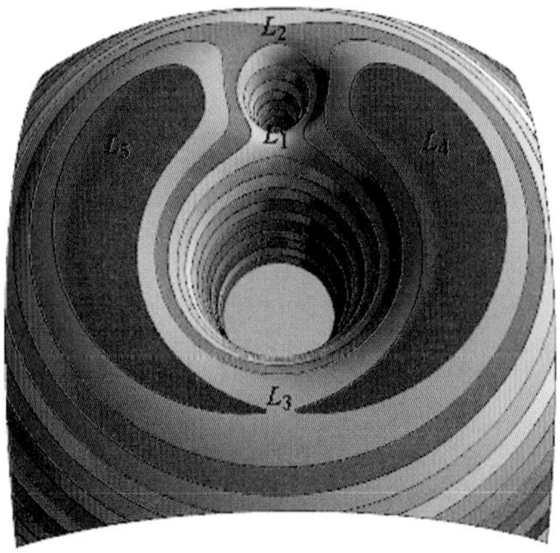


**Figure 6.6** Potential in 3D for $\mu = 0.103$ bottom view.

ately get

$$\dot{x}^2 + \dot{y}^2 = 2\Omega - C_J \qquad (6.13)$$

where $C_J$ is the integration, called Jacobi constant. From this equation we obtain the condition, that $2\Omega \geq C_J$ for all times since the speed of the third body $v = \sqrt{\dot{x}^2 + \dot{y}^2}$ is not allowed to become negative. In other words the curve given by

$$f(x, y, C_J) \equiv v^2 = 2\Omega(x, y) - C_J \qquad (6.14)$$

gives a bound on the allowed motion of the body. The curves defined by (6.14) are usually called zero velocity or Hill's curves. In the rotating coordinate system the massless body with initial conditions $(x_0, y_0)$ defines the Jacobi constant $C_J = 2\Omega(x_0, y_0)$ together with a Hill curve $f(x, y, C_J) = 0$. From the restriction of the velocity $f(x, y, C_J) > 0$ we find a region of forbidden motion (defined by $f(x, y, C_J) < 0$) together with the turning point of the third body at $f(x, y, C_J) = 0$, as it is shown for example in Figure 6.7b. A closed Hill's curve

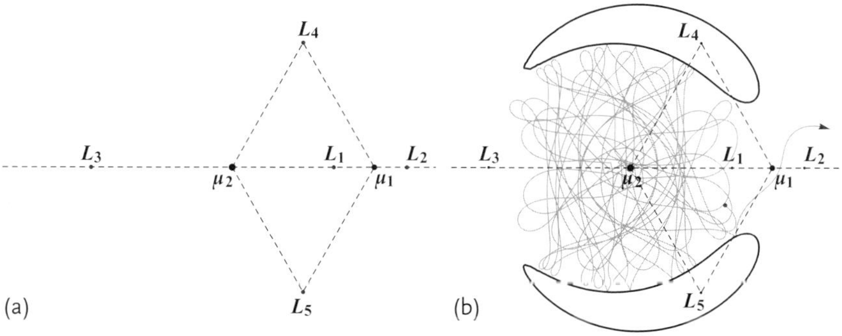


**Figure 6.7** Hill's curve for $\mu = 0.05$ and $C_J = 2\Omega_4 = 2\Omega_5 = 3$ (a) and $3 < C_J < 2\Omega_3$ (b). (a) orbits degenerate to the two triangular equilibria $L_4$ or $L_5$. (b) orbits turn at the boundary of the zero velocity curves but may leave the system.

already provides a stability criterion on the motion of the third body: if the massless body starts within a closed zero velocity curve the motion will be bounded within forever. The topology of the zero velocity curves can be investigated by looking on the critical values of $C_J = 2\Omega_k$ at the equilibrium points $L_1, \ldots, L_5$ with $k = 1, \ldots, 5$, respectively. For this reason we first write the potential as:

$$2\Omega = (1-\mu)\left(r_1^2 + \frac{2}{r_1}\right) + \mu\left(r_2^2 + \frac{2}{r_2}\right)$$

which is true in our system of units, since from $\mu_1 + \mu_2 = 1$ we have $\mu_1 r_1^2 + \mu_2 r_2^2 = x^2 + y^2 + \mu_1\mu_2$ also. In our units and the case of the triangular points $L_4$, $L_5$ with $r_1 = r_2 = 1$ we immediately find $2\Omega_4 = 2\Omega_5 = 3$. To obtain $\Omega_k$ with $k = 1, 2, 3$ we evaluate the potential at the position of the remaining equilibria and find for $\mu = 0.05$ the approximate numerical values[5] $\Omega_1 = 3.468$, $\Omega_2 = 3.402$ and $\Omega_3 = 3.097$. With these values we are able to investigate the topology of the Hill's curves as it is shown in Figures 6.7–6.10:

**Case $C_J = 2\Omega_4 = 2\Omega_5$** For $C_J < 3$ no zero velocity curves exist, for $C_J = 3$ they degenerate to the Lagrangian points $L_4$, $L_5$ (Figure 6.7a). Orbits starting at $L_4$ or $L_5$ degenerate to points and are resting at the equilibria positions forever.

**Case $3 < C_J < 2\Omega_3$** Increasing the Jacobi constant within the interval $3 < C_J < 2\Omega_3$ we find two disconnected curves (Figure 6.7b) enclosing the triangular points $L_4$ and $L_5$. Orbits starting outside the forbidden region are bounced back at the zero velocity curves but may exit the system after a while (the direction of the orbit is indicated by the arrow and finally leaves the system).

**Case $C_J = 2\Omega_3$** The tadpole like forbidden regions become more and more elongated until they touch at the unstable equilibrium $L_3$ with the value of the Jacobi constant $C_J = 2\Omega_3$. The case is shown in Figure 6.8a: an orbit enters the region which is encircled by the inner boundary of the forbidden region. It may be trapped within for a while until it leaves the system again along the direction to the second mass $\mu_1$.

**Case $2\Omega_3 < C_J < 2\Omega_2$** The two tadpole like regimes are merged to one connected forbidden region only. The boundary curve of it resembles the one of a horse-shoe, which encloses the three equilibria $L_4$, $L_3$ and $L_5$, respectively. An orbit starting within the inner boundary of the forbidden region may be trapped within for a while but is allowed to leave the system like in the former case (see Figure 6.8b).

**Case $C_J = 2\Omega_2$** With increasing Jacobi constant the outer boundaries are close the horse-shoe like forbidden region at the unstable equilibrium $L_2$ with $C_J = \Omega_2$ (see Figure 6.9a). We show the case of an orbit starting within the inner boundary

5) The location of the equilibria were obtained by a numerical root finding algorithm and checked for consistency with simplified analytical formulae.

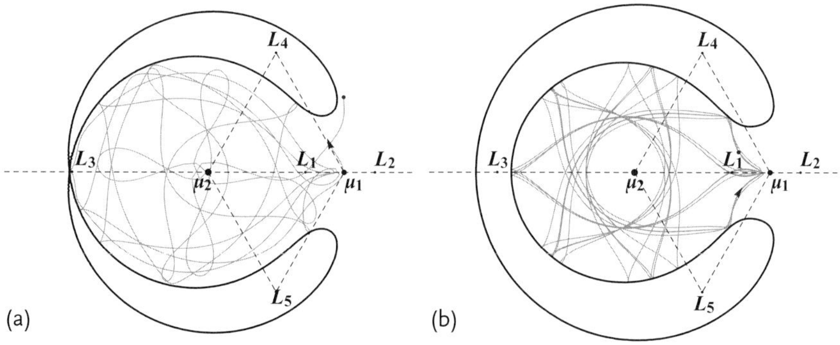


**Figure 6.8** Hill's curve for $\mu = 0.05$ and $C_J = 2\Omega_3$ (a) and $2\Omega_3 < C_J < 2\Omega_2$ (b). (a) and (b) orbits starting around the central mass $\mu_1$ are bounced back at the zero velocity curves but may exit the system into the direction to $L_3$.

of the forbidden region, which is chaotic and with irregular shape. It may become arbitrarily close to both masses $\mu_1$ and $\mu_2$ and is unable to leave the system but through the stable and unstable manifolds originating in $L_2$.

**Case $2\Omega_2 < C_J < 2\Omega_1$** The case $2\Omega_2 < C_J < 2\Omega_1$ is shown in Figure 6.9b. The forbidden region becomes again one connected zone, the inner boundary of it starts to stick at the unstable equilibrium $L_1$. An orbit starting in a satellite orbit around $\mu_2$ stays close to it at the beginning, while after times long enough it may enter the regime close to $\mu_1$. The example clearly visualizes the behavior of the orbit close to the zero-velocity curve, where the orbit turns the direction at the inner boundaries encircling the mass $\mu_1$.

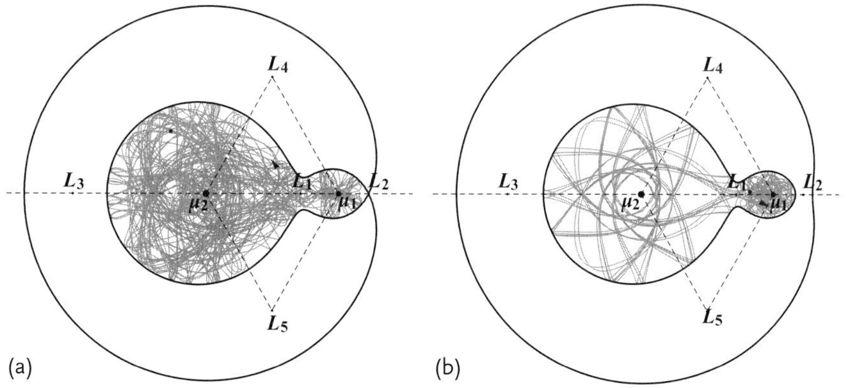


**Figure 6.9** Hill's curve for $\mu = 0.05$ and $C_J = 2\Omega_2$ (a) and $2\Omega_2 < C_J < 2\Omega_1$ (b). (a) and (b) orbits starting around $\mu_1$ or $\mu_2$ stay bounded within the inner boundary of the zero velocity curves for eternal times.

**Case $C_J = 2\Omega_1$** At the value of the Jacobi constant $C_J = \Omega_1$ the inner boundary of the forbidden region touches at the unstable equilibrium $L_1$ and an orbit starting in the circular region around $\mu_2$ stays there for very long times. The only connection between the two circular regimes around $\mu_1$ and $\mu_2$ is the one through the unstable and stable manifolds originating in $L_1$. We show a typical situation in Figure 6.10a: the orbit starting at the inner boundary of the circular region around $\mu_1$ becomes arbitrary close to $L_1$ but does not reach it for very long integration times.

**Case $C_J > 2\Omega_1$** We are left to investigate the case $C_J > 2\Omega_1$. The forbidden region completely disconnects the two regimes of allowed motion around $\mu_1$ and $\mu_2$. An orbit starting in either the circular region around the former or the latter is trapped around the two primary bodies forever. We show the case, where the orbit is trapped in the allowed regime of motion around $\mu_2$ in Figure 6.10b.

As we have seen, the zero velocity curves prevent the third body to escape from the primaries even on chaotic orbits. Moreover, a closed curve implies a forbidden region for given Jacobi constant $C_J$. If the forbidden region separates the two primaries the third body will stay close to one of the primaries forever. It is known, since [140] that transfer orbits exist between the two primaries and that they can be constructed by making use of the stable and unstable manifolds close to the equilibria.

In Figures 6.7–6.10 we fixed the mass ratio to $\mu = 0.05$, the form and topology of the zero velocity curves was solely due to the change of the Jacobi constant $C_J$. The topology of the Hill's curves also depends on the mass ratio of the primaries as it is shown in Figure 6.11. We plot the structure of the potential for the mass ratios of the primaries $\mu = 0$, $\mu = 0.082$, $\mu = 0.12$ and $\mu = 0.32$, respectively: for $\mu = 0$ the equipotential curves degenerate to concentric circles around the primary $\mu_1$ (Figure 6.11a); increasing the mass to $\mu = 0.082$ we find a situation of similar

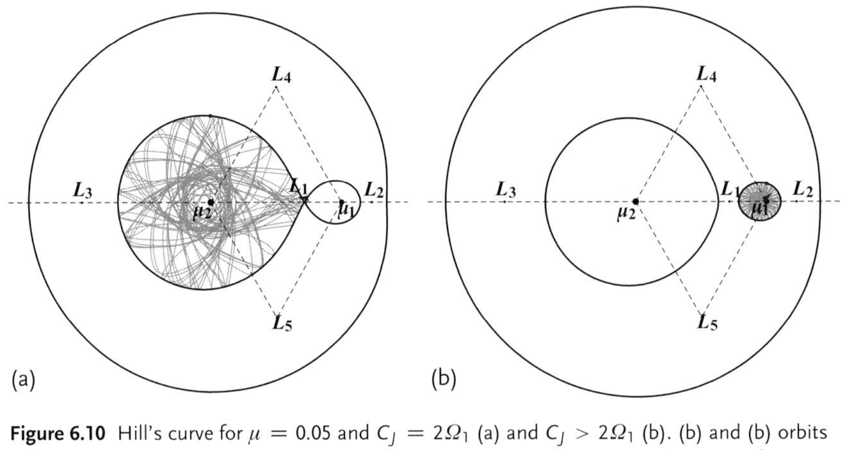


**Figure 6.10** Hill's curve for $\mu = 0.05$ and $C_J = 2\Omega_1$ (a) and $C_J > 2\Omega_1$ (b). (b) and (b) orbits starting close to $\mu_1$ (a) or close to $\mu_2$ (b) stay in satellite type motion around $\mu_1$ or $\mu_2$ forever.

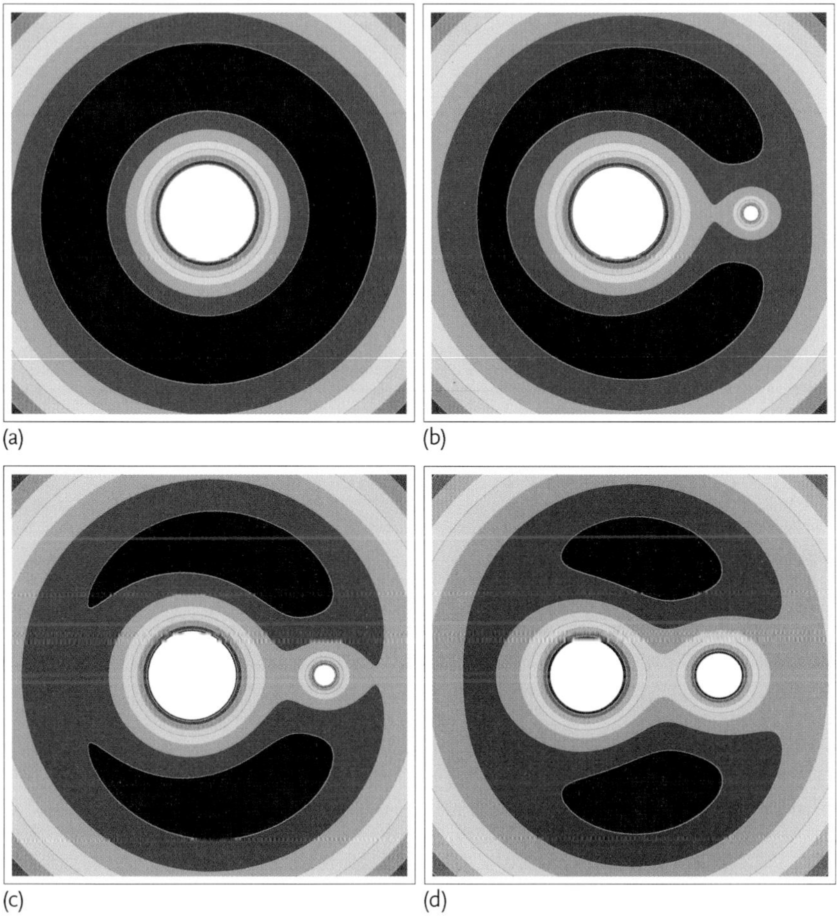

Figure 6.11 Dependency of the structure of the equipotential lines of $\Omega$ on the mass ratio of the primary bodies: $\mu = 0$, (a); $\mu = 0.08$ (b); $\mu = 0.12$ (c); $\mu = 0.32$ (d).

geometry as the one described above (Figure 6.11b). The case $\mu = 0.12$ is shown in Figure 6.11c, while the case $\mu = 0.32$ is given in Figure 6.11d.

## 6.6
## Spatial Restricted Three-Body Problem

The circular restricted three-body problem is the simplest case of the family of restricted three-body problems. If we allow the massless body to move also off the invariable plane of motion of the two primary bodies and define the inclination $i$ and the ascending node $\Omega$ of the third body with respect to this plane we get the spatial restricted three-body problem. The equations of motion can be derived in a similar way as in the planar case and turn out to be of the form:

$$\begin{aligned}\ddot{x} - 2\dot{y} &= \frac{\partial \Omega}{\partial x}\\ \ddot{y} + 2\dot{x} &= \frac{\partial \Omega}{\partial y}\\ \ddot{z} &= \frac{\partial \Omega}{\partial z}\end{aligned} \tag{6.15}$$

where the potential is given by:

$$\Omega = \frac{1}{2}(x^2 + y^2) + \frac{\mu_1}{r_1} + \frac{\mu_2}{r_2}$$

and the $r_1$, $r_2$ follow from the Euclidean distance:

$$\begin{aligned}r_1^2 &= (x - \mu_2)^2 + y^2 + z^2\\ r_2^2 &= (x + \mu_1)^2 + y^2 + z^2\end{aligned}$$

The Jacobi constant can be derived following the same procedure as in the two dimensional case: we multiply the first of (6.15) with $\dot{x}$, the second with $\dot{y}$ and the third with $\dot{z}$. Next, we sum the three equations and integrate with respect to time to get:

$$\dot{x}^2 + \dot{y}^2 + \dot{z}^2 = 2\Omega\,(x, y, z) - C_J \tag{6.16}$$

From the condition $\dot{x}^2 + \dot{y}^2 + \dot{z}^2 \geq 0$ we are able to define zero velocity surfaces defined by the equation:

$$2\Omega\,(x, y, z) - C_J = 0$$

which separates the space $(x, y, z)$ into regimes of allowed and forbidden motion like in the planar case. If we define the function:

$$f(x, y, z) = 2\Omega - C_J$$

the former are defined by $f(x, y, z) > 0$, the latter by the condition $f(x, y, z) < 0$, while the surfaces are defined as the set:

$$\{(x, y, z) : f(x, y, z) = 0\}$$

Typical geometries of zero-velocity surfaces are shown in Figure 6.12 and Figure 6.13. The plots should also be compared with Figures 6.7–6.10 (the mass ratio of the primary bodies was set again to $\mu = 0.05$). We show the surfaces for the critical values $C_J = 2\Omega_3$, $C_J = 2\Omega_2$ and $2\Omega_1$ in Figure 6.12.

**3D Case $C_J = 2\Omega_3$** With $C_J = 2\Omega_3$ (Figure 6.12a) the surface touches at the location of the equilibrium $L_3$, an analogous situation as seen in Figure 6.8a.

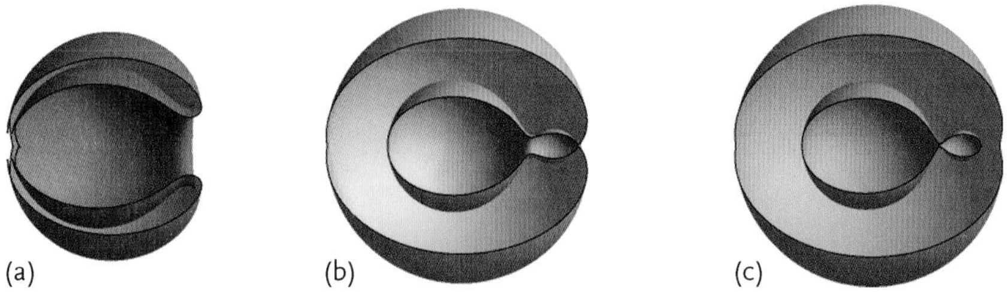


**Figure 6.12** Zero velocity surfaces for $\mu = 0.05$, excluding the part with negative $y$, for $C_J = 2\Omega_3, 2\Omega_2, 2\Omega_1$, respectively. Compare with Figures 6.7–6.10.

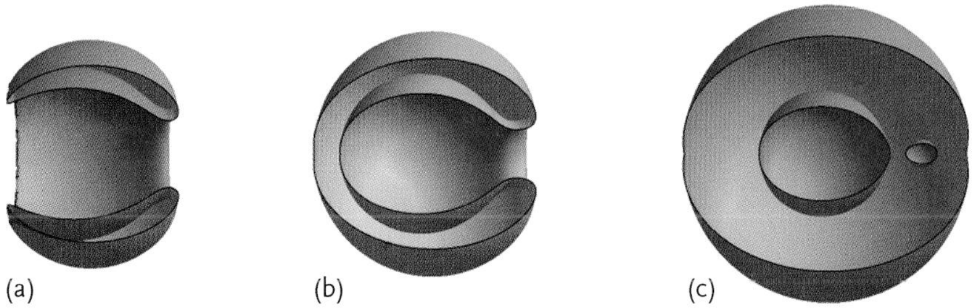


**Figure 6.13** Zero velocity surfaces for $\mu = 0.05$, excluding the part with negative $y$. (a) a case with $3 < C_J < 2\Omega_3$. (b) a case with $2\Omega_3 < C_J < 2\Omega_2$. (c) a case with $C_J > 2\Omega_1$. Compare with Figures 6.7–6.10.

**3D Case $C_J = 2\Omega_2$** For $C_J = 2\Omega_2$ the surface separates the 3D-space into three spheres of three different diameter (see Figure 6.12b). The inner two spheres define a connected regime of allowed motion, while the smaller inner sphere touches the outer largest sphere, from within, at the location of the equilibrium position $L_2$. Compare with Figure 6.9a.

**3D Case $C_J = 2\Omega_1$** The case $C_J = 2\Omega_1$ is shown in Figure 6.12d: the inner two spheres, of the former case, become disconnected from the outer sphere but touch at the location of the equilibrium position $L_1$ (see also Figure 6.10a).

We also provide typical examples of zero-velocity surfaces inbetween the critical values in Figure 6.13.

**3D Case $3 < C_J < 2\Omega_3$** For $3 < C_J < 2\Omega_3$ (as shown in Figure 6.13a) the surface forms an elongated tube around the ring formed by the equilibrium positions $L_4$ and $L_5$. Compare with Figure 6.7b.

**3D Case $2\Omega_3 < C_J < 2\Omega_2$** Within the interval $2\Omega_3 < C_J < 2\Omega_2$ the surface resembles a sphere within a sphere of similar shape (see Figure 6.13b), like a revolution plot of the curve shown in Figure 6.8, right around the $x$-axis.

**3D Case $C_J > 2\Omega_1$** We provide the case which generalizes the one shown in Figure 6.10b in Figure 6.13c. The surface consists of two disconnected spheres within a third sphere of larger diameter.

## 6.7
## Tisserand Criterion

An interesting application in our Solar system of the Jacobi integral of the circular spatial restricted three-body problem exists and is widely known as the Tisserand's criterion[6]: let $m_1$ be the mass of the Sun, $m_2$ be the mass of Jupiter and $r_1$, $r_2$ be the distance of a massless asteroid or comet from the Sun and Jupiter, respectively. Let furthermore $(x, y, z)$ be the position of the massless body and $(\dot{x}, \dot{y}, \dot{z})$ its velocity. If one takes the origin of the restricted three-body problem at the barycenter of the Sun–Jupiter system and in a fixed coordinate system the Jacobi constant takes the form:

$$C = \frac{1}{2}\left(\dot{x}^2 + \dot{y}^2 + \dot{z}^2\right) - \left(\frac{m_1}{r_1} + \frac{m_2}{r_2}\right) - (x\dot{y} - y\dot{x}) \tag{6.17}$$

where we assume that the time unit is such that the mass unit and distance unit equals one. To express the Jacobi constant in terms of osculating orbital elements we use the relations

$$\dot{x}^2 + \dot{y}^2 + \dot{z}^2 = \frac{2}{r_1} - \frac{1}{a}$$
$$x\dot{y} - y\dot{x} = \sqrt{m_1}\sqrt{p}\cos(i)$$

where $a$ is the semimajor axis, the parameter $p = a(1 - e^2)$, $e$ is the eccentricity, and $i$ is the inclination of the orbit of the massless particle. From this we get:

$$C = -\frac{1}{2}\left[\frac{1}{a} + 2m_2\left(\frac{1}{r_2} - \frac{1}{r_1}\right) + 2\sqrt{m_1}\sqrt{p}\cos(i)\right]$$

which can be arranged to get:

$$\frac{1}{2a} + \sqrt{m_1}\sqrt{a(1 - e^2)}\cos(i) = -\left[C + m_2\left(\frac{1}{r_2} - \frac{1}{r_1}\right)\right]$$

It is true that for $r_2$ large enough the left hand side is constant up to the order of the perturbing mass $m_2$. The left hand side does just depend on $(a, e, i)$ while the mass $m_1$ is constant. If we denote by $(a_1, e_1, i_1)$ the reduced set of orbital elements of the asteroid or comet before a close encounter with $m_1$ and by $(a_2, e_2, i_2)$ the orbital elements after it we find

$$\frac{1}{2a_1} + \sqrt{a_1\left(1 - e_1^2\right)}\cos(i_1) \simeq \frac{1}{2a_2} + \sqrt{a_2\left(1 - e_2^2\right)}\cos(i_2)$$

which is one formulation of Tisserand's criterion. It is useful to decide if a comet or asteroid observed before it is hidden by the Sun coincides with the observation of it after the event.

6) We follow the exposition as given in [138]. See also [143, 144] for a different exposition.

## 6.8 Elliptic Restricted Three-Body Problem

Another generalization of the circular restricted three-body problem is to allow the primaries to have nonzero eccentricity. In a first attempt we again restrict the third body to move in the same plane as the primaries and implement a coordinate system that can be fixed relative to the motion of the primaries, as we did in the circular case. Since in the elliptic problem the distance $l$ between the primaries is no longer constant and, moreover, the velocities of the primaries depend on time, the only way is to perform a time-dependent transformation that rotates and pulsates with the same periods as the primary bodies orbit around their common barycenter. In the rotating pulsating coordinate system the position of the primaries are again fixed, and the motion of the third body is determined by a similar set of equations of motion as in the circular case. If we define the position and velocity in the rotopulsating coordinate system as $(\xi, \eta)$ the equations of motion can be written [137]:

$$\begin{aligned} \frac{d^2\xi}{dE^2} - 2\frac{d\eta}{dE} &= \frac{\partial \Omega_E}{\partial \xi} \\ \frac{d^2\eta}{dE^2} + 2\frac{d\xi}{dE} &= \frac{\partial \Omega_E}{\partial \eta} \end{aligned} \tag{6.18}$$

where $\Omega_E$ is the potential in the elliptic case related to the potential of the circular case by:

$$\Omega_E = \Omega(1 + e\cos\nu)^{-1}$$

and $\Omega$ in terms of $(\xi, \eta)$ has the usual form:

$$\Omega(\xi, \eta) = \frac{1}{2}(\xi^2 + \eta^2) + \frac{\mu_1}{r_1} + \frac{\mu_2}{r_2}$$

If we repeat the calculation to obtain the Jacobi constant $C_J$ we get the equation:

$$\left(\frac{d\xi}{d\nu}\right)^2 + \left(\frac{d\eta}{d\nu}\right)^2 = 2\int\left(\frac{d\omega}{d\xi} + \frac{d\omega}{d\eta}\right) = 2\omega - 2e\int\frac{\Omega\sin\nu}{(1 + e\cos\nu)^2}d\nu - C_J \tag{6.19}$$

The integral in the right-hand side of the preceding equation can not be solved. It is not possible to find the additional constant of motion as in the circular problem. It is a remarkable fact that, formally, the equations of motion of the elliptic problem remind of the equations of motion of the circular problem. The difference, of course, is the definition of the potential, which in the elliptic case is time dependent due to the presence of the true anomaly $\nu$.

The above form of the equations of motion goes back to [145], and was implemented in the case of the Sun–Jupiter system. The planar case was reduced to

Hill's form of the equations of motion in [146], and on asymptotic solution with application to the Trojan case was found in [147]. A detailed numerical survey of quasi-periodic orbits in the elliptic restricted three-body problem was done in [148] based on the Lie-integration method derived in [26]. A symplectic mapping (first derived in [149]) instead of the equations of motion was used to derive analytical estimates based on Nekhoroshev-like theory in [150]. The problem of integrability of the problem was investigated by analytical means in [151] two integrals of motion have been found in [152] related to the so-called 'third integral' of motion. The disturbing function for the planar elliptic restricted three-body problem was obtained in [153]. The equilibria of the elliptic and spatial problem were analyzed in [154]. See also [137, 139] and references therein for further information.

## 6.9 Dissipative Restricted Three-Body Problem

In presence of dissipative forces the problem can still be treated in a near Hamiltonian framework, that is, it is possible to investigate the problem

$$\begin{aligned} \ddot{x} - 2\dot{y} &= \frac{\partial \Omega}{\partial x} + F_x \\ \ddot{y} + 2\dot{x} &= \frac{\partial \Omega}{\partial y} + F_y \end{aligned}$$

where $F_x$, $F_y$ are dissipative forces. We summarize three types relevant for Solar system dynamics. The simple nebular drag:

$$\begin{aligned} F_x &= k\dot{x}r^{\nu} \\ F_y &= k\dot{y}r^{\nu} \end{aligned}$$

where $k$, $\nu$ are parameters. The Poynting–Robertson light drag:

$$\begin{aligned} F_x &= k\frac{1}{r^2}\left[\dot{x} - y + \frac{x}{r^2}\left(x\dot{x} + y\dot{y}\right)\right] \\ F_y &= k\frac{1}{r^2}\left[\dot{y} + x + \frac{y}{r^2}\left(x\dot{x} + y\dot{y}\right)\right] \end{aligned}$$

and the nebular (or Stokes) gas drag (see [155]):

$$\begin{aligned} F_x &= -K\rho v_{\text{rel}}\left(\dot{x} - y + \Omega_{\text{g}} y\right) \\ F_y &= K\rho v_{\text{rel}}\left(\dot{y} + x - \Omega_{\text{g}} x\right) \end{aligned}$$

with

$$v_{\text{rel}} = \left[\left(\dot{x} - y + \Omega_{\text{g}} y\right)^2 + \left(\dot{y} + x - \Omega_{\text{g}} x\right)^2\right]^{1/2}$$

where the parameter $K$ is a constant and the density $\rho = \rho(r)$ and angular velocity $\Omega_{\text{g}} = \Omega_{\text{g}}(r)$ of the gas are functions depending on the radial component $r$ also.

The problem was investigated under the influence of Stokes and Poynting–Roberston drag in [156, 157] see also [158, 159] (and references therein) for further information.

# 7
# The Sitnikov Problem

The Sitnikov problem (= SP) serves as one of the best known models for a dynamical system which is chaotic. The name goes back to a publication some 50 years ago by K.A. Sitnikov [160]. We can define the SP in the following way: Two equally massive bodies (called primary bodies) are in a Keplerian orbit around their common barycenter; perpendicular to the plane of motion of the primaries a third massless body (called planet) is moving through the center of mass (see Figure 7.1). Whereas the motion of the primaries is regular by definition (it is the two body problem) whatever their eccentricity is, the motion of the planet is regular with all initial conditions when the primaries are in circular orbits (MacMillan problem). In the case of eccentric orbits of the primaries the motion of the third body depends sensitively on the initial conditions which is the characterization for chaotic motion. To demonstrate it the best is to cite one of the great mathematicians of the second half of the twentieth century [117]:

> "... we consider a solution of $z(t)$ with infinitely many zeroes $t_k(k = 0, \pm 1, \pm 2, \ldots)$ which are ordered according to size, $t_k < t_{k+1}, z(t_k) = 0$. Then we introduce the integers
>
> $$s_k = \left( \frac{t_{k+1} - t_k}{2\pi} \right) \tag{7.1}$$
>
> which measures the number of complete revolutions of the primaries between two zeroes of $z(t)$. This way we can associate to every such solution a double infinite sequence of integers. The main result can be expressed as the converse statement:
> Theorem. Given a sufficiently small eccentricity $\varepsilon > 0$ there exists an integer $S_{\text{lim}} = S_{\text{lim}}(\varepsilon)$ such that any sequence s with $s_k \geq m$ corresponds to a solution ..."

This very strong statement will be explained in the following with the aid of mathematical considerations but also with results of numerical experiments.

*Celestial Dynamics*, First Edition. R. Dvorak and C. Lhotka.

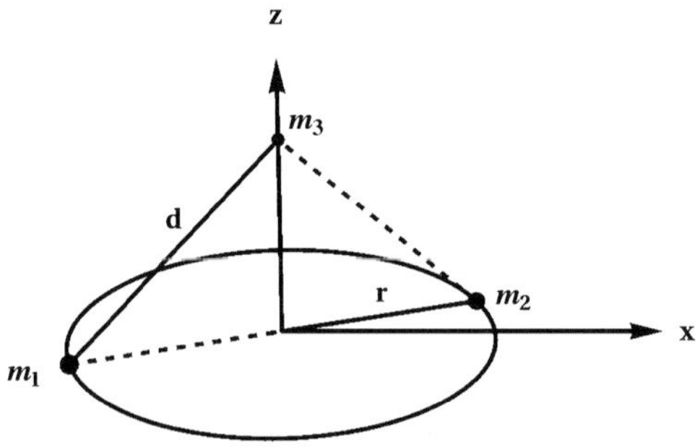


**Figure 7.1** Configuration of the MacMillan problem.

## 7.1 Circular Case: the MacMillan Problem

The case, when the primaries move on circular orbits, the MacMillan problem (=CSP), was already known to Leonhard Euler who treated it as a special case of the two fixed center problem. Under the gravitational force of two fixed masses ($m_1 = m_2$) a third massless body $m_3$ moves perpendicular to the plane of orbit of the two primaries through their common barycenter. Euler succeeded to solve the two fixed center problem for any motion using elliptic functions, where the CSP can be regarded as one special case. Two centuries later, in 1907, G. Pavanini [161] found solutions of the CSP with the aid of Weierstrass elliptic functions. Shortly after MacMillan [162] derived a solution in form of a Fourier series expansion; the coefficients are power series depending on the distance of the planet from the barycenter $z$ (sufficiently small), which is the reason that sometimes the CSP is named after MacMillan.

### 7.1.1 Qualitative Estimates

In the following we will develop simple analytical considerations for the CSP to get a first insight into the dynamics of this problem. With $m_1 = m_2 = m$, $d^2 = r^2 + z^2$ and $r =$ constant. In this case the constant Hamiltonian $H$ of the systems is

$$H = \frac{1}{2}\dot{z}^2 - \frac{2m}{d} \tag{7.2}$$

The velocity can be derived from the constant Hamiltonian ($H = h$)

$$\frac{dz}{dt} = \sqrt{2h + \frac{4m}{d}} \tag{7.3}$$

and the equation of motion for the planet is easily found by differentiating (7.2) with respect to the time

$$\ddot{z} = -\frac{2mz}{d^3} \tag{7.4}$$

Setting $2m = 1$ and $r = 1$ leads to the equation of motion in the following form:

$$\ddot{z} = -\frac{z}{\left(\sqrt{1+z^2}\right)^3} \tag{7.5}$$

We find a restriction for the square of the velocity from the Hamiltonian

$$\frac{1}{2}\dot{z}^2 = h + \frac{1}{\sqrt{1+z^2}} = h + \frac{1}{d} \leq h + 1 \tag{7.6}$$

because $\sqrt{1+z^2} = d$ and $d \geq 1$ is always fulfilled. As long as $h \geq -1$ the velocity is $\dot{z} \geq 0$ and the former mentioned condition valid, which means that there exists point of return ($z_{\max} = \pm\gamma$) for the planet when $\dot{z} = 0$

$$h = -\frac{1}{\sqrt{1+\gamma^2}} \tag{7.7}$$

As a consequence the planet has a bounded motion only for negative $h$ which is limited by two extreme values of $z$ symmetric to the plane of motion of the primaries.

$$z_{\max} = \pm\gamma = \pm\sqrt{\frac{1}{h^2} - 1} \tag{7.8}$$

We can distinguish three qualitatively different cases:

- A $\gamma = 0$ in the case of $h = -1$, the planet is exactly in the barycenter and does not move.
- B For a vanishing Hamiltonian we see that there is the point of reversion shifted to infinity.
- C For $h > 0$ the velocity is always finite.

B and C can be called *parabolic case* and *hyperbolic case*. The most interesting motion is when $-1 < h < 0$ because then it is always symmetric to $z = 0$ and reaches its maximum velocity in the barycenter

$$\dot{z}_{\max} = \pm\sqrt{2(h+1)} = \pm 2k\,; \quad \text{with } 0 \leq k = \sqrt{\frac{1+h}{2}} < \frac{1}{\sqrt{2}} \tag{7.9}$$

This purely periodic motion can be represented with the aid of a Fourier series

$$z = a_1 \sin \nu\tau + a_3 \sin 3\nu\tau + a_5 \sin 5\nu\tau + \ldots \tag{7.10}$$

where $\nu = 2\pi/P$ and $P$ is the period of the oscillation which can be derived using (7.3)

$$dt = \pm\frac{dz}{\sqrt{2h + \frac{2}{d}}} \tag{7.11}$$

Integration leads to

$$t - t_0 = \tau = \frac{1}{\sqrt{2}} \int\limits_{z_0}^{z} \frac{dz}{\sqrt{h + \frac{1}{\sqrt{1+z^2}}}} \tag{7.12}$$

which can be used to determine the period of oscillation

$$\frac{p}{4} = \frac{1}{\sqrt{2}} \int\limits_{z_0}^{z} \frac{dz}{\sqrt{h + \frac{1}{\sqrt{1+z^2}}}} \tag{7.13}$$

The solution (7.10) leads to an elliptic integral of the third kind. After rather lengthy transformations[1)] finally one ends up with a time series for $z$ where we use the parameter $\mu = \gamma^2/(1 + \gamma^2)$

$$z = \gamma \left[ \sin \nu\tau + \frac{3}{64}\mu(\sin \nu\tau + \sin 3\nu\tau) + \right. \tag{7.14}$$

$$\left. + \frac{1}{4096}\mu^2 \left(79 \sin \nu\tau + 108 \sin 3\nu\tau + 29 \sin 5\nu\tau\right) + \ldots \right] \tag{7.15}$$

In Figure 7.2 we show the periodic motion of the planet in the CSP for three different initial conditions $z_{\text{ini}}$ and $\dot{z}_{\text{ini}} = 0$. The frequency decrease with the amplitudes and converges to a final value for infinitesimal oscillations to $P = 2\sqrt{2}$, a value which will be derived in the next section.

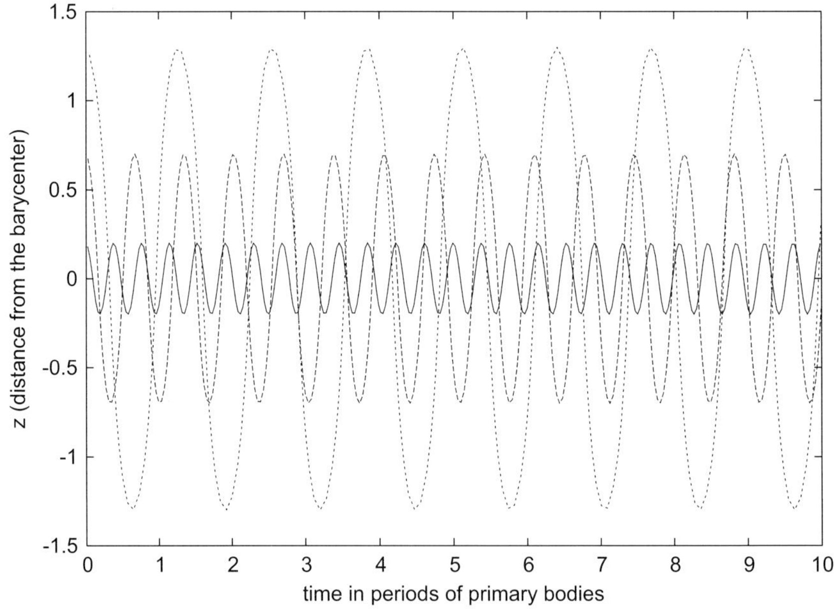


**Figure 7.2** Oscillations of $m_3$ for $z_{\text{ini}} = 0.2, 0.7, 1.3$ for 10 revolutions of the primary bodies.

1) For details see Stumpff [133, p. 75 ff.]

## 7.2 Motion of the Planet off the z-Axes

In the CSP Soulis *et al.* [163] found an infinite sequence of stability intervals on the $z$-axis with growing width for larger distances from the barycenter, which seems to tend versus a fixed nonzero value. The first interval is between $5.044 < z < 5.052$ and from the respective Table 7.1 we can see that the width decreases with the initial distance $z$ from the barycenter and seems to converge to a maximum value. These stable intervals were established with the chaos indicator SALI[2], which distinguishes quite well between stable and chaotic motion for a specific orbit. According to the definition of the SALI (bounded orbits for SALI $= 0$) certain small stable intervals in $z_{\text{ini}}$ (compare Table 7.1) are found and in between these stable intervals negative values of the SALI indicate chaotic (here unbounded) motion. In the respective Figure 7.3 we see the computations of the SALI along the $z$-axis between $5 < z_{\text{ini}} < 6$ where the first three stable windows are visible.

In Figures 7.4 and 7.5 we can see how the number of stable intervals along the initial distance to the barycenter $z$ changes with increasing $z_{\text{ini}}$. Whereas the length of the interval is increasing with $z$ up to $z = 15$, where it reaches is maximum extension $\delta z \sim 0.055$, it is then decreasing slightly. This is a different behavior as found in the former mentioned paper [163] where it is claimed that $\delta z \sim 0.05$ is a final value. Computations up to $z = 100$, which have been undertaken recently

**Table 7.1** Selected stable intervals on the z-axis; the convergence to a final value is expected (for more see text)

| Interval | Lower limit | Upper limit | Width |
|---|---|---|---|
| 1 | 5.044 | 5.052 | 0.008 |
| 2 | 5.453 | 5.470 | 0.017 |
| 3 | 5.848 | 5.872 | 0.024 |
| 4 | 6.230 | 6.259 | 0.029 |
| 5 | 6.600 | 6.634 | 0.034 |
| 6 | 6.961 | 6.999 | 0.037 |
| 7 | 7.314 | 7.353 | 0.039 |
| 8 | 7.657 | 7.699 | 0.042 |
| ⋮ | ⋮ | ⋮ | ⋮ |
| 12 | 8.965 | 9.010 | 0.045 |
| ⋮ | ⋮ | ⋮ | ⋮ |
| 20 | 11.332 | 11.381 | 0.049 |
| ⋮ | ⋮ | ⋮ | ⋮ |
| 24 | 12.425 | 12.475 | 0.050 |
| 25 | 12.691 | 12.741 | 0.050 |

2) Which was introduced briefly in Chapter 3.

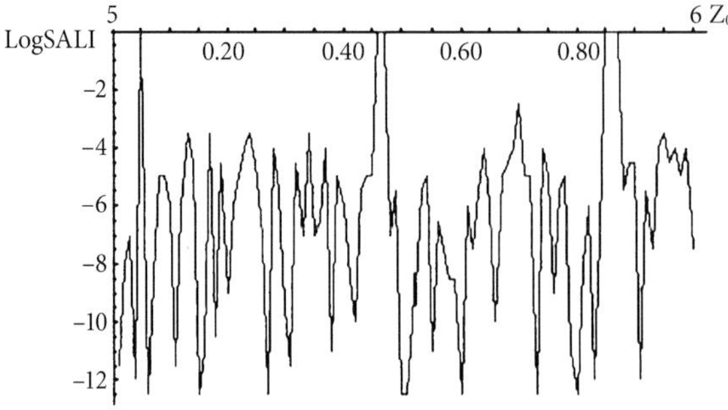


**Figure 7.3** The graph shows the numerical value of the SALI along the $z$-axis between $5 < z_{\text{ini}} < 6$ ($x$-axis) versus the numerical value of the SALI (after [163]).

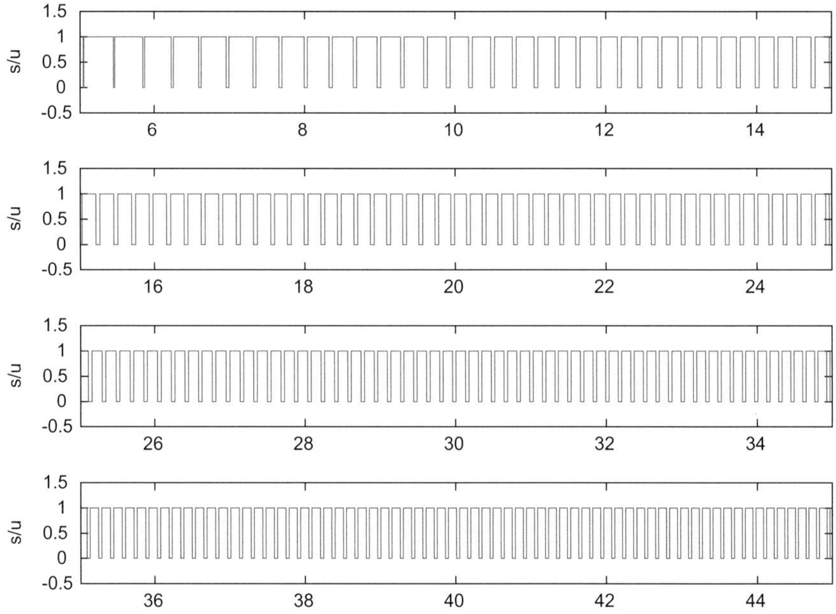


**Figure 7.4** The graphs show the stable ($y = 0$) and unstable ($y = 1$) intervals versus $z_{\text{ini}} < 45$.

indicate that for very large distances on the $z$-axis the ratio $\rho$ between stable and unstable intervals reaches asymptotically $\rho_\infty \sim 0.04$.

According to the stability analysis very small initial displacement from the $z$-axes for such intervals do not lead to escapes. Without the possibility to determine the radius of such small 'allowed' deviations analytically numerical simulations showed the possible displacements which still lead to stable motion. The quite interesting dynamical behavior is shown in Figure 7.6 where we plot the projection of such an orbit started inside the stable interval number 12 onto the plane of motion of the two primaries. A projection on the x-z plane (Figure 7.7) and a three dimensional plot (Figure 7.8) show how small the region of motion is when the

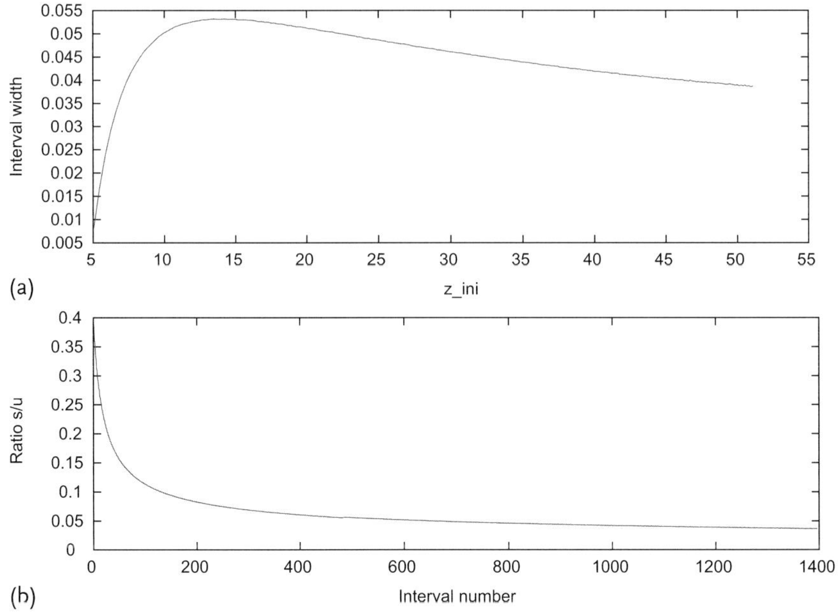


**Figure 7.5** (a) shows the length of the stable intervals on the $y$-axis versus the initial distance to the barycenter up to $z_{\text{ini}} = 50$; (b) depicts the ratio $\rho$ between stable and unstable intervals ($y$-axis) versus the interval number (up to $z_{\text{ini}} = 100$)

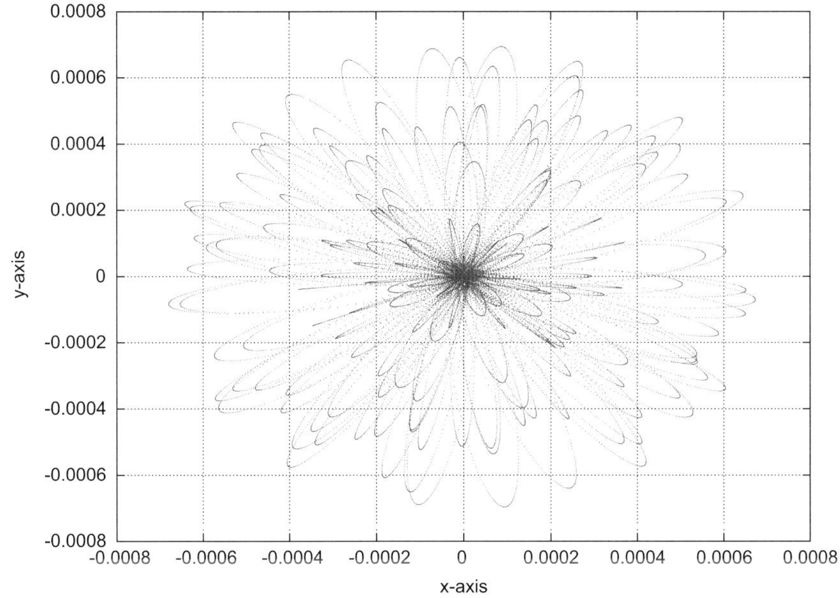


**Figure 7.6** A projection of a stable orbit in interval 12 on the $x$–$y$ plane. It is evident that the orbit lies inside a circle with $r \sim 0.0007$.

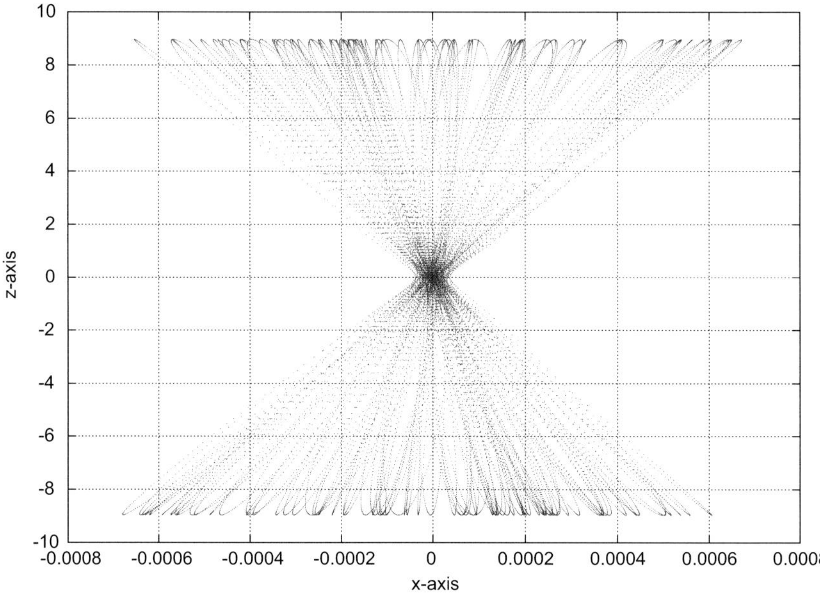


**Figure 7.7** A projection of a stable orbit in interval 12 on the $x$–$z$ plane

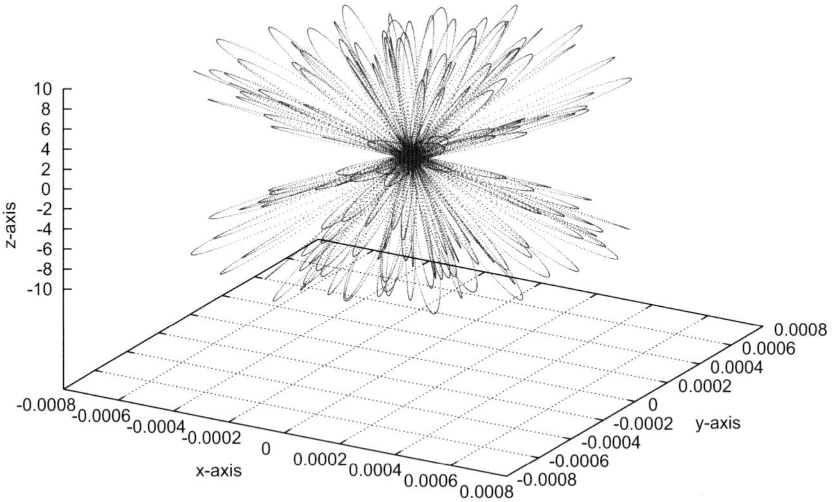


**Figure 7.8** A three dimensional view of an orbit starting in the stable interval 12 in the $x$–$y$–$z$ space.

planet passes through the barycenter (note the different scaling in the two axis). We emphasize that the projection on the $y$–$z$ plane is qualitatively quite the same; this means that this *double bouquet* orbit is inside a double cone extended symmetrically from the $x$–$y$ plane for $z = 0$.

Another property of these three dimensional orbits is that with increasing distance $z_{\text{ini}}$ from the barycenter the $x$–$y$ region, which leads to bounded motion, is also increasing; this is in contrast to the length of the stable intervals in the $z$-axis as it was shown in Figure 7.5a. It would be interesting to determine the volume of the double-cone where inside stable orbits exist: is it decreasing or increasing with larger with $z_{\text{ini}}$?

## 7.3
## Elliptic Case

In the elliptic case ($e > 0$) the distance of the primaries from the barycenter $r$ is not constant anymore (see Figure 7.9). For this reason the Hamiltonian explicitly depends on time; the equation of motion reads

$$\ddot{z} = -\frac{2mz}{\left(\sqrt{z^2 + r(t)^2}\right)^3} \tag{7.16}$$

No further integrals of motion are known – it is not possible to solve (7.16) in terms of analytic functions and only approximate solutions for small parameter $e$ can be derived (see end of this chapter). No global solution exists, only in the vicinity of a regular solution it is possible to use lengthy power series in time $t$ (in most of the cases Fourier series) to find – up to a certain time $T$ – solutions for a specific set of initial conditions. The appropriate tool is the numerical integration of the respective equations of motion that means that a solution (only ONE) can be computed step by step up to a time $T$. The reliability of the solution in both cases (analytical and numerical) up to $T$ depends where we are looking for solutions; in the chaotic regime the Lyapunov time tells us after what time 'the solution' forgets where it comes from, in a regular regime the validity of the solution is longer or, at least, it stays longer in the vicinity of the 'real' orbit. Nevertheless both methods, the 'analytical' approach with the aid of perturbation theory and the numerical one can be used complementary and provide interesting insights in the dynamical behavior

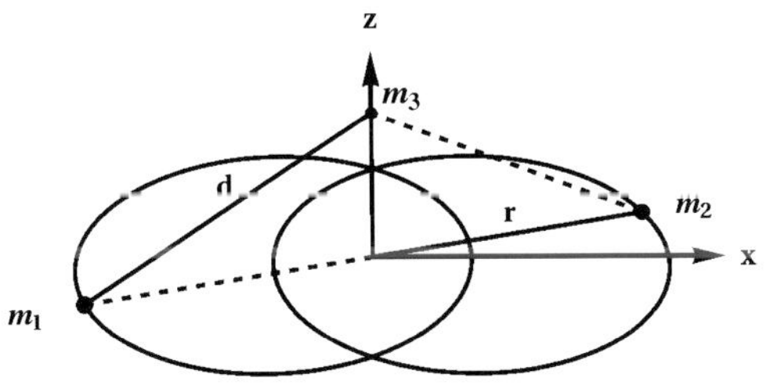


**Figure 7.9** Configuration of the Sitnikov problem.

of a system like the Sitnikov problem. In the case of the SP an appropriate way to study the dynamics with the aid of numerical results is to construct surfaces of section (SOS) where we compute orbits for various initial conditions.

### 7.3.1
### Numerical Results

In the Sitnikov problem with $e \neq 0$ the situation is very different: the structure of phase space is the one which we know from the study of nonlinear dynamical systems with periodic orbits (= PO), quasi-periodic orbits and chaotic ones. As example for the difference we show the orbital behavior of the third body for the same initial conditions ($z_{\text{ini}} = 0.05$ and the primaries in periastron) but for the eccentricities ($e = 0.0, 0.3, 0.6, 0.9$) of the primary bodies (Figure 7.10): from these plots it is well visible how irregular the orbits behave with increasing $e$.

The structure of the phase space for different eccentricities is shown in Figures 7.11 and 7.12. Every orbit in the SOS starts with different initial conditions for $z$ with $\dot{z} = 0$ and $r = r_{\min}$ (= periastron); one can imagine that we "drop" the third mass from this position. Points on this subspace of the three dimensional phase space are plotted for the next crossing of the orbit with the SOS when the primaries are again in their periastron position. We thus have chosen the positional

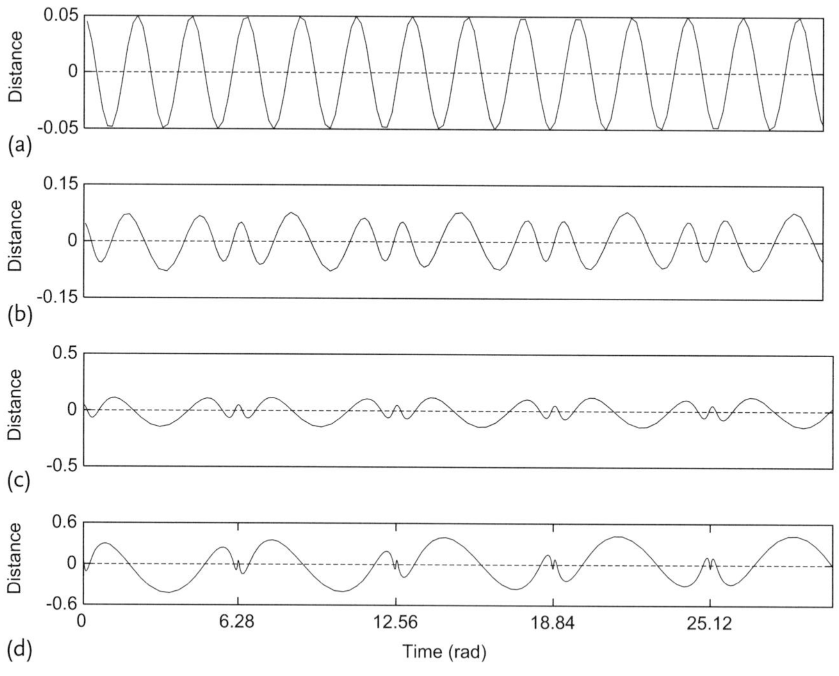


**Figure 7.10** Time evolution of the distance $z$ of the massless body ($y$-axes) from the in the Sitnikov problem for $e = 0, 0.3, 0.6$, and $0.9$ (a–d) where time ($x$-axes) is measured in radians

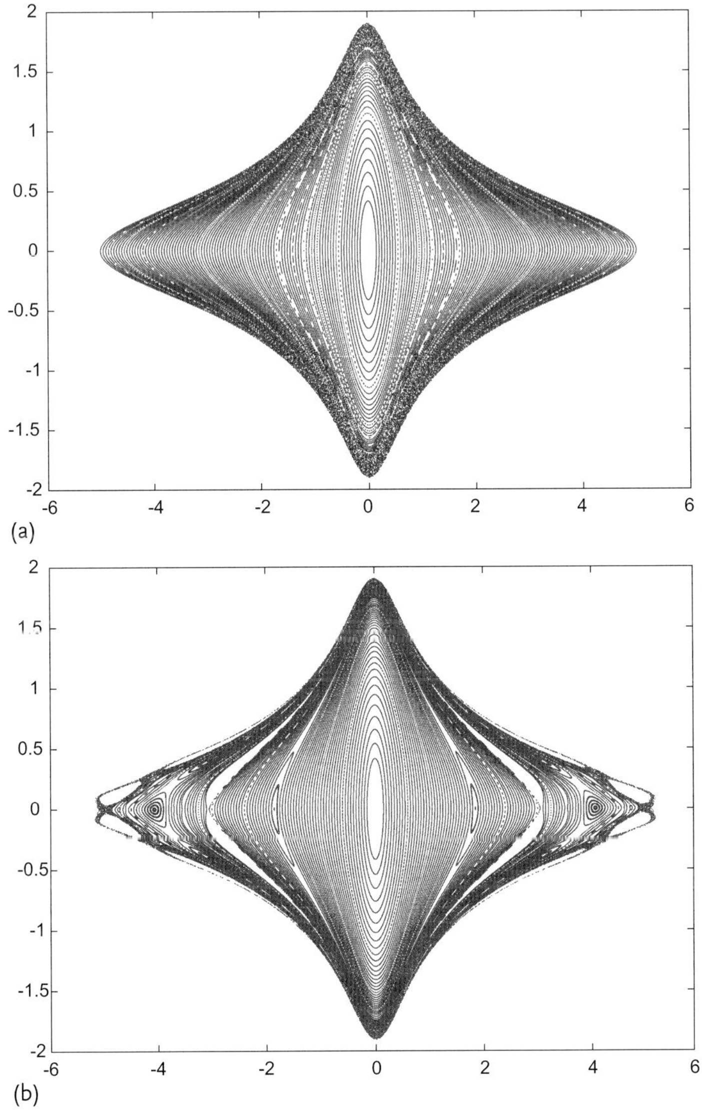


**Figure 7.11** Surfaces of Section ($z$ ($x$-axis) versus $\dot{z}$) for slightly different eccentricities of the primaries: $e = 0$ (a) $e = 0.005$ (b); a detailed description is given in the text.

angle of the primaries being zero as initial condition and as defining the surface of section:

- The SOS for $e = 0$ Figure 7.11a shows the well known behavior of an integrable dynamical system without any chaotic motion: all consecutive points for one initial condition are on one invariant curve.

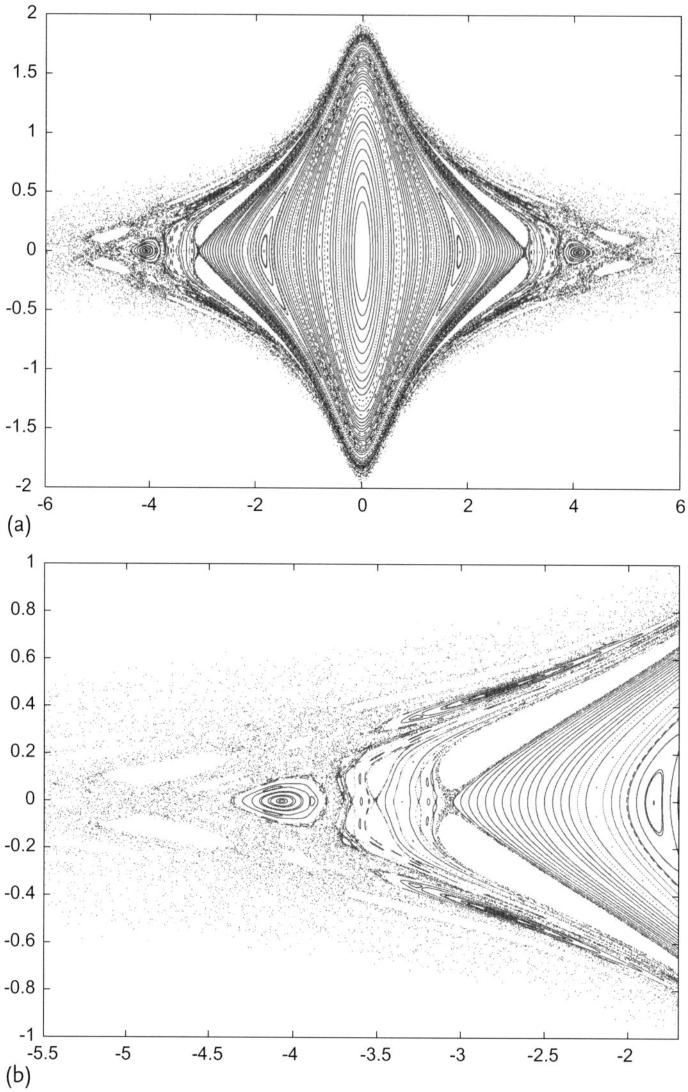


**Figure 7.12** Surfaces of Section ($z$ ($x$-axis) versus $\dot{z}$) for $e = 0.01$ (a) and a zoom of this graph (b); a detailed description is given in the text.

- The SOS for $e = 0.005$ Figure 7.11b shows already a very different picture, namely closed invariant curves around the center and periodic orbits represented by points, which are surrounded by 'islands', closed invariant curves around such a PO (e.g. close to $z = 4$ and $\dot{z} = 0$). The whole structure is explained in the chapter on the KAM theory. I emphasize one special island of regular motion visible through several small invariant curves around a PO at

$z \sim 1.848$ and $\dot{z} = 0$ which is the 1 : 2 resonance[3]. When we discuss the whole structure for $0 < e < 0.94$ we will see that this resonance is of outstanding importance.

- The SOS for $e = 0.01$ Figure 7.12a has again a different appearance especially for large $z$-values. The 1 : 2 PO is still surrounded by stable islands, but close to the island at $z = 4$ and $\dot{z} = 0$ many scattered points indicate chaotic motion. Whereas in the former picture this island is enclosed by invariant curves – making any orbit in this region confined to stay there forever – in the new plot we can see that with only a slightly higher eccentricity ($\delta e = 0.05$) an orbit there can escape to infinity. In this sense the small chaotic layer from before is not connected to the 'large chaotic sea'.
- A zoom of the SOS for $e = 0.01$ Figure 7.12b shows it even better and reveals the complicated structure of phase space with chains of small island (close to $z = -3.3$ and $\dot{z} = 0$ and another tiny chain of islands for $z = -3.6$ and $\dot{z} = 0$

The points in the middle of the islands are periodic orbits (PO) related to resonances between the oscillation of the planet and the period of the primaries, which are characterized by the ratio of two rational numbers. There are an infinite number of POs in a dynamical system; they are – as already Poincaré pointed out – the skeleton of the phase space. A very detailed discussion on this topic concerning POs, KAM-Tori, asymptotic curves and sticky orbits is given in several papers (e.g. [164]). Just for completeness we show in Table 7.2 the initial conditions $z_{\text{ini}}$ for some of the orbits shown in the preceding figures.

The outstanding role of the 1 : 2 resonance is evident when we discuss the complete structure of the phase space of the Sitnikov configuration using Figures 7.13 and 7.14. There we show in detail the phase space for eccentricities $0.0 < e < 0.94$ for a specially chosen SOS. These pictures are derived with the aid of cuts of $0.5 < z_{\text{ini}} < 3$ along the line $\dot{z}_{\text{ini}} = 0$; the integrations started for the primaries in their periastron. Stable orbits, defined as nonescaping up to a distance in the SOS $(z^2 + \dot{z}^2) < 10$ within the integration time of 1000 revolutions of the primaries were marked in yellow; the different color codes (from orange to blue) mark the escape time. Orbits starting in the dark blue region immediately lead to an escape.

**Table 7.2** Initial conditions for the distance $z$ for selected periodic orbits in the Sitnikov problem; the primaries are in periastron position and $\dot{z} = 0$; $n$ is the mean motion

| $n_{\text{prim}} : n_{\text{m3}}$ | $e = 0$ | $e = 0.2$ | $e = 0.4$ | $e = 0.6$ | $e = 0.8$ |
|---|---|---|---|---|---|
| 1 : 1 | 1.043 698 | 0.872 719 | 0.691 427 | 0.496 293 | 0.278 653 |
| 1 : 2 | 1.848 460 | 1.836 787 | 1.816 155 | 1.786 429 | 1.747 073 |
| 1 : 3 | 2.495 394 | 2.924 811 | 3.024 252 | 3.048 240 | 3.040 044 |
| 1 : 4 | 3.064 713 | 3.071 739 | 3.071 158 | 3.062 020 | 3.042 885 |
| 1 : 5 | 3.584 274 | 3.137 770 | 3.090 335 | 3.067 454 | 3.043 343 |

3) Whereas the planet makes one full oscillation through the barycenter, the primaries make two revolutions.

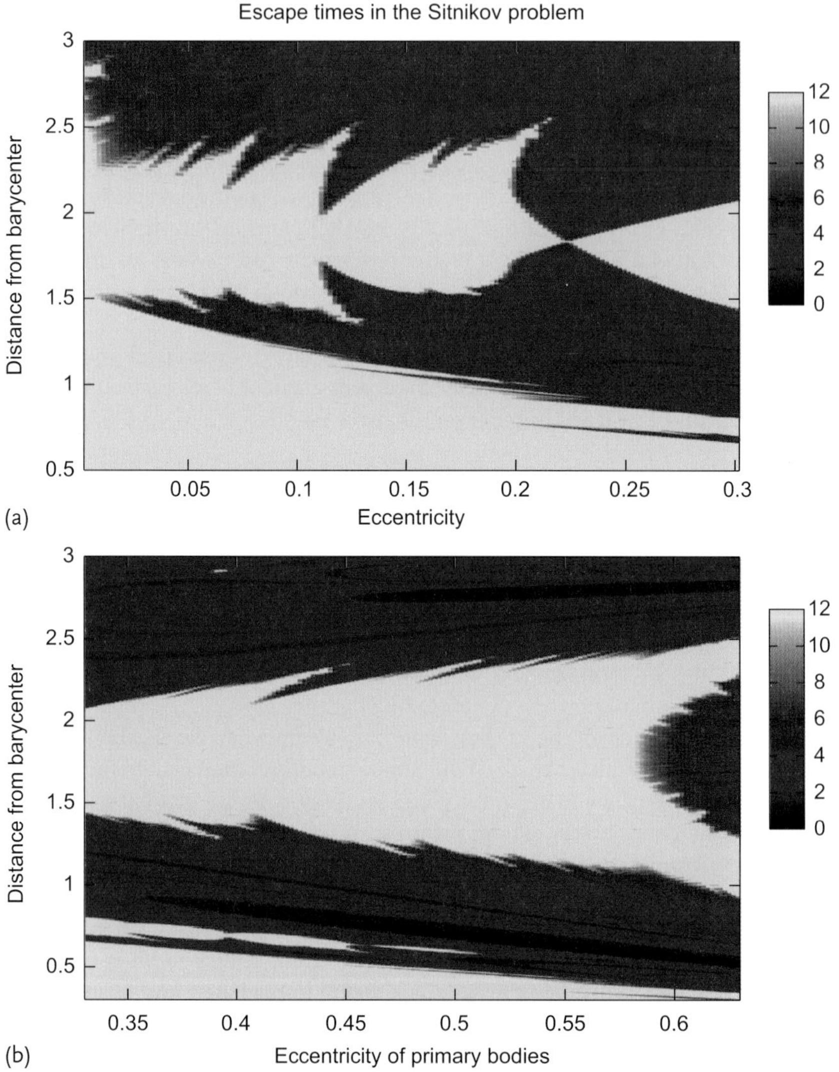


**Figure 7.13** Parameter space depending on the initial distance of the planet to the barycenter $0.5 < z_{\text{ini}} < 3.0$ ($y$-axis) and on the eccentricity $0 < e < 0.3$ ($x$-axis) of the primaries (a) and for $0.33 < e < 0.2$ (b). A detailed description is given in the text

- *The parameter space from* $e = 0$ *to* $e = 0.3$
  In Figure 7.13a for small values of $z_{\text{ini}}$ at the bottom a large yellow region is visible which characterize the 'main land' of stable orbits around the barycenter which diminishes from $z_{\text{ini}} = 1.5$ on to $z_{\text{ini}} \sim 1$ with increasing $e$. For small eccentricities the yellow region between $1.5 < z_{\text{ini}} < 2.5$ is the large $1:2$ island showing higher-order PO on the upper edge and the lower edge, which

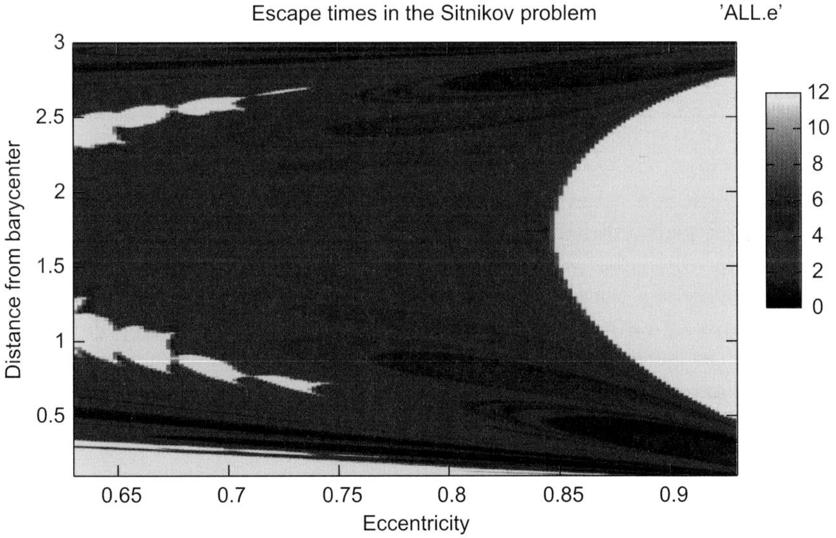


**Figure 7.14** Caption like in Figure 7.13 but for the eccentricity $0.6 < e < 0.94$ of the primaries.

split from this island (in form of spikes) with increasing eccentricity. Whenever such an island separates from the big 1 : 2 island, it suffers from a sudden decrease. Close to $e \sim 0.225$ it disappears almost completely, but then increases constantly. The red lines emanating from the edges of the island mark 'sticky' orbits[4], which stay long time close to a stable region but then escape (into the chaotic sea).

- *The parameter space from $e = 0.3$ to $e = 0.6$*
  In Figure 7.13b we can see again how the 'main land' of stable motion shrinks with increasing $e$ because of the separation of small islands which then disappear completely. The 1 : 2 island is steadily growing (since its disappearance mentioned before at $e \sim 0.225$) up to $e \sim 0.57$ where it splits into two islands[5]. Again we observe the red edges of all yellow regions, which is present at the moment of the splitting of the 1 : 2 island. The spikes of sticky orbits emanating from here form a whole area with lines parallel to the lower $e$ values from $e \sim 0.57$ on with an extension in $1.5 < z_{\text{ini}} < 2.2$. The island which split from the main land at $e = 0.2$ for small $z$-values disappears completely after several splittings at $e = 0.45$ but reappears as a very narrow stable island between $0.46 < e < 0.52$. Characteristic are also the small deep blue valleys of immediate escape slightly inclined with increasing eccentricity.

4) *Stickiness* is a special feature in dynamical system. It is closely connected to the breakup of the last KAM-torus into a Cantorus consisting of a chain of PO with with holes. Through this holes along the unstable manifold going through a homoclinic point – an unstable PO – orbits can escape into the large chaotic region.

5) Which is the well known phenomenon of period doubling in dynamical systems and simple mappings like in the logistic map.

- *The parameter space from $e = 0.6$ to $e = 0.94$*
  From Figure 7.14 it is obvious that the name 'crab-diagram' is a good description. The two fingers of the high-order islands emanating from the 1 : 2 island continue up to $e \sim 0.75$ and then only a tiny main land around the fixed point in the center indicates stable orbits up to $e \sim 0.85$; but in fact it is continuously present up to $e = 1$ with small intervals in $e$ for which the center turns out to be unstable (see below). Then the 1 : 2 island starts to grow continuously up to very large eccentricities. At the bottom from $e \sim 0.5$ on two inclined dark blue lines show that for this initial conditions the orbits escape immediately. Two boomerang shaped escape channel are also visible 'above' and 'below' the 1 : 2 island for $e > 0.75$. An interesting feature is the 'red straight line' at the location of the 1 : 2 island from the disappearance at ($e \sim 0.57$) on the former plot (Figure 7.13b) up to its reappearance. Again it shows a sticky region and can – just descriptively – regarded as an 'island just below the ocean of unstable motion'.

For even larger eccentricities than $e = 0.94$ the 'main land' is more and more shrinking whereas the big island for the 2 : 1 resonance is still growing.

#### 7.3.1.1 The Unstable Center

A very interesting behavior of the SP was discovered by Alfaro and Chiralt [165] for selected intervals for large parameters $e$: the motion in the close vicinity of the barycenter is unstable. They found the respective values for this very small unstable windows with respect to the eccentricity with a very accurate stability analysis. In Table 7.3 we show their results for the first five intervals; note how the extension of the unstable intervals in eccentricity $e$ diminish: the last one is only $\Delta e = 3\times10^{-15}$!

In Figure 7.15 we show the instability of the center for a special value of the eccentricity inside the first unstable window (compare Table 7.3). 9 orbits were computed numerically with a specially designed computer code; to derive a complete invariant curve takes several $10^6$ points on the SOS so close to the center. It is evident that from the lower value of $e$ listed in the respective table with increasing eccentricities inside this unstable window the period doubling formed two POs

**Table 7.3** Intervals in the eccentricity of the primaries where the center ($z = 0, \dot{z} = 0$) is an unstable hyperbolic fixed point.

| Interval | $e_{lower}$ | $e_{upper}$ |
|---|---|---|
| 1 | 0.855 861 796 455 97 | 0.855 863 313 749 44 |
| 2 | 0.977 521 503 363 75 | 0.977 721 898 154 92 |
| 3 | 0.996 021 712 161 88 | 0.996 021 786 342 71 |
| 4 | 0.999 275 593 816 15 | 0.999 275 607 435 21 |
| 5 | 0.999 867 194 118 188 | 0.999 867 196 617 65 |
| 6–12 | ... | ... |
| 13 | 9 999 999 999 827 872 | 9 999 999 999 827 875 |

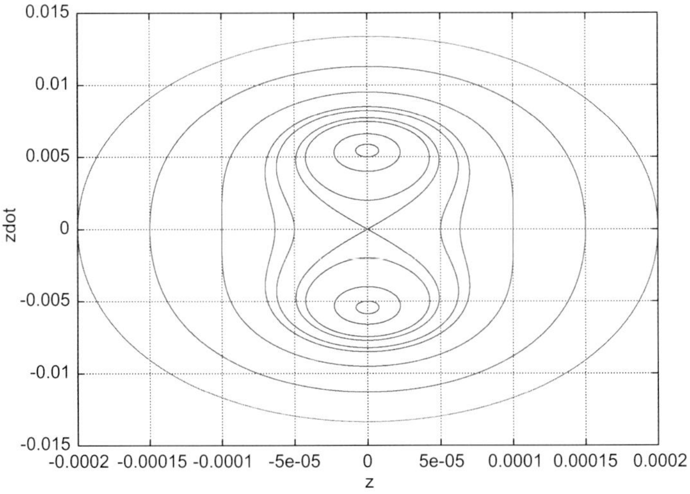


**Figure 7.15** Invariant curves close to the hyperbolic point at the center for $e = 0.855\,862\,5$ derived via numerical integrations. The two stable PO are approximately at $z = 0$ and $\dot{z} = \pm 0.006$; they correspond to a 4 : 7 resonance orbit.

(close to the 7 : 4 resonance). It is expected that these two PO will suffer from an infinite number of period doubling up to the moment when the center is again stable.

## 7.3.2
## Analytical Results

Since the circular case is integrable in terms of elliptic functions the elliptic case can be seen as the perturbed circular system. Here the small parameters are the eccentricity of the primary bodies as well as the initial distance of the massless body from the barycenter. If both are small enough one attempts to find series expansions in the small parameters which approximate the real solution for given but finites times. A symplectic mapping was derived in [166], Hamiltonian perturbation theory was used in [167, 168]. In the former the author introduced the action-angle variables of the harmonic oscillator and found the solution together with an approximate integral of motion. In the later the authors implemented a nonresonant normal form of high orders in the small parameters and estimated the domain of exponential stability in the $z(0)$, $e$-plane. Since the equation of motion is of one and a half dimension it is also possible to implement the perturbation techniques directly on the equation of motion. In a first attempt one tries to find a form of the equations of motion suitable for perturbation theory. In an interesting work [169] the author found a transformation for the low energy case by using as the independent variable to the angle $\zeta = z/a(\varphi)$ – which the tangent of the angle between the barycenter and one primaries' position seen from the mass $m_3$. Then the equation of motion for small eccentricities and small z reads

$$\zeta'' + \frac{e\cos\varphi + \left(\frac{1}{4} + \zeta^2\right)^{-\frac{3}{2}}}{1 + e\cos\varphi}\zeta = 0 \tag{7.17}$$

In this equation $'$ is the derivation with respect to the true anomaly $\varphi$. [167, 170] developed the $\Xi(\zeta)$ into a polynomial differential equation:

$$\zeta'' + g_1(\varphi)\zeta + \sum_{k=2}^{N} g_k(\varphi)\zeta^k = 0 \tag{7.18}$$

with $N$ chosen such that the results are accurate for $\zeta$ for small values of $e$. One can expand the expression $\Xi(\zeta) = (1/4 + \zeta^2)^{-3/2}$ into Chebycheff polynomials

$$\Xi(\zeta) = 8 - 46\zeta^2 + 240\zeta^4 - 1120\zeta^6 + 5040\zeta^8 + O(\zeta^{10}) \tag{7.19}$$

for $-0.8 < \zeta < 0.8$. With this the authors could improve the domain of convergence of the various series expansions compared to the classical Taylor series approach. In the vicinity of the barycenter the linearized equation becomes:

$$\zeta'' + g_1(\varphi)\zeta = 0 \tag{7.20}$$

One can show that the equation is of Hill's type with periodic coefficients defined by the function $g_1(\varpi)$. In this form it is possible to derive analytical solutions for the frequencies and the amplitudes using Floquet theory. The approach can also be used to transform the nonlinear equation to the one of the perturbed harmonic oscillator. With this it is possible to derive a high-order perturbation theory as it was worked out in [170]. The idea was later used by [171] but implemented in case of the original set of variables. In [172, 173] the method was extended to high orders in $e$ and $z(0)$. We briefly summarize the main results in the following discussion:

#### 7.3.2.1 Analytical Solution of the Sitnikov Problem

For small oscillations of the third body $|z| < 1$ the equations of motion (7.16) can be expanded into Taylor-series around $z = 0$ in the form:

$$\ddot{z} + \sum_{k=0}^{\infty} \binom{-\frac{3}{2}}{k} \frac{z^{2k+1}}{r(t)^{2k+3}} = 0 \tag{7.21}$$

The eccentricity of the primaries enters into the equations through the distance $r = r(t)$. Neglecting the influence of the third mass on the motion of the primaries it is determined by the equation:

$$\ddot{r} = \frac{1-e^2}{16r^3} - \frac{1}{8r^2}$$

The solution of the unperturbed Kepler problem can be expressed in terms of the eccentric anomaly $E = E(t)$ via

$$r(t) = \frac{1}{2}\left[1 - e\cos(E(t))\right]$$

where $E$ is related to the mean anomaly $M = M(t) = n(t - t_0)$ by:

$$M(t) = E(t) - e\sin(E(t))$$

In our units of time $n = 1$ and for $t_0 = 0$ Kepler's equation reduces to

$$t = E(t) - e\sin(E(t))$$

which implicitly defines $E(t)$. Assuming $e \ll 1$ we may solve the above equation and substitute in the definition of $r(t)$ to get:

$$r(t) = \frac{1}{2}\left(\sum_{k=1}^{\infty} r_k(t)e^k + 1\right) \tag{7.22}$$

Here $r_k = r_k(t)$ are trigonometric functions in the time variable $t$. Up to second order we get:

$$\begin{aligned} r_0(t) &= \frac{1}{2}\,, \\ r_1(t) &= -\cos(t)\,, \\ r_2(t) &= \frac{1}{2}\left\lfloor 1 - \cos(2t)\right\rfloor\,. \end{aligned}$$

For $e = 0$ the distance of the primaries becomes just

$$r(t) = r_0(t) = \frac{1}{2}$$

and the equations of motion reduce to the well known equations of motion of the MacMillan problem (see (7.5)) with the Hamiltonian

$$H(z, \dot{z}) = \frac{\dot{z}^2}{2} - \frac{1}{\sqrt{z^2 + \frac{1}{4}}}$$

In the following subsections we are interested in the motion of the primary for nonzero $e$ and small but nonzero amplitudes $z$.

#### 7.3.2.2 The Linearized Solution

If we substitute the expression (7.22) into the Taylor series expansion (7.21) the linearized equations of motion reduce to the simple harmonic oscillator:

$$\ddot{z} + 8z = 0$$

with the linearized solution:

$$z_0(t) = z_0\cos(\omega_0 t) + \frac{v_0}{\omega_0}\sin(\omega_0 t)$$

where $z_0 = z(0)$ and $v_0 = \dot{z}(0)$ are the initial conditions and $\omega_0 = 2\sqrt{2}$ defines the fundamental frequency of motion of the unperturbed problem. In a first step we are interested to the solution for small $|z| \ll 1$ but nonvanishing eccentricity of the primaries. The linearized equation of motion according to (7.21) becomes:

$$\ddot{z} + \frac{z}{r(t)^3} = 0$$

where we replace $r(t)$ with the expression (7.22) to get:

$$\ddot{z} + zg(t; e) = 0 \tag{7.23}$$

The Fourier representation of the function $g = g(t; e)$ is given by:

$$g(t, e) = \sum_{k>0} g_k(t) e^k$$

where $g_k = g_k(t)$ up to second order are given by:

$$\begin{aligned} g_0(t) &= 8 \,, \\ g_1(t) &= 24 \cos(t) \,, \\ g_2(t) &= 12 \left[ 3 \cos(2t) + 1 \right] \,. \end{aligned}$$

We notice that (7.23) is an ordinary equation of motion of Hill's type with a time periodic coefficient in front of the linear term in $z$. From Floquet theory [174] it follows that the solution can be written as

$$z(t) = a w(t) \cos \left[ b + \psi(t) \right] \tag{7.24}$$

where $a, b$ are constants and $w = w(t)$, $\psi = \psi(t)$ are the time dependent amplitude and phase function, respectively. Substituting the above solution into the linearized equation of motion (7.23) we get the coupled system of equations:

$$\begin{aligned} \ddot{w} + w g(t, e) - w \dot{\psi}^2 &= 0 \,, \\ 2\dot{w}\dot{\psi} + w \ddot{\psi} &= 0 \,, \end{aligned}$$

or

$$\begin{aligned} \ddot{w} + w g(t, e) - \frac{1}{w^3} &= 0 \,, \\ \dot{\psi} &= \frac{1}{w^2} \,. \end{aligned} \tag{7.25}$$

The constants $a, b$ are related to the initial conditions and are determined by the set of equations:

$$\begin{aligned} a &= \frac{\sec(b) z_0}{w_0} \,, \\ \tan(b) &= w_0^2 \left( \frac{\dot{w}_0}{w_0} - \frac{v_0}{z_0} \right) \end{aligned}$$

being $w_0 = w(0)$ and $\dot{w}_0 = \dot{w}(0)$ the solution of (7.25) for $t = 0$. Since we are interested in the solution close to $e = 0$ we may expand $w = w(t)$ into a Taylor series of the form:

$$w(t) = w_0(t) + w_1(t)e + w_2(t)e^2 + O(e^3)$$

We get order by order the system of equations:

$$w_0'' + 8w_0 = \frac{1}{w_0^3} ,$$

$$w_1'' + 8w_1 = -24\cos(t)w_0 - \frac{3w_1}{w_0^4} ,$$

$$w_2'' + 8w_2 = -36\cos(2t)w_0 - 24\cos(t)w_1 + \frac{6w_1^2 + w_0\left(-12w_0^5 - 3w_2\right)}{w_0^5}$$

which can be solved recursively for the $w_j$ for known $w_k$ with $0 \le k < j$. At zeroth order we get:

$$w_0(t) = \frac{1}{2^{\frac{3}{4}}}$$

We notice that the form of the equations is of the type of driven harmonic oscillators of the form:

$$\ddot{w}_i + 8w_i = h_i(w_k, t)$$

with $0 \le k < i$ and the solution can be easily found. Up to second order in the eccentricity $e$ we get:

$$w_1(t) = -\frac{12}{31}\sqrt[4]{2}\cos(t) ,$$

$$w_2(t) = \frac{2805}{7688 \cdot 2^{\frac{3}{4}}} - \frac{423}{961 \cdot 2^{\frac{3}{4}}}\cos(2t) .$$

Once, the system of ordinary equations determining the amplitude $w$ of the linearized solution order by order is found the time dependent phase function $\psi$ can be obtained by direct integration of the second of (7.25):

$$\psi(t) = \int_{t_0}^{t} \frac{1}{w(s)^2} ds$$

Since $w = w_0 + ew_1 + \ldots$ the phase $\psi$ must have the form

$$\psi(t) = \psi_0(t) + \psi_1(t)e + \psi_2(t)e^2 + O(e^3)$$

We get up to $O(e^3)$:

$$\psi_0(t) = 2\sqrt{2}t ,$$

$$\psi_1(t) = \frac{96}{31}\sqrt{2}\sin(t) ,$$

$$\psi_2(t) = \frac{3\sqrt{2}\left[217t + 1140\sin(2t)\right]}{1922} .$$

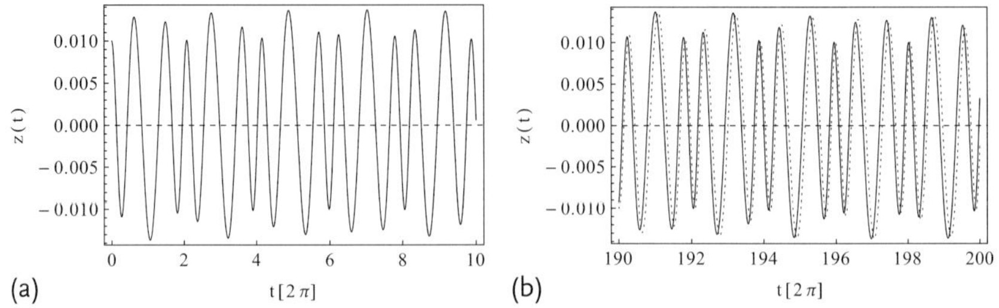


**Figure 7.16** Comparison of analytical (black) and numerical (dotted) obtained solution in the Sitnikov problem. The time unit is given in revolution periods of the primary bodies. See text.

Since the amplitude and phase is time dependent no strict period exists. Nevertheless it is possible to define a mean period $\bar{T}$ which can be identified with the constant part of the Fourier series representation:

$$\tilde{T} = \frac{1}{2\pi} \int_0^{2\pi} \frac{1}{w(s)^2}\, ds = \frac{\psi(2\pi)}{2\pi}$$

To low orders $O(e^3)$ we get:

$$\tilde{T} = \sqrt{2}\left[2 + \frac{21e^2}{62} + O(e^3)\right]$$

In Figure 7.16 we plot the analytical solution (black) for $z_0 = 0.01$, $v_0 = 0$ and $e = 0.2$ and compare with a numerically obtained solution (dotted). The analytical solution is based on $w$, $\psi$ expanded up to order $O(e^{14})$ and allows high values in the eccentricity of the primaries but small initial amplitudes $z_0$. The agreement is up to machine precision within the first revolution periods of the primaries (Figure 7.16a) and still quite good for longer times (Figure 7.16b): after 200 revolution periods of the primaries the approximate solution still recovers the correct amplitudes and frequencies with a small shift due to the cumulative error of the approximation.

#### 7.3.2.3 **Linear Stability**

In this subsection we use the solution of the linearized problem to obtain the stability of the motion based on the transfer matrix of the problem defined by (7.23). The transfer matrix is defined as:

$$R = \begin{pmatrix} c(T) & s(T) \\ c'(T) & s'(T) \end{pmatrix} \tag{7.26}$$

where $c, s$ denote the cosine and sine like solution of an ordinary differential equation of Hill's type and $T$ is the period of the solution. The notion comes from the fact that we require that $c, s$ match the following "initial conditions" $c(0) = \dot{s}(0) = 1$ and $\dot{c}(0) = s(0) = 0$ at $t = 0$. Although $c = c(t)$ and $s = s(t)$ are time dependent solutions for fixed period $T = 2\pi$ the monodromy matrix $R$ is constant with

$\det(R) = 1$. It defines the symplectic change of coordinates $Z_n \equiv (z(nT), \dot{z}(nT))$ to $Z_{n+1}$ through the relation:

$$Z_{n+1} = RZ_n$$

The solution of the transformation can be written as

$$Z_n = A_1\lambda_1^n + \lambda_2^n A_2$$

where $\lambda_1, \lambda_2$ are the eigenvalues and $A_1, A_2$ are the corresponding eigenvectors of $R$. The eigenvalues can be obtained from the characteristic polynomial

$$\lambda^2 - \mathrm{Tr}(R)\lambda + 1 = 0$$

where $\mathrm{Tr}(R)$ denotes the trace of the matrix $R$. From

$$\lambda_{1,2} = \frac{\mathrm{Tr}(R)}{2} \pm \sqrt{\frac{\mathrm{Tr}(R)^2}{4} - 1}$$

we clearly see that for bounded motions $|\lambda_1|, |\lambda_2| < 1$ the trace of the transfer matrix is $\mathrm{Tr}(R) < 2$. Using the linearized solution (7.24) to obtain (7.26) we have

$$\mathrm{Tr}(R) = 2\cos\left[\sqrt{2}\pi\left(4 + 0.677\,419e^2\right)\right]$$

which also depends on the eccentricity $e$. We plot the $\mathrm{Tr}(R)$ for different values of $e$ in Figure 7.17. The linearized stability analysis based on Floquet theory shows that critical parameters are $e = \pm 0.544\ldots, \pm 0.859\ldots, \pm 0.966\ldots$ (a negative value of the eccentricity denotes that the primaries start at their apocenter instead of their pericenter.

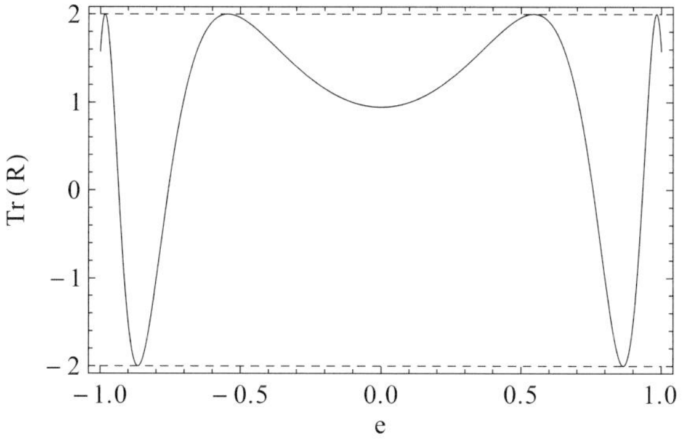


**Figure 7.17** Trace of the transfer matrix of the linearized problem. For $\mathrm{Tr}(R) = 2$ the motion is close to an unstable situation. See text.

#### 7.3.2.4 Nonlinear Solution

The linearized solution obtained from Floquet theory can also be used to investigate the motion of the third body for larger amplitudes of oscillation. An ordinary equation of motion of the type:

$$\ddot{z} + g(t)z + F(z, t, e) = 0$$

can be transformed into

$$y''(\psi) + y(\psi) + F\big(w(t(\psi))\, y(\psi),\, e\big)\, w(t(\psi))^3 = 0$$

by the change of variables

$$y = \frac{z}{w(t)} ,$$
$$\psi' = \frac{1}{w(t)^2} .$$

Here $w = w(t)$ defines the time dependent amplitude function obtained from Floquet theory and $'$ denotes the derivative with respect to the phase taken as the new independent variable. The function $F = F(y, e)$ can be explicitly constructed. From:

$$z(t) = y(t)w(t)$$

and its derivatives:

$$\dot{z}(t) = y(t)\dot{w}(t) + w(t)\dot{y}(t) ,$$
$$\ddot{z}(t) = w(t)\ddot{y}(t) + 2\dot{w}(t)\dot{y}(t) + y(t)\ddot{w}(t) ,$$

we find

$$w(t)\ddot{y} + 2\dot{w}(t)\dot{y} + g(t)w(t)y + y\ddot{w}(t) + F(w(t)y(t), e) = 0$$

From

$$y(t) = y(\psi(t))$$

we get

$$\dot{y}(t) = y'(\psi)\dot{\psi}(t) = \frac{y'(\psi)}{w(t)^2}$$

and also

$$\ddot{y}(t) = y''(\psi) - 2y'(\psi)w(t)\dot{w}(t)$$

Putting all together we find

$$\frac{y''(\psi)}{w(t)^3} + g(t)w(t)y(\psi) + y(\psi)w''(t) + F(w(t)\, y(\psi), e) = 0$$

which can be simplified to

$$y''(\psi) + y(\psi) + F\big(w(t(\psi))\, y(\psi), e\big)\, w(t(\psi))^3 = 0$$

since by definition

$$\ddot{w}(t) = \frac{1}{w(t)^3} - g(t)w(t)$$

We give the results of the transformed equation of motion of the Sitnikov problem up to order $O(e^m z^n)$ with $m + n = 5$. Defining the new independent variable

$$\tau = \frac{\psi}{\sqrt{8}}$$

we get:

$$\begin{aligned} y'' + 8y - 12\sqrt{2}y^3 - \frac{132}{31}\sqrt{2}e\cos(\tau)y^3 + 30y^5 + \\ e^2\left[-\frac{2784}{961}\sqrt{2}\cos(2\tau) - \frac{825\sqrt{2}}{961}\right]y^3 = 0\,. \end{aligned} \tag{7.27}$$

Equation 7.27 approximates the motion of the third body in the Sitnikov problem for small amplitudes $z$ (in fact $y$) and small eccentricities $e$. For vanishing $e$ the equation is again that of the perturbed harmonic oscillator:

$$y'' + 8y = F(y)$$

for nonzero $e$ an additional time dependency is introduced:

$$y'' + 8y = F(y, t; e)$$

The perturbation is polynomial in the amplitudes $y$ and eccentricities $e$ and periodic in time $t$. Equations of motion of the type (7.27) can be treated by means of Poincaré–Lindstedt methods [175, 176] strongly related to the concepts of KAM-theory. The solution may therefore be written in the form:

$$y(\tau) = y_0(\tau) + y_1(\tau) + y_2(\tau) + O(e^m y^n) \quad \text{with } m + n = 5$$

where the independent variable $\tau$ is expressed itself as a Taylor series of the type:

$$\tau(\sigma) = \sigma + \tau_1(\sigma) + \tau_2(\sigma) + O(e^m y^n) \quad \text{with } m + n = 5$$

and a coefficient function $y_k$, $\tau_k$ denotes a function proportional to $y^m e^n$ where $m + n = k$. The substitution in (7.27) gives the recursive system of equations up to second order in the small parameters:

$$\begin{aligned} &y_0'' + 8y_0 = 0\,, \\ &y_1'' + 8y_1 = 2\tau_1' y_0'' + y_0' \tau_1''\,, \\ &y_2'' + 8y_2 = 2\tau_2' y_0'' + \tau_1'\left(2y_1'' - 3\tau_1' y_0''\right) + y_1'\tau_1'' + y_0'\left(\tau_2'' - 3\tau_1'\tau_1''\right) \end{aligned}$$

where the $'$ denotes the derivative with respect to the new independent variable $\sigma$ from now on. At zeroth order we get again the solution of the harmonic oscillator:

$$y_0 = y_0 \cos(\omega_0 \sigma) + \frac{y_0'}{\omega_0} \sin(\omega_0 \sigma)$$

where $y_0 = y(0)$ and $y_0' = y'(0)$ are the initial conditions. We notice that the generic form of the above equations is again of the form of perturbed harmonic oscillators:

$$y_i'' + 8y_i = F_i(y_k, \sigma)$$

where $0 \le k < i$. The constants of integration are already defined from the initial conditions at the zeroth-order solution. The solution at order $i$ can therefore be written formally as:

$$y_i(\sigma) = \left[\int_0^\sigma -\frac{\sin\left(2\sqrt{2}\rho\right) F_i(\ldots,\rho)}{2\sqrt{2}}\, d\rho\right] \cos\left(2\sqrt{2}\sigma\right) + \left[\int_0^\sigma \frac{\cos\left(2\sqrt{2}\rho\right) F_i(\ldots,\rho)}{2\sqrt{2}}\, d\rho\right] \sin\left(2\sqrt{2}\sigma\right)$$

up to any order in $i$. Once the $y_k$ are found the transformation from $z \to y$ and $t \to \tau \to \sigma$ has to be inverted to get the solution $z = z(t)$. It is of the type:

$$z(t) = \sum_{k=1}^{N} \left[A_k(t) \cos(\Psi_k(t)) + B_k(t) \sin(\Psi_k(t))\right] \tag{7.28}$$

where $A_k = A_k(t; z_0, v_0, e)$, $B_k = B_k(t, z_0, v_0, e)$ are the time dependent amplitudes and $\Psi_k = \Psi_k(t;, z_0, v_0, e)$ are the time dependent frequencies of the system of order $k$ in $e \cdot z$. The form of the solution clearly shows that in the nonlinear problem the frequencies and amplitudes are not fixed anymore. They depend on the parameter $e$ the initial conditions $(z_0, v_0)$ and on the time also. We plot in Figure 7.18 the solution obtained from (7.28) with $N = 7$ for $z_0 = 0.2$ and $e = 0.2$ and

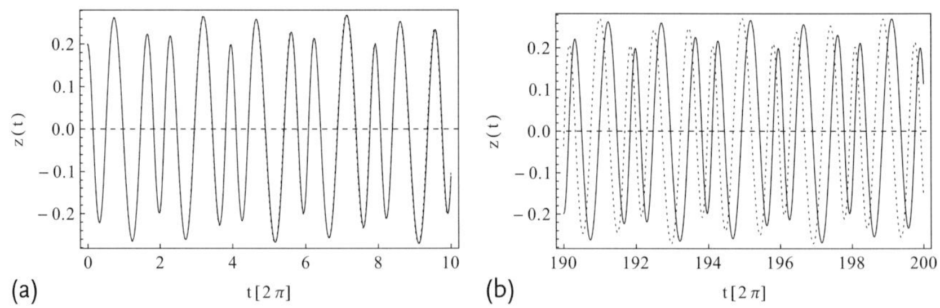


**Figure 7.18** Comparison of analytical (black) and numerical (dotted) obtained solution in the Sitnikov problem. The time unit is given in revolution periods of the primary bodies. See text.

compare it with the solution for same initial conditions obtained from a numerical integration. The two solutions perfectly agree within the first revolution periods of the primaries (Figure 7.18a) and still reproduce the qualitative aspects of the solution after longer integration time (Figure 7.18b). The shift in phase and amplitude is only due to the accumulation of the approximation errors.

We compare the numerical and analytical obtained solution in the grid $(e, z_0)$ with $0 \leq e \leq 0.6$ and $0 \leq z_0 \leq 0.2$ in Figure 7.19. The plot shows the integration time when the absolute difference of the numerical found solution $z_{\text{num}}(t)$ and the analytical solution $z_{\text{ana}}(t)$ is of the order of 1% of the initial condition $z_0$. As we can see the approximate solution is valid within a large domain in $e$ and $z(0)$. We also

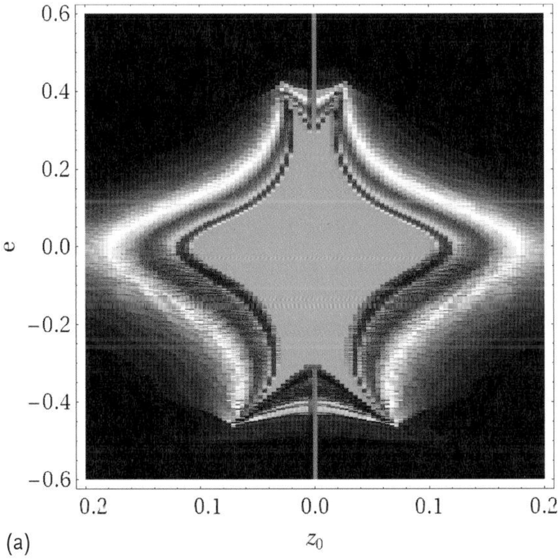


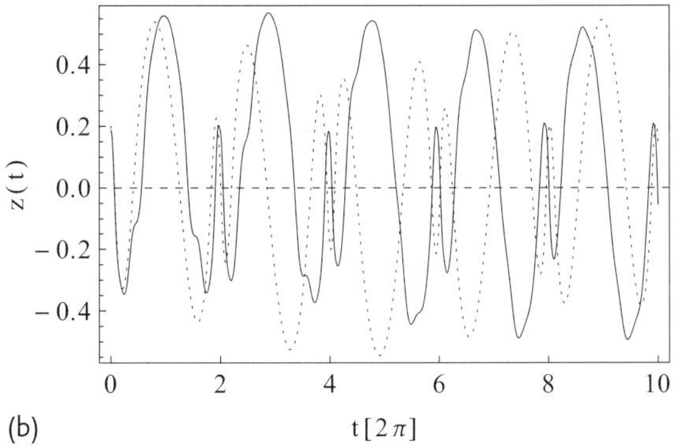


**Figure 7.19** (a) validity of the approximate analytical solution on the grid $(e, z_0)$ with $-0.6 \leq e \leq 0.6$ and $-0.2 \leq z_0 \leq 0.2$ versus the time where the difference of the numerical and analytical solution is less than 1% of the initial condition $z_0$. Green corresponds to 100, blue to 60, red to 25, white to 10 and black to less than 10 revolution periods of the primary bodies. An orbit of the limiting case, for $e = 0.6$ and $z_0 = 0.2$, is shown in (b).

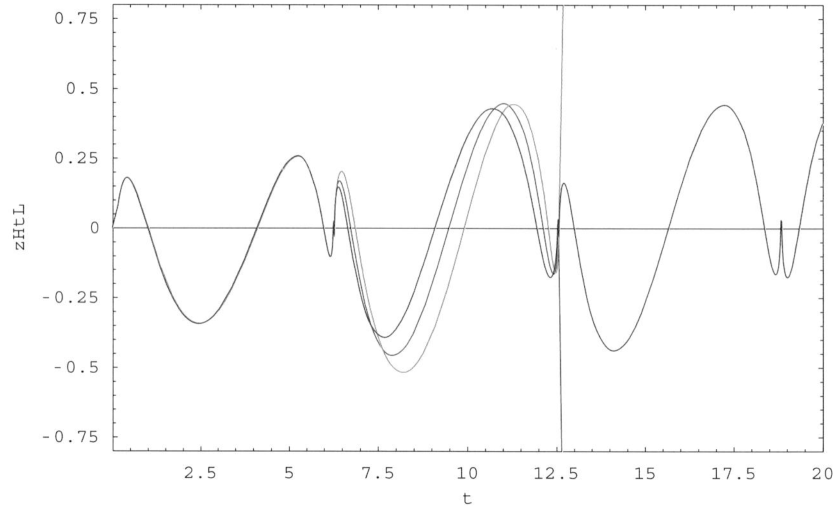


**Figure 7.20** For a large eccentricity of the primaries ($e = 0.9$) we see the different behavior of three orbits initially only separated by $S_Z = 10^{-6}$ in position (for details see text).

demonstrate a case for which the nonlinear solution (7.28) does not reproduce the numerical obtained solution anymore. The nonlinear effects become too strong and higher-order terms cannot be neglected. We demonstrate in Figure 7.20 the sensitivity of an orbit with respect to a very small change in the initial conditions ($\delta \dot{z} = 10^{-6}$). After only a few revolutions of the primaries the orbit is completely different:

1. for $z(0) = 10^{-3}$ $m_3$ leaves after 2 primary revolutions along the negative $z$-axis (red line),
2. for $z(0) = 10^{-3} + 10^{-6}$ $m_3$ leaves the system in the other direction (green line),
3. for $z(0) = 10^{-3} - 10^{-6}$ $m_3$ the orbit stays bounded for long time but nevertheless escapes.

The results of the present section are produced on the basis of [173]. A detailed description of the methods and calculations involved can be found within.

## 7.4
## The Vrabec Mapping

We now discuss how one can realize the Moser theorem cited at the beginning, namely whatever sequence of integer numbers $s_k$ (which corresponds to revolutions of the primary bodies), one can imagine, there is always one 'real' orbit having these successive crossings of the barycenter (with the restriction $s_k \geq S_{\text{lim}}$[6]).

6) $S_{\text{lim}}$ depends on the eccentricity of the primaries.

Following an idea of F. Vrabec described in [177] one can visualize the complexity of the phase space of the SP in the following way: setting a fixed value of the eccentricity $e$ for a grid of initial conditions for the true anomaly $0^\circ < \nu < 360^\circ$ of the primaries and the velocity $\dot{z} > 0$ of the planet one starts the computations exactly in the barycenter ($z = 0$). The first crossing of the massless body **A** gives the first number in the sequence depending on the number of revolutions of the primaries between the starting point and the first crossing of the massless body. In Figure 7.21 this occurs during the second revolution – as a consequence $s_1 = 2$. We then ask for the second crossing **B**, which happens for this special orbit during the third revolution, which means that $s_2 = 3$. The number of revolutions between the third and the fourth crossing is 1, between the fourth and fifth it is 0, between the fifth and the sixth it is 2 and so on. Consequently for this orbit we get the sequence

$$2/3/4/6/7/7/10\ldots \tag{7.29}$$

We mark these numbers with the aid of a color code: after the first crossing in an initial condition diagram for a great number of orbits with different initial conditions for $\nu$ and $\dot{z}$ which now defines different regions. We proceed to the next crossing, count again the number of revolutions between the last crossing and this one which defines now the regions in a new diagram. This has been done up to 100(!) crossings for many different eccentricities. The grid for initial conditions for the numerical integration of the equations of motion was set to $2000 \times 2000$.

The results presented for the Vrabec mapping are derived in a master thesis by N. Lang [178]; especially the color figures are taken from this work. In Figure 7.22 the black region (top of (a)) for $\dot{z} > 1.7$ marks the initial conditions which lead

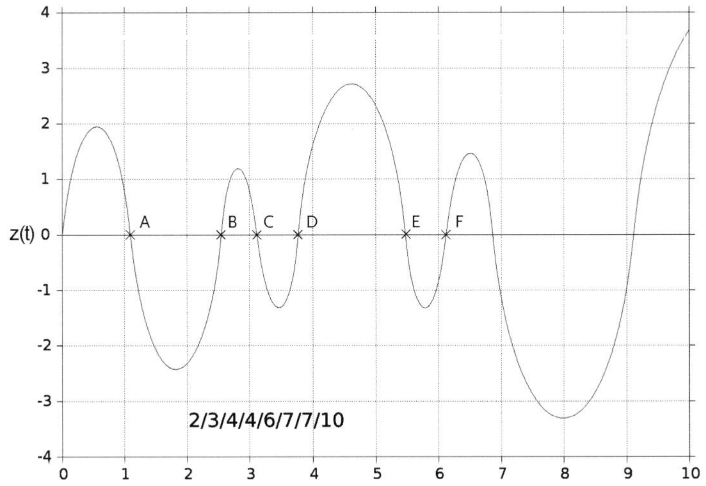


**Figure 7.21** Time evolution of an orbit which defines a sequence of numbers (here 2/3/4/4...) which characterizes the number of periods between two consecutive crossings of the planet through the barycenter of the system. $x$-axis denotes the number of revolutions of the primaries.

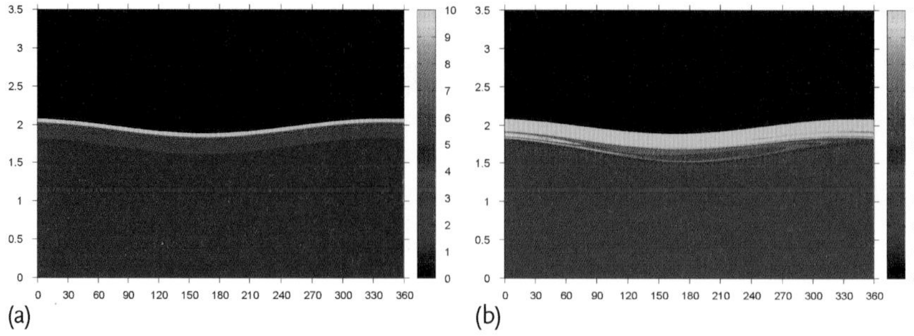


**Figure 7.22** This plot initial true anomaly $\nu$ of the primaries ($x$-axes) versus the initial velocity ($\dot{z} > 0$) of the planets indicates the number of periods of the primaries until the next crossing of the planet through the barycenter. The significance of the color code is given in the text. In (a) the diagram of initial conditions is shown for $e = 0.1$ after the crossing **A**, in (b) the diagram is shown after the crossing **F**; for details see text.

to an escape without coming back to the barycenter[7]. The yellow region is the one with initial conditions where the planet returns and crosses the barycenter after more than 10 periods of the primaries; the tiny orange to blue regions (small horizontal stripes) stand for the first crossing of the barycenter between zero and 9 respective periods of the motion of the primaries. In this first return (marked '**A**') in Figure 7.21 almost all initial velocities with $\dot{z} \leq 1.7$ cross the barycenter still during the first period of the primaries.

In Figure 7.22b again (as in all other graphs) the black region marks the escape from the system after the fifth crossing, the other colors have the same significance like in the last figure; for example yellow marks that after the fifth crossing the next crossing happens after more than 10 periods of the primaries.

The whole structure is better visible when we follow the increase of complexity for an initial eccentricity of the primaries of $e = 0.6$. In Figure 7.23 we can see how it develops: In the diagram (Figure 7.23a) representing the first crossing there is a big black region visible $1.7 \leq \dot{z} \leq 3$ with a minimum close to $\nu = 180°$. In fact there is a slight asymmetry of the structures which appears quite clear in the following graphs where more and more black regions appear especially for $\nu \geq 180°$. An interesting feature is the growing black triangle like structure in Figure 7.23c,d which is located close to $\dot{z} \sim 2$ and $\nu \sim 240°$. Another property is the asymmetry of the lobes between $0° < \nu < 180°$ and $180° < \nu < 360°$ which can be understood by the different distances of the planet to the primaries causing different accelerations before and after the apoastron ($\nu = 180°$). Small black stripes appear also inside the regions which in the figures before marked the recurrence of the planet after several periods. These orbits stay more or less in the vicinity of the barycenter for several periods of the primaries but then (black) they escape to infinity. Always the black region is enclosed by yellow regions (more than 10 periods after returning) and the yellow region by an orange one and so on.

7) $H(t, z, \dot{z}) > 0$.

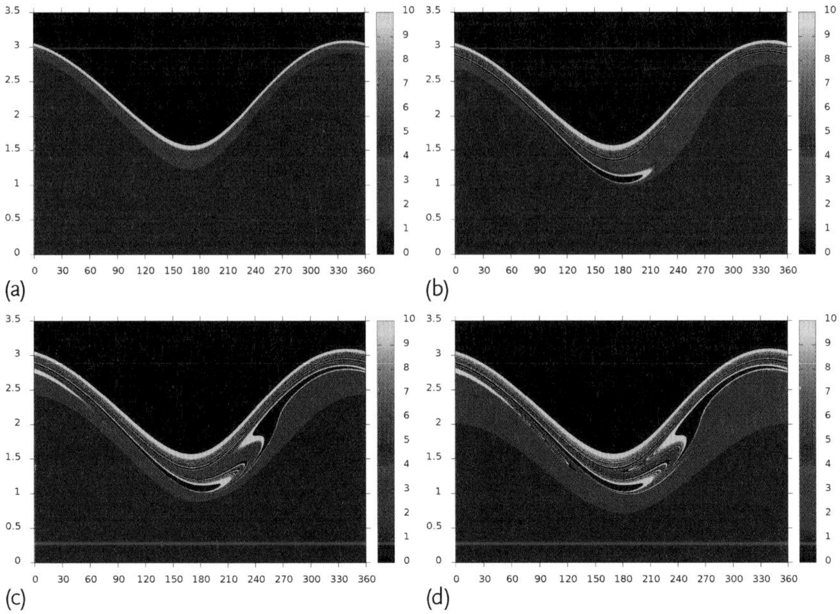


**Figure 7.23** Description like in Figure 7.22 but for $e = 0.6$ and the first four crossings **A**, **B**, **C** and **D**.

The dark blue regions marking immediate return within 1 period retreats gradually with the larger number of returns well visible in all figure parts of Figure 7.23.

In a separate picture Figure 7.24 we show the fine structure after the fifth return: tiny black regions appear everywhere between $1 \leq \dot{z} \leq 3$. To underline the complexity of the SP with all the known structures of self-similarity we show in Figure 7.25 a zoom of the diagram for $e = 0.6$ after several tenths of recurrences.

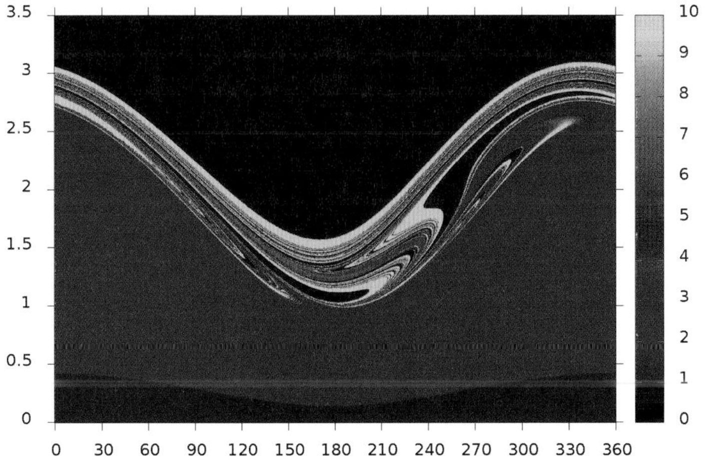


**Figure 7.24** Description like in Figure 7.22 but for $e = 0.6$ and the crossing **F**

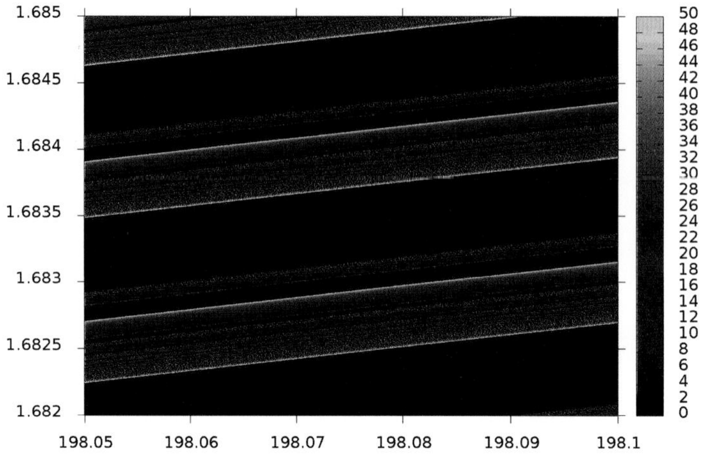


**Figure 7.25** Zoom of the diagram Figure 7.24 $\nu$ ($x$-axis) versus $\dot{z}$ showing the fine fractal structure and the self-similarity of the recurrence plots for $e = 0.6$

## 7.5 General Sitnikov Problem

In the generalized SP (GSP) one studies the motion of three massive bodies in the Sitnikov configuration. The perturbed motion of the primaries is still on a plane perpendicular to the $z$-axis where the planet is confined to move; this plane is in oscillatory motion around the barycenter of the system. When the three masses are comparable large (e.g. $m_1 = m_2 = m_3$) we call it 'extended Sitnikov Problem' (= ESP)) ; we will primarily deal with this configuration, but we will also show one interesting numerical result when the planet is more massive ($m_1 + m_2 = m_3$).

### 7.5.1 Qualitative Estimates

Let us start with the equations of motion when all three masses are equal ($m_1 = m_2 = m_3 = m$). According to the Sitnikov configuration we have to take into account several symmetries and find that we just need three second differential equations

$$\begin{aligned}
\frac{d^2x}{dt^2} &= -\frac{2mx}{(4x^2+4y^2)^{\frac{3}{2}}} - \frac{mx}{(x^2+y^2+9z^2)^{\frac{3}{2}}} \\
\frac{d^2y}{dt^2} &= -\frac{2my}{(4x^2+4y^2)^{\frac{3}{2}}} - \frac{my}{(x^2+y^2+9z^2)^{\frac{3}{2}}} \\
\frac{d^2z}{dt^2} &= -\frac{3mz}{(x^2+y^2+9z^2)^{\frac{3}{2}}}
\end{aligned} \tag{7.30}$$

where the position in the plane of the primaries is $x$ and $y$ and the distance of the planet to the barycenter is $z$ [8]. After a transformation to $x = r\cos\theta$ and $y = r\sin\theta$ we find a simplified formulation of the equations of motion

$$\frac{d^2 r}{dt^2} = \frac{c^2}{4m^2 r^3} - \frac{mr}{(r^2+9z^2)^{\frac{3}{2}}} - \frac{m}{4r^2} \tag{7.31}$$

$$\frac{d^2 z}{dt^2} = -\frac{3mz}{(r^2+9z^2)^{\frac{3}{2}}} \tag{7.32}$$

with the angular momentum integral $r^2\dot{\theta} = c/2m$, where $c$, the angular momentum integral as well as the energy integral are constant

$$m\left(\dot{r}^2 + \frac{c^2}{4m^2 r^3} + \frac{3}{4}\dot{z}^2\right) - \frac{m^2}{2r} - \frac{2m^2}{\sqrt{r^2+9z^2}} = H = \text{const} \tag{7.33}$$

For a given value for the angular momentum $c$ we find an upper limit for $H$

$$H \geq \frac{c^2}{4mr^2} - \frac{5m^2}{2r} \equiv f(r) \tag{7.34}$$

and an minimum value

$$H_{\min} = f(r_0) = -\frac{25m^5}{4c^2} \equiv H_0 \tag{7.35}$$

The value of $H$ depends on the minimum $H_1$

$$\left(\frac{c^2}{4mr^2} - \frac{m^2}{2r}\right)_{r=\frac{c^2}{m^3}} = -\frac{m^5}{4c^2} \equiv H_1 \tag{7.36}$$

For $H_0 \leq H < H_1$ one sees that

$$H \geq H_1 - \frac{2m^2}{\sqrt{r^2+9z^2}} \tag{7.37}$$

from which follows that the motion is bounded

$$\sqrt{r^2+9z^2} \leq \frac{2m^2}{H_1 - H} \tag{7.38}$$

From the energy relation an important condition defining regions of permitted motion in the SOS $r$ versus $\dot{r}$ can be derived

$$\dot{r}^2 + \frac{c^2}{4m^2r^2} - \frac{5m}{2r} \leq \frac{H}{m} \tag{7.39}$$

8) Note that in this special case the distance of the plane of motion of the primaries is always $z/2$.

### 7.5.2
### Phase Space Structure

Because of the possible reduction two a 2-degree of freedom system and the existence of the energy-integral it is possible to represent the motion on a SOS. A very illustrative representation turned out to be on the plane $r/2$, the distance between the two primaries, versus the velocity $\dot{r}$. We show just two examples: the first SOS is for 3 equally massive planets, the second example for the already mentioned case where the planet is massive 'planet' than the primary bodies.

In Figure 7.26 we show the structure of the phase space $r$ versus $\dot{r}$ for three equally massive bodies. The outermost curve is the zero-velocity curve, inside most of the motion is regular. Close to $r = 1.4$ around a hyperbolic fixed point – which corresponds to the $1:2$ resonance – a small chaotic layer is visible. On both sides of this unstable motion symmetric to the $\dot{r} = 0$ axes two stable elliptic points surrounded by invariant curves are visible, which are the outcome of a pitchfork bifurcation of the former stable $1:2$ PO on the $x$-axis.

In Figure 7.27 one can see how the 'simple' phase space structure changes with a bigger 'planet'. The motion close to the stable $1:2$ PO (located at approximately $r = 2.4$ and $\dot{r} = 0$ is regular; off the $x$-axes two symmetric pear shaped islands around another stable fixed point are visible. These islands are surrounded by high-order chains of islands in the SOS. They are all lying in a big sea of chaotic motion which extends almost to the border, the zero velocity curve. This problem of three massive bodies always in the isosceles configuration[9] deserves many more

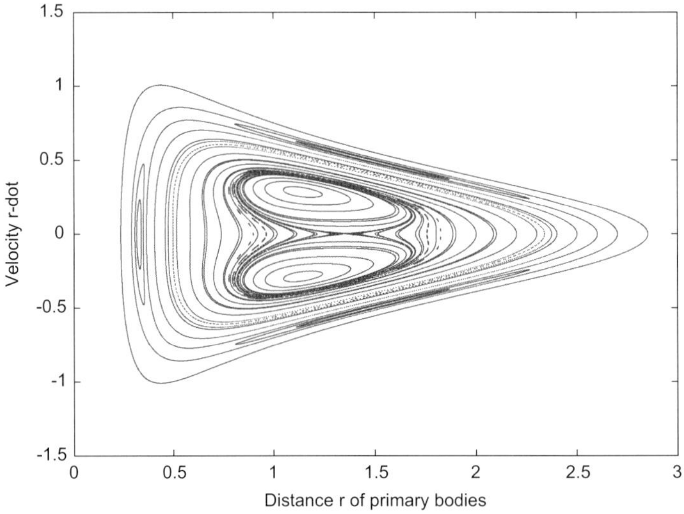


**Figure 7.26** SOS for three equally massive bodies $r$ versus $\dot{r}$.

9) An isosceles triangle is a triangle with two equal sides. In the Sitnikov configuration the three masses are always on the vertices of such a triangle, but the angles are continuously changing. When the three masses are aligned two angles are zero and one is 180°.

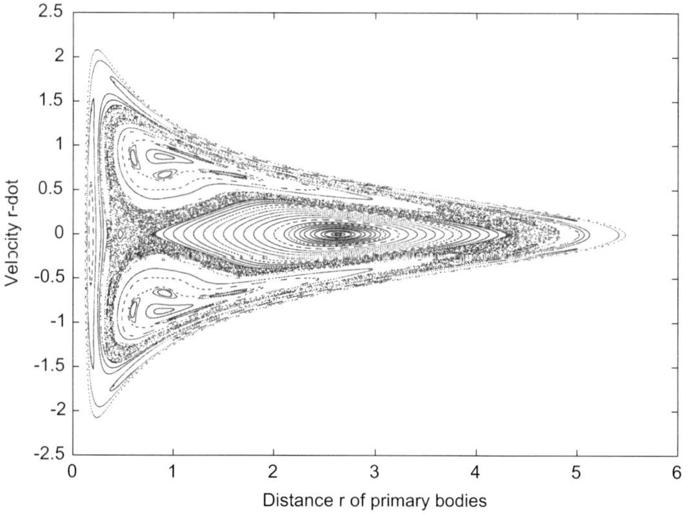


**Figure 7.27** SOS ($r$ versus $\dot{r}$) for three bodies with $m_3 = m_1 + m_2$.

profound investigation. Let us think of a very massive central body $m_3$ where two small masses are in orbits on opposite sides of this body – a somewhat strange extrasolar planetary system?

# 8
# Planetary Theory

After Kepler's discovery of the ellipticity of orbits of the planets it took almost a hundred years before Newton (1643–1727) found the 'Law of Gravitation'. Although Gravitation is the weakest of the 4 fundamental forces[1] it is the reason for the hierarchical structuring of our universe from planets to superclusters of galaxies. After almost another 100 years astronomers and mathematicians[2], making use of the gravitational forces acting between all planets in the Solar system found, that the Keplerian orbits are only an approximation to the real motion. In fact the Keplerian elements are not constant but undergo small variations with time and need to be modeled by differential equations.

## 8.1
## Planetary Perturbation Theory

### 8.1.1
### A Simple Example

To understand the principles of perturbation theory in celestial mechanics we will explain the method using a simple example which is taken from [179]

$$\frac{d^2x}{dt^2} + k^2 x = -m_1 \left(\frac{dx}{dt}\right)^3 + m_2 \cos \lambda t \tag{8.1}$$

$k^2$, $m_1$ and $m_2$ as well as $\lambda$ are constants; with $m_1 = m_2 = 0$ the right hand side vanish and the solution is the one of the harmonic oscillator. Equation (8.1) is the equation of motion (the Duffing oscillator[3]) of a perturbed oscillator which we will now solve in the same way we will later use to solve the Lagrange equation of

1) If we set the strength of the *gravitational interaction* to 1 then the *weak interaction* is $10^{25}$ times stronger, the *electromagnetic interaction* $10^{36}$ and the *strong interaction* $10^{38}$ stronger.
2) Joseph-Louis Lagrange (1736–1813), Pierre-Simon Laplace (1749–1827), Leonard Euler (1707–1783) and Jean d'Alembert (1717–1783).
3) An example of a periodically forced oscillator with a nonlinear elasticity.

*Celestial Dynamics*, First Edition. R. Dvorak and C. Lhotka.

planetary theory. After separating (8.1) into two first-order differential equations

$$\begin{aligned} \frac{dx_1}{dt} - x_2 &= 0 \\ \frac{dx_2}{dt} + k^2 x_1 &= -m_1 x_2^3 + m_2 \cos \lambda t \end{aligned} \tag{8.2}$$

For the unperturbed case ($m_1 = m_2 = 0$) the solution is simply

$$\begin{aligned} x_1 &= a \cos kt + b \sin kt \\ x_2 &= -ka \sin kt + kb \cos kt \end{aligned} \tag{8.3}$$

with $a$ and $b$ as integration constants which may be regarded as osculating elements in planetary theory. The problem is to find the variations of these two 'elements' – which will not be constant in the presence of perturbations – such that (8.2) is satisfied. Inserting (8.3) into (8.2) we have

$$\begin{aligned} &\frac{da}{dt} \cos kt + \frac{db}{dt} \sin kt = 0 \\ &\frac{db}{dt} \cos kt - \frac{da}{dt} \sin kt = m_1 k^2 (a \sin kt - b \cos kt)^3 + \frac{m_2}{k} \cos \lambda t \end{aligned} \tag{8.4}$$

Now we find differential equations for the 'elements'

$$\begin{aligned} \frac{da}{dt} &= -m_1 k^2 (a \sin kt - b \cos kt)^3 \sin kt - \frac{m_2}{k} \cos \lambda t \sin kt \\ \frac{db}{dt} &= m_1 k^2 (a \sin kt - b \cos kt)^3 \cos kt + \frac{m_2}{k} \cos \lambda t \cos kt \end{aligned} \tag{8.5}$$

What is the gain of these transformations into new variables? Already the name 'perturbation theory' tells us that the solution will be close to one of an integrable equation. And in fact the two parameters $m_1$ and $m_2$ we have chosen as perturbing the oscillator will be small quantities, namely the masses of the planets compared to the Sun's mass in planetary theory [4]. This means that $a$ and $b$ in our example, the osculating elements in planetary theory, will change slowly! We find the first-order solution with respect to the small quantities when we keep $a$ and $b$ constant. Introducing the constants $c_1 - c_7$

$$\begin{aligned} &c_1 = \frac{3}{8}\left(a^2 + b^2\right), \quad c_2 = \frac{1}{8k}\left(3a^2 + b^2\right), \quad c_3 = \frac{1}{32k}\left(3a^2 - b^2\right) \\ &c_4 = \frac{1}{32k}\left(a^2 - 3b^2\right), \quad c_5 = \frac{1}{8k}\left(a^2 + 3b^2\right) \\ &c_6 = \frac{m_2}{2k\,(\lambda + k)}, \quad c_7 = \frac{m_2}{2k\,(\lambda - k)} \end{aligned} \tag{8.6}$$

4) The largest planet Jupiter has only 1/1000 of the mass of the Sun!

In this solution there are constant terms, secular terms, pure periodic terms and mixed terms[5] like $[\cos(\lambda - k)]t$. The solutions for $a$ and $b$ read

$$a = a_0 - m_1 k^2 \left[ a c_1 t + b\,(c_2 \cos 2kt - c_3 \cos 4kt) - \frac{a^3}{4k} \sin 2kt \right.$$
$$\left. + a c_4 \sin 4kt \right] + c_6 \cos t\,(\lambda + k) - c_7 \cos t\,(\lambda - k)$$

respectively

$$b = b_0 - m_1 k^2 \left[ b c_1 t + a(c_5 \cos 2kt - c_4 \cos 4kt) + \frac{b^3}{4k} \sin 2kt \right.$$
$$\left. - b c_3 \sin 4kt \right] + c_6 \sin\left((\lambda + k)\,t\right) - c_7 \sin\left((\lambda - k)\,t\right)$$

The appearance of terms proportional to the time $t$ seems to blow up the solution for large $t$. But in fact to higher order in the theory there appear terms in $t^2$, $t^3$ and so on. Therefore the series' behavior is not necessarily unbound, as the example of $\sin t = t - t^3/3! + t^5/5! \dots$ shows; despite these terms in $t$ there is an upper and lower limit for $\sin t$!

## 8.1.2
## Principles of Planetary Theory

In planetary theory we proceed in the same way. Without going into details we outline the principal steps. Well written introductions can be found for example in [143, 180–183]. Starting from the equations of motion in rectangular coordinates for a planet $m_1, \boldsymbol{q}_1 = (x_1, y_1, z_1)^{\mathrm{T}}$ perturbed by another planet $m_2, \boldsymbol{q}_2 = (x_2, y_2, z_2)^{\mathrm{T}}$, both orbiting the Sun ($m_0$) we can write for $i = 1, 2,\ i \neq j$ and $r_i = |\boldsymbol{q}_i|$

$$\frac{d^2 \boldsymbol{q}_i}{dt^2} + k^2 \frac{m_0 + m_i}{r_i^3} \boldsymbol{q}_i = m_j \nabla F_i \tag{8.7}$$

and separate into four differential equations of first order for the couple of two planets

$$\begin{aligned} \frac{d\boldsymbol{q}_i}{dt} - \boldsymbol{q}_i' &= 0, \\ \frac{d\boldsymbol{q}_i'}{dt} + k^2 \frac{m_0 + m_i}{r_i^3} \boldsymbol{q}_i &= m_j \nabla F_i \end{aligned} \tag{8.8}$$

with

$$\nabla F_i = k^2 \left( \frac{\boldsymbol{q}_j - \boldsymbol{q}_i}{r_{ji}^3} - \frac{\boldsymbol{q}_i}{r_i^3} \right) \tag{8.9}$$

5) In planetary theory called Poisson terms, after Siméon-Denis Poisson, a french astronomer (1781–1840).

for $i = 1, 2$ and $i \neq j$. In the next chapter we will see how $F_i$ can be developed into Legendre Polynomials. For the unperturbed problem we can write $\nabla F_i = 0$ and we know the solutions are the Kepler elements. So – as we have done for (8.2) – we use a transformation of the coordinates from $\boldsymbol{q}_1, \boldsymbol{q}_2 \rightarrow \boldsymbol{a}, \boldsymbol{b}$ so that we can write

$$\begin{aligned}
\boldsymbol{q}_1 &= \boldsymbol{f}(a_1, a_2, a_3, a_4, a_5, a_6, t),\\
\boldsymbol{q}_1' &= \boldsymbol{g}(a_1, a_2, a_3, a_4, a_5, a_6, t),\\
\boldsymbol{q}_2 &= \boldsymbol{f}(b_1, b_2, b_3, b_4, b_5, b_6, t),\\
\boldsymbol{q}_2' &= \boldsymbol{g}(b_1, b_2, b_3, b_4, b_5, b_6, t)
\end{aligned} \tag{8.10}$$

We conduct this transformation by building the derivatives of (8.10)

$$\begin{aligned}
\frac{d\boldsymbol{q}_1}{dt} &= \frac{\partial \boldsymbol{q}_1}{\partial t} + \sum_{k=1}^{6} \frac{\partial \boldsymbol{q}_1}{\partial a_k}\frac{da_k}{dt},\\
\frac{d\boldsymbol{q}_1'}{dt} &= \frac{\partial \boldsymbol{q}_1'}{\partial t} + \sum_{k=1}^{6} \frac{\partial \boldsymbol{q}_1'}{\partial a_k}\frac{da_k}{dt},\\
\frac{d\boldsymbol{q}_2}{dt} &= \frac{\partial \boldsymbol{q}_2}{\partial t} + \sum_{k=1}^{6} \frac{\partial \boldsymbol{q}_2}{\partial b_k}\frac{db_k}{dt},\\
\frac{d\boldsymbol{q}_2'}{dt} &= \frac{\partial \boldsymbol{q}_2'}{\partial t} + \sum_{k=1}^{6} \frac{\partial \boldsymbol{q}_2'}{\partial b_k}\frac{db_k}{dt}
\end{aligned} \tag{8.11}$$

which we substitute in (8.8). We have

$$\begin{aligned}
&\frac{\partial \boldsymbol{q}_1}{\partial t} - \boldsymbol{q}_1' + \sum_{k=1}^{6} \frac{\partial \boldsymbol{q}_1}{\partial a_k}\frac{da_k}{dt} = 0,\\
&\frac{\partial \boldsymbol{q}_1'}{\partial t} + k^2 \frac{m_0 + m_1}{r_1^3}\boldsymbol{q}_1 + \sum_{k=1}^{6} \frac{\partial \boldsymbol{q}_1'}{\partial a_k}\frac{da_k}{dt} = m_2 \nabla F_1
\end{aligned} \tag{8.12}$$

and the same equations hold for the other planet with mass $m_2$ and the position vector $\boldsymbol{q}_2$. When these equations are solutions of the unperturbed problem then for the coordinates and the velocities the partial derivatives with respect to time need to be equivalent to the total derivative. It follows that

$$\sum_{k=1}^{6} \frac{\partial \boldsymbol{q}_1}{\partial a_k}\frac{da_k}{dt} = 0$$

$$\sum_{k=1}^{6} \frac{\partial \boldsymbol{q}_2}{\partial b_k}\frac{db_k}{dt} = 0$$

and we have the additional properties

$$\sum_{k=1}^{6} \frac{\partial \boldsymbol{q}_1'}{\partial a_k}\frac{da_k}{dt} = m_2 \nabla F_1$$

$$\sum_{k=1}^{6} \frac{\partial \boldsymbol{q}_2'}{\partial b_k}\frac{db_k}{dt} = m_1 \nabla F_2$$

which provides us with equations to be solved for the $a_i, b_i, i = 1, 6$ for which we can write

$$\begin{aligned} \frac{da_i}{dt} &= m_2 \Phi_i(a_r, b_r, t), \, r = 1, 6 \\ \frac{db_i}{dt} &= m_1 \Psi_i(a_r, b_r, t), \, r = 1, 6 \end{aligned} \tag{8.13}$$

Now we need to remember that in planetary dynamics the perturbing fixed parameters, namely the masses are constant values with at most $m_i = m_{\text{Sun}}/1000$. Thus we can develop the solutions as convergent power series in the masses $m_1, m_2$

$$\begin{aligned} a_i &= \sum_{j=0}^{\infty} \sum_{k=0}^{\infty} a_i^{(j,k)} m_1^j m_2^k \\ b_i &= \sum_{j=0}^{\infty} \sum_{k=0}^{\infty} b_i^{(j,k)} m_1^j m_2^k \end{aligned} \tag{8.14}$$

In the case of the planetary system Poincaré proved that given the nonintersecting planetary orbits for an initial time $t_0$ then for sufficiently small perturbing parameters $m_1$ and $m_2$ the solutions given in (8.14) converge for any finite time interval. The 'problem' in our Solar System is that the masses of the planets are not parameters which can be chosen arbitrarily small – they are given by nature! Consequently the theorem by Poincaré is only valid for a time interval $\Delta t \leq T$ dependent on the masses of the planets (Cauchy theorem[6]) – which is in the order of several $10^5$ years. Given $T$ small enough (8.14) are solutions where we now have to determine the values of the $a_i^{(j,k)}$ and the $b_i^{(j,k)}$. We therefore insert the solution (8.14) in (8.13) where we develop the right hand side of (8.13) into a Taylor series:

$$\begin{aligned} &\frac{da_i^{(0,0)}}{dt} + m_2 \frac{da_i^{(0,1)}}{dt} + m_1 \frac{da_i^{(1,0)}}{dt} + m_1 m_2 \frac{da_i^{(1,1)}}{dt} + m_2^2 \frac{da_i^{(0,2)}}{dt} + m_1^2 \frac{da_i^{(2,0)}}{dt} \\ &\quad + m_2^3 \frac{da_i^{(0,3)}}{dt} + m_1 m_2^2 \frac{da_1^{(1,2)}}{dt} + m_1^2 m_2 \frac{da_i^{(2,1)}}{dt} + m_1^3 \frac{da_i^{(3,0)}}{dt} + \ldots \\ &= m_2 \Phi_i \left( a_r^{(0,0)}, b_r^{(0,0)}, t \right) + m_2 \sum_{j=1}^{6} \frac{\partial \Phi_i}{\partial a_j} \left( m_2 a_j^{(0,1)} + m_1 a_j^{(1,0)} \right) \\ &\quad + m_2 \sum_{j=1}^{6} \frac{\partial \Phi_i}{\partial b_j} \left( m_2 b_j^{(0,1)} + m_1 b_j^{(1,0)} \right) + \ldots \end{aligned} \tag{8.15}$$

A similar equation can be derived for the parameter $b_i$; the next step is a comparison of the respective coefficients of $m_1, m_2, m_1^2, m_2^2, m_1 m_2, \ldots$ in (8.15) from which follows for both parameters $a_i, b_i$

$$\frac{da_i^{(0,0)}}{dt} = 0 \,, \quad \frac{db_i^{(0,0)}}{dt} = 0 \,, \quad i = 1, 6 \tag{8.16}$$

6) Which states criteria for a series to be convergent.

After integration we get the constant terms $a_i^{(0,0)}, b_i^{(0,0)}$ which we replace in the terms having $m_1$ or $m_2$ as factors

$$
\begin{aligned}
\frac{da_i^{(0,1)}}{dt} &= \Phi_i\left(a_i^{(0,0)}, b_i^{(0,0)}, t\right), \quad i = 1,6, \\
\frac{da_i^{(1,0)}}{dt} &= 0, \qquad \frac{db^{(0,1)}}{dt} = 0 \\
\frac{db_i^{(1,0)}}{dt} &= \Psi_i\left(a_i^{(0,0)}, b_i^{(0,0)}, t\right), \quad i = 1,6
\end{aligned}
\tag{8.17}
$$

Consequently the parameters $a_i^{(0,1)}, a_i^{(1,0)}, b_i^{(0,1)}, b_i^{(1,0)}$ can also be determined by integrating (8.17). As next step we compare for the second order in the masses and get

$$
\begin{aligned}
\frac{da_i^{(1,1)}}{dt} &= \sum_{j=1}^{6} \frac{\partial \Phi_i}{\partial a_j} a_j^{(1,0)} + \sum_{j=1}^{6} \frac{\partial \Phi_i}{\partial b_j} b_j^{(1,0)} \\
\frac{da_i^{(0,2)}}{dt} &= \sum_{j=1}^{6} \frac{\partial \Phi_i}{\partial a_j} a_j^{(0,1)} + \sum_{j=1}^{6} \frac{\partial \Phi_i}{\partial b_j} b_j^{(0,1)} \\
\frac{da_i^{(2,0)}}{dt} &= 0\,, \qquad \frac{db_i^{(0,2)}}{dt} = 0 \\
\frac{db_i^{(1,1)}}{dt} &= \sum_{j=1}^{6} \frac{\partial \Psi_i}{\partial a_j} a_j^{(0,1)} + \sum_{j=1}^{6} \frac{\partial \Psi_i}{\partial b_j} b_j^{(0,1)} \\
\frac{db_i^{(2,0)}}{dt} &= \sum_{j=1}^{6} \frac{\partial \Psi_i}{\partial a_j} a_j^{(1,0)} + \sum_{j=1}^{6} \frac{\partial \Psi_i}{\partial b_j} b_j^{(1,0)}
\end{aligned}
\tag{8.18}
$$

which we will deal with after the determination of the constants of integration.

### 8.1.3
### The Integration Constants – the Osculating Elements

Now we need to determine the constants of integration for the $a_i^{(j,k)}$ and $b_i^{(j,k)}$. In the following we denote the respective constants with $\bar{a}_i^{(j,k)}$ and $\bar{b}_i^{(j,k)}$. In (8.15) $m_2$ is always present as factor in the right hand sides; in the equations for the $b_i^{(j,k)}$ $m_1$ is the corresponding factor. It is evident that

$$
a_i^{(j,0)} = \bar{a}_i^{(j,0)} \quad \text{and} \quad b_i^{(0,k)} = \bar{b}_i^{(0,k)}
\tag{8.19}
$$

for all indices $j$ and $k$. According to the ansatz in Eq. (8.14) we find that all the new terms $a_i^{(j,k)}$ as well as $b_i^{(j,k)}$ can be derived by quadratures and therefore we just added two functions $f_i^{(j,k)}(t)$ and $g_i^{(j,k)}(t)$. They have the following form

$$
a_i^{(j,k)} = f_i^{(j,k)}(t) - \bar{a}_i^{(j,k)}\,, \quad b_i^{(j,k)} = g_i^{(j,k)}(t) - \bar{b}_i^{(j,k)}
\tag{8.20}
$$

and (8.14) reads

$$a_i = \sum_{j=0}^{\infty} a_i^{(j,0)} m_1^j + \sum_{j=0}^{\infty}\sum_{k=1}^{\infty} \left( f_i^{(j,k)}(t) - \bar{a}_i^{(j,k)} \right) m_1^j m_2^k$$
$$b_i = \sum_{k=0}^{\infty} b_i^{(0,k)} m_2^k + \sum_{j=1}^{\infty}\sum_{k=0}^{\infty} \left( g_i^{(j,k)}(t) - \bar{b}_i^{(j,k)} \right) m_1^j m_2^k \tag{8.21}$$

When the initial values for the time $t = t_0$ are $a_i = a_i^{(0)}$ and $b_i = b_i^{(0)}$ then in (8.21) we just replace these values and get immediately that

$$a_i^{(0,0)} = a_i^{(0)}\,, \quad a_i^{(j,0)} = 0\,, \quad (j \geq 1)$$
$$b_i^{(0,0)} = a_i^{(0)}\,, \quad b_i^{(0,k)} = 0\,, \quad (k \geq 1) \tag{8.22}$$

and the second terms in the right hand sides of (8.21) vanish for $j, k = 1, \ldots, \infty$ because the powers on both sides are equal. Consequently all terms except the first one disappear and these values are the osculating elements $a_i$ and $b_i$ of the two orbits and we need to integrate from $t = t_0$ to $t = T$ to find the solutions of the differential equations for the elements for a time $T$.

## 8.1.4
## First-Order Perturbation

In a first-order theory we assume that the perturbing planets move on fixed Keplerian orbits. The perturbations caused by those planets can simply be added to the orbit of the perturbed planet due to the linearity of the first order problem. As a consequence the orbit is not any more a fixed ellipse, but a slightly distorted one. We can write for the first-order perturbation in a 3 planet system for the orbital elements $a_i(m_1)$, $b_i(m_2)$, $c_i(m_3)$, (for $i = 1, \ldots, 6$)[7]

$$\begin{aligned}
\frac{d a_i^{(0,1,0)}}{dt} &= \Phi_i \left( a_r^{(0)}, b_r^{(0)}, t \right), \quad i = 1, 6,\\
\frac{d a_i^{(0,0,1)}}{dt} &= \Phi_i \left( a_r^{(0)}, c_r^{(0)}, t \right), \quad i = 1, 6,\\
\frac{d b_i^{(1,0,0)}}{dt} &= \Psi_i \left( a_r^{(0)}, b_r^{(0)}, t \right), \quad i = 1, 6,\\
\frac{d b_i^{(0,0,1)}}{dt} &= \Psi_i \left( b_r^{(0)}, c_r^{(0)}, t \right), \quad i = 1, 6,\\
\frac{d c_i^{(1,0,0)}}{dt} &= \Xi_i \left( a_r^{(0)}, c_r^{(0)}, t \right), \quad i = 1, 6,\\
\frac{d c_i^{(0,1,0)}}{dt} &= \Xi_i \left( b_r^{(0)}, c_r^{(0)}, t \right), \quad i = 1, 6
\end{aligned} \tag{8.23}$$

7) The index inside the functions $\Phi$, $\Psi$ and $\Xi$ stands for $r = 1, \ldots, 6$.

and

$$\frac{da_i^{(1,0,0)}}{dt} = 0,$$
$$\frac{db_i^{(0,1,0)}}{dt} = 0 \qquad (8.24)$$
$$\frac{dc_i^{(0,0,1)}}{dt} = 0$$

It is evident that in this approximation every planet disturbs the other ones such that their ellipses become distorted; thus for determining the perturbations as example for the planet $m_1$ the first and second equations in (8.23) need to be integrated as if the elliptic elements of the perturbing planets would be constant. In this approximation we can just add the perturbations from any pair of planets $m_i - m_j$ for $j = 1, 8;\ i \neq j$.

### 8.1.5
### Second-Order Perturbation

In a higher-order theory the perturbed ellipse of the perturbing planet acts on the already perturbed orbit of the regarded planet. This means a correction of the perturbation of the first order stemming from all the planets (when we consider more than two planet in the system). These second-order corrections in the case of two planets $m_1$ and $m_2$ can be determined with the aid of the following equations

$$\frac{da_i^{(1,1)}}{dt} = \sum_{j=1}^{6} \frac{\partial \Phi_i\left(a_r^0, b_r^0, t\right)}{\partial b_j} b_j^{(1,0)}$$
$$\frac{da_i^{(0,2)}}{dt} = \sum_{j=1}^{6} \frac{\partial \Phi_i\left(a_r^0, b_r^0, t\right)}{\partial a_j} a_j^{(0,1)}$$
$$\frac{db_i^{(1,1)}}{dt} = \sum_{j=1}^{6} \frac{\partial \Psi_i\left(a_r^0, b_r^0, t\right)}{\partial a_j} a_j^{(0,1)} \qquad (8.25)$$
$$\frac{db_i^{(2,0)}}{dt} = \sum_{j=1}^{6} \frac{\partial \Psi_i\left(a_r^0, b_r^0, t\right)}{\partial b_j} b_j^{(1,0)}$$

because we have already seen that $a_i^{(1,0)}, a_i^{(2,0)}, b_i^{(0,1)}, b_i^{(0,2)}$ vanish and therefore in (8.18) the terms with these parameters vanish also. This procedure can be continued and the $a_i$ and $b_i$ determined with any desired precision as far as the series expansions are convergent in the regarded time span $T$.

For three planets $m_1$, $m_2$ and $m_3$ the determination of the second order perturbations is more complicated. In the following we will deal only with the perturbations on $m_1$. For the elements $a_i$ we then use the ansatz

$$a_i = \sum_{j=0}^{\infty}\sum_{k=0}^{\infty}\sum_{l=0}^{\infty} a_i^{(j,k,l)} m_1^j m_2^k m_3^l \qquad (8.26)$$

where for the Taylor expansion which in principle is an expansion for three arguments of the functions but we only need to use pairs of variables

$$\frac{da_i}{dt} = m_2 \Phi_i(a_r, b_r, t) + m_3 \Phi_i(a_r, b_r, t) \tag{8.27}$$

because in the comparison of the coefficients the other masses $m_3$ for the first term in (8.27) and $m_2$ for the second term are not present. As a consequence the equations for the perturbation of $m_1$ to the second order are the following ones:

$$\begin{aligned}
\frac{da_i^{(1,1,0)}}{dt} &= \sum_{j=1}^{6} \frac{\partial \Phi_i\left(a_i^{(0)}, b_i^{(0)}, t\right)}{\partial b_j} b_j^{(1,0,0)} \\
\frac{da_i^{(1,0,1)}}{dt} &= \sum_{j=1}^{6} \frac{\partial \Phi_i\left(a_i^{(0)}, c_i^{(0)}, t\right)}{\partial c_j} c_j^{(1,0,0)} \\
\frac{da_i^{(0,2,0)}}{dt} &= \sum_{j=1}^{6} \frac{\partial \Phi_i\left(a_i^{(0)}, b_i^{(0)}, t\right)}{\partial a_j} a_j^{(0,1,0)} \\
\frac{da_i^{(0,0,2)}}{dt} &= \sum_{j=1}^{6} \frac{\partial \Phi_i\left(a_i^{(0)}, c_i^{(0)}, t\right)}{\partial a_j} a_j^{(0,0,1)} \\
\frac{da_i^{(0,1,1)}}{dt} &= \sum_{j=1}^{6} \frac{\partial \Phi_i\left(a_i^{(0)}, b_i^{(0)}, t\right)}{\partial a_j} a_j^{(0,0,1)} \\
&+ \sum_{j=1}^{6} \frac{\partial \Phi_i\left(a_i^{(0)}, b_i^{(0)}, t\right)}{\partial b_j} b_j^{(0,0,1)} \\
&+ \sum_{j=1}^{6} \frac{\partial \Phi_i\left(a_i^{(0)}, c_i^{(0)}, t\right)}{\partial a_j} a_j^{(0,1,0)} \\
&+ \sum_{j=1}^{6} \frac{\partial \Phi_i\left(a_i^{(0)}, c_i^{(0)}, t\right)}{\partial c_j} c_j^{(0,1,0)}
\end{aligned} \tag{8.28}$$

In (8.28) the following terms one by one arise:

- terms coming from the perturbation of $m_2$ caused by the perturbations of the first order of $m_2$ by $m_1$ (factor $b_j^{(1,0,0)}$)
- terms coming from the perturbation of $m_3$ caused by the perturbations of the first order of $m_3$ by $m_1$ (factor $c_j^{(1,0,0)}$)
- terms coming from the perturbation of $m_2$ caused by the perturbations of the first order of $m_1$ by $m_2$ (factor $a_j^{(0,1,0)}$)
- terms coming from the perturbation of $m_3$ caused by the perturbations of the first order of $m_1$ by $m_3$ (factor $a_j^{(0,0,1)}$)

- terms coming from the perturbation of $m_2$ caused by the perturbations of the first order of $m_1$ by $m_3$ (factor $a_j^{(0,0,1)}$)
- terms coming from the perturbation of $m_2$ caused by the perturbations of the first order of $m_2$ by $m_3$ (factor $b_j^{(0,0,1)}$)
- terms coming from the perturbation of $m_3$ caused by the perturbations of the first order of $m_1$ by $m_2$ (factor $a_j^{(0,1,0)}$)
- terms coming from the perturbation of $m_3$ caused by the perturbations of the first order of $m_3$ by $m_2$ (factor $c_j^{(0,1,0)}$)

The complicated structure shows that already for the perturbation of the second order it is not possible to separate into pairs of two planets.

## 8.2 Equations of Motion for n Bodies

In an inertial coordinate system the force acting on a celestial body $m_i$ is given by the summation of the forces of all other masses $m_j$ $(j = 0, n, j \neq i)$[8] in an inertial frame (Figure 8.1)

$$\boldsymbol{F}_i \equiv m_i \ddot{\boldsymbol{p}}_i = -k^2 m_i \sum_{j=0, j\neq i}^{n} \frac{m_j}{r_{ij}^3} (\boldsymbol{p}_i - \boldsymbol{p}_j) \tag{8.29}$$

When we now add all the forces in the system we get

$$\sum_{i=0}^{n} m_i \ddot{\boldsymbol{p}}_i = -k^2 \sum_{i=0}^{n} \sum_{j=0, j\neq i}^{n} \frac{m_i m_j}{r_{ij}^3} (\boldsymbol{p}_i - \boldsymbol{p}_j) \equiv 0 \tag{8.30}$$

because in the summation the vectors $\boldsymbol{p}_i - \boldsymbol{p}_j$ and $\boldsymbol{p}_j - \boldsymbol{p}_i$[9] cancel each other. Integrating twice $\sum_{i=0}^{n} m_i \cdot \ddot{\boldsymbol{p}}_i$ leads to two constant vectors which define the barycenter $\boldsymbol{S}$ of the system which is moving on a straight line

$$\boldsymbol{S} = \boldsymbol{c}_1 t + \boldsymbol{c}_2 \tag{8.31}$$

and defines six integrals of motion which is the principle of linear momentum. We now can move the origin to the barycenter from which follows $\ddot{\boldsymbol{S}} = \dot{\boldsymbol{S}} = \boldsymbol{S} = 0$. The angular momentum vector $\mathbf{g}$ defines another three constants

$$\sum_{i=0}^{n} \boldsymbol{p}_i \times m_i \dot{\boldsymbol{p}}_i = \mathbf{g} \tag{8.32}$$

and finally for conservative systems the energy $h$ is constant.

$$h = \frac{1}{2} \sum_{i=0}^{n} m_i \left(\dot{\boldsymbol{p}}_i \dot{\boldsymbol{p}}_i\right) - U \tag{8.33}$$

8) Thought to be mass points.
9) $|\boldsymbol{p}_j - \boldsymbol{p}_i| = r_{ij}$.

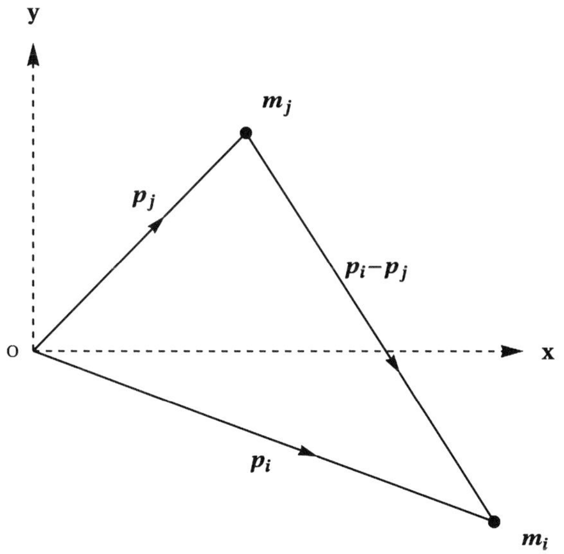


**Figure 8.1** Configuration in the inertial frame.

where the potential can be expressed

$$U = k^2 \sum_{i=0}^{n} \sum_{j=0, i \neq j}^{n} \frac{m_i m_j}{r_{ij}} \tag{8.34}$$

### 8.2.1
### The Virial Theorem

Although the results of the virial theorem are more of statistical value it is a good point of discussion in connection with the planetary system. In the case of n bodies we can compute the quantity

$$G = \sum_{i=0}^{n} (\boldsymbol{p}_i \cdot m_i \dot{\boldsymbol{p}}_i) \tag{8.35}$$

where we have to build the sum over all bodies in the system. After differentiation with respect to the time

$$\frac{dG}{dt} = \sum_{i=0}^{n} (\dot{\boldsymbol{p}}_i \cdot m_i \dot{\boldsymbol{p}}_i) + \sum_{i=0}^{n} (\boldsymbol{p}_i \cdot m_i \ddot{\boldsymbol{p}}_i) \tag{8.36}$$

where we can write for the first term

$$\sum_{i=0}^{n} (\dot{\boldsymbol{p}}_i \cdot m_i \dot{\boldsymbol{p}}_i) = \sum_{i=0}^{n} m_i v_i^2 = 2T$$

and for the second term

$$\sum_{i=0}^{n}(\boldsymbol{p}_i \cdot m_i \ddot{\boldsymbol{p}}_i) = \sum_{i=0}^{n}(\boldsymbol{F}_i \cdot \boldsymbol{p}_i)$$

and as a consequence for (8.36)

$$\frac{dG}{dt} = 2T + \sum_{i=0}^{n}(\boldsymbol{F}_i \cdot \boldsymbol{p}_i) \tag{8.37}$$

Now we build the time average over a time interval $\tau$ by integrating both sides from 0 to $\tau$

$$\frac{1}{\tau}\int_0^{\tau} \frac{dG}{dt}\,dt \equiv \bar{G} = 2\overline{T} + \overline{\sum_{i=0}^{n}(\boldsymbol{F}_i \cdot \boldsymbol{p}_i)} \tag{8.38}$$

When $\tau = P$, the period, which is chosen such that the coordinates repeat which means that (8.38) vanishes; this expression vanishes also when $G$ has an upper bound. When $\tau$ is chosen very large, the quantity

$$2\bar{T} + \overline{\sum_{i=0}^{n}(\boldsymbol{F}_i \cdot \boldsymbol{p}_i)} = \frac{1}{\tau}[G(\tau) - G(0)]$$

vanishes also and the result is

$$\bar{T} = -\frac{1}{2}\overline{\sum_{i=0}^{n}(\boldsymbol{F}_i \cdot \boldsymbol{p}_i)} \tag{8.39}$$

Because $\boldsymbol{F}_i \cdot \boldsymbol{p}_i$ can be written as $\nabla_i U.\boldsymbol{p}_i$ we can formulate the so-called

Theorem (Virial Theorem)

For self-gravitating masses the total kinetic energy of the objects is equal to minus 1/2 of the total gravitational potential energy.[10]

This theorem is of special importance for many body systems like globular clusters as well as galaxies but also inside a star these relations hold.[11]

## 8.2.2
## Reduction to Heliocentric Coordinates

The situation in the Solar system is a special one because more than 99.9% of the mass in the Solar system is in the Sun ($m_0$) itself. Consequently we take the Sun

10) Note that the potential needs to be of the form $\alpha r^n$.

11) It goes back until 1870 when Rudolf Clausius (1822–1888) formulated it in one of his lectures in connection with his fundamental thermodynamical principles. Today it also applies to estimate the amount of dark matter in a galaxy.

as center of the coordinate system which means we use heliocentric coordinates $\boldsymbol{q}_i = \boldsymbol{p}_i - \boldsymbol{p}_0$. For the motion of the planet $m_i$ in the inertial coordinate system we can write

$$\ddot{\boldsymbol{p}}_i = k^2 \left[ \frac{m_0}{r_{0i}^3}(\boldsymbol{p}_0 - \boldsymbol{p}_i) + \sum_{j=1, j\neq i}^{n} \frac{m_j}{r_{ij}^3}(\boldsymbol{p}_j - \boldsymbol{p}_i) \right] \tag{8.40}$$

whereas the equation of motion for the Sun reads

$$\ddot{\boldsymbol{p}}_0 = k^2 \left[ \frac{m_i}{r_{0i}^3}(\boldsymbol{p}_i - \boldsymbol{p}_0) + \sum_{j=1, j\neq i}^{n} \frac{m_j}{r_{j1}^3}(\boldsymbol{p}_j - \boldsymbol{p}_0) \right] \tag{8.41}$$

The following subtraction leads to the equations for the planet $m_i$ in the heliocentric frame

$$\ddot{\boldsymbol{q}}_i = k^2 \left[ -\frac{m_0}{r_{0i}^3}(\boldsymbol{p}_i - \boldsymbol{p}_0) - \frac{m_i}{r_{0i}^3}(\boldsymbol{p}_i - \boldsymbol{p}_0) \right] + k^2 \sum_{j=1, j\neq i}^{n} \left[ \frac{m_j}{r_{ji}^3}(\boldsymbol{p}_j - \boldsymbol{p}_i) - \frac{m_j}{r_{j0}^3}(\boldsymbol{p}_j - \boldsymbol{p}_0) \right]$$

In a different form we can write it as

$$\ddot{\boldsymbol{q}}_i = -k^2 \frac{m_0 + m_i}{r_{0i}^3} \boldsymbol{q}_i + \boldsymbol{P}_i \tag{8.42}$$

where we introduced the perturbation $\boldsymbol{P}_i$

$$\boldsymbol{P}_i = k^2 \sum_{j=1, j\neq i}^{n} m_j \left( \frac{\boldsymbol{q}_j - \boldsymbol{q}_i}{r_{ji}^3} - \frac{\boldsymbol{q}_j}{r_{j0}^3} \right) \tag{8.43}$$

This differential equation (8.42) of second order (Figure 8.2) is very practical to describe the motions in the Solar system where, for the planets, the comets and asteroids the Sun is dominating. But it can also be used for the Moon or Near Earth Asteroids (then we may take the Earth (now $m_0$) as the center for geocentric coordinates). In (8.43) we see that there may be a problem when the distance between two bodies $r_{ij}$ is small, because this quantity is in the denominator! Although the masses $m_j$ are small the 'perturbation $\boldsymbol{P}$' may be larger than the first term on the right hand side in (8.42) which is nothing else than the two-body motion of a planet $m_i$ around the Sun $m_0$. Therefore, when using perturbation theory, one has to be sure that the 'perturbation' stays small, otherwise it does not make sense to use the equations in this form. In the next chapters we will introduce the different analytical methods of solving these differential equations. Another way to solve them is a numerical integration method for which many efficient ones are available; most of them are extensively tested and in all details described in various textbooks and articles; we will shortly explain it later in this section.

## 8.3
## Lagrange Equations of the Planetary n-Body Problem

The vector $\boldsymbol{P}_i$ can be described by a scalar function $F_i$ in form of

$$\nabla F_i = k^2 \sum_{j=1, j\neq i}^{n} m_j \left( \frac{\boldsymbol{q}_j - \boldsymbol{q}_i}{r_{ji}^3} - \frac{\boldsymbol{q}_j}{r_{j0}^3} \right) \tag{8.44}$$

where $F_i$ is the so-called perturbation function describing the perturbations of a planet $m_i$ perturbed by $n-1$ planets

$$F_i = k^2 \sum_{j=1, j\neq i}^{n} m_j \left( \frac{1}{r_{ij}} - \frac{\boldsymbol{q}_i \cdot \boldsymbol{q}_j}{r_{j0}^3} \right) \tag{8.45}$$

which has a direct part $1/\rho$ with $\rho = r_{ij}$ and an indirect part; the latter is the product of the two heliocentric position vectors of the planets. We will in detail explain how one gets the Fourier expansion of the direct part for a pair of two planets, which can be extended to any pair of planets. But we will also explain the principle of perturbation theories of higher orders. Just to mention shortly that in these theories we also need to take care of the fact that the perturbed planet is also perturbing the other planets which were regarded in the first-order theory as moving on unperturbed ellipses. It turns out that in such theories products of all the masses involved are present so that in $P_i$ in a second-order theory factors of the form $m_i m_j$ for $j = 1, n$ with $j \neq i$ appear and so on.

### 8.3.1
### Legendre Polynomials

In the triangle shown in Figure 8.2 we can compute the distance between the two planets $\rho = \sqrt{r^2 + r'^2 - 2rr' \cos\phi}$ so that the direct part is simply derived by the trigonometric relation

$$\frac{1}{\rho} = \frac{1}{r'} \left(1 - 2\alpha \cos\phi + \alpha^2\right)^{-\frac{1}{2}} \tag{8.46}$$

As depicted in Figure 8.2 we deal with an inner planet perturbed by an outer one ($r/r' = \alpha < 1$). A development with respect to a parameter x of the form $(1-x)^m$ requires $|x| < 1$. Unfortunately in (8.46) the term $x = -2\alpha\cos\phi + \alpha^2$ does not fulfill this requirement because in several cases for a pair of planets this term is large (e. g. for Venus perturbed by the Earth $\alpha \sim 0.72$; depending on the angle $\phi$ $x$ achieves values up to almost 2).[12] To avoid this problem we divide the expression to develop into two parts and use the following relation:

$$2\cos\phi = \sigma + \sigma^{-1} \quad \text{for} \quad \sigma = e^{i\phi} = \cos\phi + i\sin\phi \tag{8.47}$$

12) The development can directly be used for the Moon orbiting the Earth perturbed by the Sun with $\alpha \sim 1/400$.

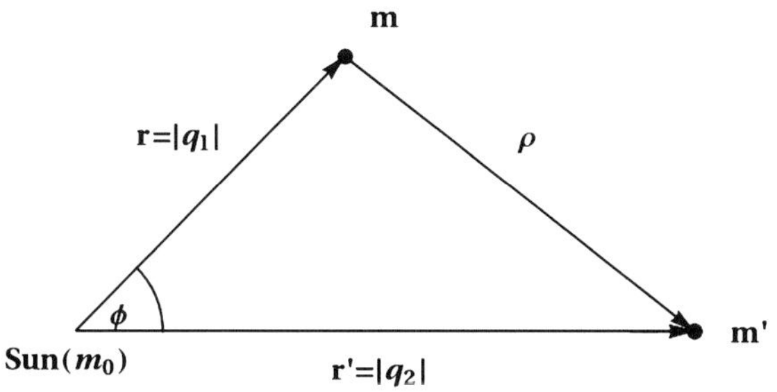


**Figure 8.2** Heliocentric position vectors.

which can now be expressed as

$$\frac{1}{\rho} = \frac{1}{r'} \left(1 - \alpha\sigma\right)^{-\frac{1}{2}} \left(1 - \alpha\sigma^{-1}\right)^{-\frac{1}{2}}$$

Both binomial series replacing the expression $1 - 2\alpha \cos\phi + \alpha^2$ are absolutely convergent because of $|\sigma| = 1$ and $|\alpha| < 1$ and we can use the development for $(1-x)^m$ for $m = -1/2$.[13] Multiplying the two series leads to the intermediate result for the development

$$1 + \frac{1}{2}\alpha\sigma + \frac{1}{2}\alpha\sigma^{-1} + \frac{3}{8}\alpha^2\sigma^2 + \frac{1}{4}\alpha^2 + \frac{3}{8}\alpha^2\sigma^{-2} + \frac{5}{16}\alpha^3\sigma^3 + \frac{3}{16}\alpha^3\left(\sigma + \sigma^{-1}\right) + \frac{5}{16}\alpha^3\sigma^{-3} + \ldots$$

which can be combined to

$$1 + \frac{1}{2}\alpha\left(\sigma + \sigma^{-1}\right) + \alpha^2\left[\frac{1}{4} + \frac{3}{8}\left(\sigma^2 + \sigma^{-2}\right)\right] + \alpha^3\left[\frac{3}{16}\left(\sigma + \sigma^{-1}\right) + \frac{5}{16}\left(\sigma^3 + \sigma^{-3}\right)\right] + \ldots$$

This can be reduced by applying the well known formula ($\cos 2\phi = 2\cos^2\phi - 1$, $\cos 3\phi = 4\cos^3\phi - 3\cos\phi$, $\cos 4\phi = \ldots$), where at the same time we have to keep in mind that $2\cos 2\phi = \sigma^2 + \sigma^{-2}$, $2\cos 3\phi = \sigma^3 + \sigma^{-3}$ and so on. Finally we get

$$\frac{1}{\rho} = \frac{1}{r'}\left[1 + \alpha\cos\phi + \alpha^2\left(-\frac{1}{2} + \frac{3}{2}\cos^2\phi\right) + \alpha^3\left(-\frac{3}{2}\cos\phi + \frac{5}{2}\cos^3\phi\right) + \ldots\right] \tag{8.48}$$

which can be expressed by the Legendre polynomials such that

$$\frac{1}{\rho} = \frac{1}{r'}\left[P_0(\cos\phi) + \alpha P_1(\cos\phi) + \alpha^2 P_2(\cos\phi) + \alpha^3 P_3(\cos\phi) + \ldots\right] \tag{8.49}$$

13) $(1-x)^{-1/2} = 1 + x/2 + (3/8)x^2 + (5/16)x^3 + (35/128)x^4 + \ldots$

The first seven Legendre polynomials read

$$\begin{aligned}
P_0(x) &= 1\\
P_1(x) &= x\\
P_2(x) &= \frac{1}{2}\left(3x^2 - 1\right)\\
P_3(x) &= \frac{1}{2}\left(5x^3 - 3x\right)\\
P_4(x) &= \frac{1}{8}\left(35x^4 - 30x^2 + 3\right)\\
P_5(x) &= \frac{1}{8}\left(63x^5 - 70x^3 + 15x\right)\\
P_6(x) &= \frac{1}{16}\left(231x^6 - 315x^4 + 105x^2 - 5\right)
\end{aligned} \tag{8.50}$$

which can be derived up to any desired Legendre Polynomial by the generating function

$$P_n(x) = \frac{1}{2^n n!}\frac{d^n}{dx^n}\left(x^2 - 1\right)^n \tag{8.51}$$

Inserting the development of the direct part of $F_i$ with Legendre polynomials into the perturbation function (8.45) yields

$$F = \frac{k^2 m'}{r'} \sum_{j=2}^{\infty} \alpha^n P_n(\cos\phi) \tag{8.52}$$

because the indirect part in (8.45) cancels with the first term in the expansion of the direct part $1/\rho$ in (8.48).

### 8.3.2
### Delaunay Elements

We now introduce the Delaunay-elements to be able to describe the motion of Solar system bodies in a simple form via canonical equations. A very instructive book in this respect is [134]. The equation of motion in form of (8.42) can now be expressed as

$$\ddot{\boldsymbol{q}}_i = -k^2 \frac{m_0 + m_i}{r_{0i}^3} \boldsymbol{q}_i + \nabla F_i \tag{8.53}$$

An appropriate way to describe the perturbations acting on a planet is to use differential equations for the change of the Keplerian elements. Unfortunately the equations which we were able to derive are in form of rectangular coordinates; consequently we have to find how we can transform the force function $F_i$ in their dependence of the Keplerian elements and how we can then develop this function into a Fourier-series with respect to time. To derive the appropriate differential equations for the perturbed Keplerian elements we introduce first the Delaunay elements (for $i = 1, \ldots, n$ planets) with $\kappa^2 = k^2(m_0 + m_i)$

$$\boldsymbol{\chi}_i = \begin{pmatrix} L_i \\ G_i \\ H_i \end{pmatrix} = \begin{pmatrix} \kappa_i \sqrt{a} \\ L_i \sqrt{1 - e_i^2} \\ G_i \cos i_i \end{pmatrix}$$

$$\overline{\boldsymbol{\chi}}_i = \begin{pmatrix} l_i \\ g_i \\ h_i \end{pmatrix} = \begin{pmatrix} M_i \\ \omega_i \\ \Omega_i \end{pmatrix}$$

which are useful as they are of canonical form obeying the equations of motion

$$\begin{aligned} \dot{\chi}_i &= \frac{\partial F_i}{\partial \overline{\chi}_i} \\ \dot{\overline{\chi}}_i &= -\frac{\partial F_i}{\partial \chi_i} \end{aligned} \tag{8.54}$$

For deriving the Lagrange equations we now omit the subscript 'i'

$$\begin{pmatrix} \dot{a} \\ \dot{e} \\ \dot{i} \end{pmatrix} - A^{-1} \begin{pmatrix} \dot{L} \\ \dot{G} \\ \dot{H} \end{pmatrix}$$

To derive the Lagrange equations for the Keplerian elements we need to compute a matrix $\boldsymbol{A}$

$$A = \begin{pmatrix} \frac{\partial L}{\partial a} & \frac{\partial L}{\partial e} & \frac{\partial L}{\partial i} \\ \frac{\partial G}{\partial a} & \frac{\partial G}{\partial c} & \frac{\partial G}{\partial i} \\ \frac{\partial H}{\partial a} & \frac{\partial H}{\partial e} & \frac{\partial H}{\partial i} \end{pmatrix}$$

$$= \begin{pmatrix} \frac{\kappa}{2\sqrt{a}} & 0 & 0 \\ \frac{\kappa}{2}\sqrt{\frac{1-e^2}{a}} & -\kappa e \sqrt{\frac{a}{1-e^2}} & 0 \\ \frac{\kappa}{2}\sqrt{\frac{1-e^2}{a}} \cos i & -\kappa e \sqrt{\frac{a}{1-e^2}} \cos i & -\kappa \sqrt{a(1-e^2)} \sin i \end{pmatrix}$$

with its inverse matrix $\boldsymbol{A}^{-1}$

$$A^{-1} = \begin{pmatrix} \frac{2\sqrt{a}}{\kappa} & 0 & 0 \\ \frac{1-e^2}{\kappa e \sqrt{a}} & -\frac{1}{\kappa e}\sqrt{\frac{1-e^2}{a}} & 0 \\ 0 & \frac{\cot i}{\kappa \sqrt{a(1-e^2)}} & -\frac{1}{\kappa \sqrt{a(1-e^2)} \sin i} \end{pmatrix}$$

When we now use the KIII we can replace $\kappa = na^{3/2}$ so that the Lagrange equations for the orbital elements $a, e, i$ read

$$\begin{aligned}
\frac{da}{dt} &= \frac{2}{na}\frac{\partial F}{\partial M} \\
\frac{de}{dt} &= \frac{-\sqrt{1-e^2}}{na^2 e}\frac{\partial F}{\partial \omega} + \frac{1-e^2}{na^2 e}\frac{\partial F}{\partial M} \\
\frac{di}{dt} &= \frac{-1}{na^2\sqrt{1-e^2}\sin i}\frac{\partial F}{\partial \Omega} + \frac{\cos i}{na^2\sqrt{1-e^2}\sin i}\frac{\partial F}{\partial \omega}
\end{aligned} \tag{8.55}$$

To derive the Lagrange equations for the other elements we will proceed differently. Taking into account that

$$a = \frac{L^2}{\kappa}, \sqrt{1-e^2} = \frac{G}{L}$$

we see that we can express $e$ and $i$ such that the respective derivatives can easily be computed

$$e = \sqrt{1 - \frac{G^2}{L^2}}\ ; \quad \cos i = \frac{H}{G}$$

and consequently the first three equations read

$$\begin{aligned}
\frac{\partial i}{\partial H} &= \frac{1}{na^2\sqrt{1-e^2}\sin i} \\
\frac{\partial e}{\partial G} &= -\frac{G}{L^2}\frac{1}{\sqrt{1-\frac{G^2}{L^2}}} \\
\frac{\partial i}{\partial G} &= -\frac{1}{\sin i}\frac{-H}{G^2}\ .
\end{aligned} \tag{8.56}$$

Using the Hamilton formalism again we find

$$\frac{dh}{dt} = \frac{d\Omega}{dt} = -\frac{\partial F}{\partial H} = -\frac{\partial F}{\partial i}\frac{\partial i}{\partial H} \tag{8.57}$$

$$\frac{dg}{dt} = \frac{d\omega}{dt} = -\frac{\partial F}{\partial G} = \frac{\partial F}{\partial e}\frac{\partial e}{\partial G} - \frac{\partial F}{\partial i}\frac{\partial i}{\partial G} \tag{8.58}$$

where we now can use (8.56) to find the Lagrange equations for the angles $\Omega$ and $\omega$.

For the sixth equation for $M$ we need some additional considerations concerning the mean motion $n$ because it is not a constant any more but defined through KIII where $a$ is not any more a constant. We cannot avoid a double integration which leads to squared small divisors; we refer to this problem in more details when we will speak about the mean motion resonances. Note that in the perturbing function we omitted that the motion is a perturbed Kepler motion which means that the equation of motion possesses an additional term of the form $\kappa^2/2a$ which stands for the unperturbed elliptic motion (compare first term in (8.42)). First we build the following derivatives:

$$\frac{\partial a}{\partial L} = \frac{2L}{\kappa} \tag{8.59}$$

$$\frac{\partial e}{\partial L} = \frac{G^2}{L^3}\left(1 - \frac{G^2}{L^2}\right)^{-\frac{1}{2}}$$

which we can introduce in the following equation

$$\frac{dl}{dt} = \frac{dM}{dt} = -\frac{\partial}{\partial L}\left(\frac{\kappa^2}{2a}\right) - \frac{\partial F}{\partial L} = -\frac{\partial}{\partial L}\left(\frac{\kappa^4}{2L^2}\right) - \frac{\partial F}{\partial a}\frac{\partial a}{\partial L} - \frac{\partial F}{\partial e}\frac{\partial e}{\partial L} \tag{8.60}$$

so that the Lagrange equations for the angles read

$$\begin{aligned}
\frac{d\Omega}{dt} &= \frac{1}{na^2\sqrt{1-e^2}\sin i}\frac{\partial F}{\partial i} \\
\frac{d\omega}{dt} &= \frac{\sqrt{1-e^2}}{na^2 e}\frac{\partial F}{\partial e} - \frac{\cos i}{na^2\sqrt{1-e^2}\sin i}\frac{\partial F}{\partial i} \\
\frac{dM}{dt} &= n - \frac{2}{na}\frac{\partial F}{\partial a} - \frac{1-e^2}{na^2 e}\frac{\partial F}{\partial e}
\end{aligned} \tag{8.61}$$

## 8.4 The Perturbing Function in Elliptic Orbital Elements

We now need to express $F(t)$ as a series expansion in terms of the orbital elements which we expressed in Cartesian coordinates. An extensive description how to make this transformation can be found for example in [181] or [143]. The desired form of the perturbing function is the following

$$F = \kappa^2 \sum C\left(a, a', e, e', i, i'\right)\cos\Phi \tag{8.62}$$

where the angle reads $\Phi = j_1\lambda' + j_2\lambda + j_3\varpi' + j_4\varpi + j_5\Omega' + j_6\Omega$. The $j_1, \ldots, j_6$ are the summation indices in (8.62) with the special property

$$q\sum_{i=1}^{6} j_i = 0 \tag{8.63}$$

known as the *d'Alembert property* which follows from the characteristics of the function $F_i$ (for details [83, p. 36 ff])

1. it is invariant under a simultaneous change of sign of all the angles involved
2. it is invariant for any rotation around the $z$-axis
3. it is invariant for a change of the signs of all inclinations

Because of the Hamilton equation for the two body problem

$$\mathcal{H}_i = \frac{1}{2}|\dot{\boldsymbol{q}}_i|^2 - \frac{\kappa_i^2}{|\boldsymbol{q}_i|} = -\frac{\kappa_i^2}{2a_i} \tag{8.64}$$

we can write the equations of motion for the Delaunay elements in a "half" canonical way, where for every planet $m_i$ the perturbation function $F_i$ looks slightly different according to the planet we are dealing with.
For a planet (elements $\chi$) perturbed by another one (elements $\chi'$) the perturbing function $F_1$ can be expressed with the aid of the Legendre polynomials as we have shown before (see (8.52)). To understand how we can derive this expression in the desired form of (8.62) we only use in the development the first terms for $j = 2, 3, 4$ so that we deal with

$$F_1 = \frac{k^2 m'}{r'^3} r^2 \left( -\frac{1}{2} + \frac{3}{2} \cos^2 \phi \right) \tag{8.65}$$

$$F_2 = \frac{k^2 m'}{r'^4} r^3 \left( -\frac{3}{2} \cos \phi + \frac{5}{2} \cos^3 \phi \right) \tag{8.66}$$

$$F_3 = \frac{k^2 m'}{r'^5} r^4 \left( \frac{3}{8} - \frac{15}{4} \cos^2 \phi + \frac{35}{8} \cos^4 \phi \right) \tag{8.67}$$

For a better understanding we go further in the simplifications and explain how we need to proceed to change from heliocentric rectangular coordinates to osculating elements: as example we chose the perturbation of a planet (mass $m$) perturbed by another planet (mass $m'$). In the corresponding Figure 8.3 we show a coordinate system centered at the Sun with the orbital plane of motion of the perturbed planet as reference plane. We use the following notation introducing the new angle true longitude $\Psi = \omega + v$ and $\Psi' = \Omega' + \omega' + v'$

$$\begin{aligned}
\widehat{\Upsilon P} &= \omega + v \\
\widehat{\Upsilon N} &= \Omega' \\
\widehat{NP'} &= \omega' + v' = \Psi' + \Omega' \\
\widehat{PN} &= \Omega' - \omega - v
\end{aligned} \tag{8.68}$$

The angle $\phi$ between the directions to the two planets is independent of the origin from which we count the angles involved; respectively they occur only in linear combinations in the perturbing function. When we increase them by an arbitrary amount there is no change in the perturbation function which means that the sum of the coefficients is zero for each term in $F$[14].

We now compute the angle between the direction to planet and the direction to

14) Which is one property of the d'Alembert rule mentioned before.

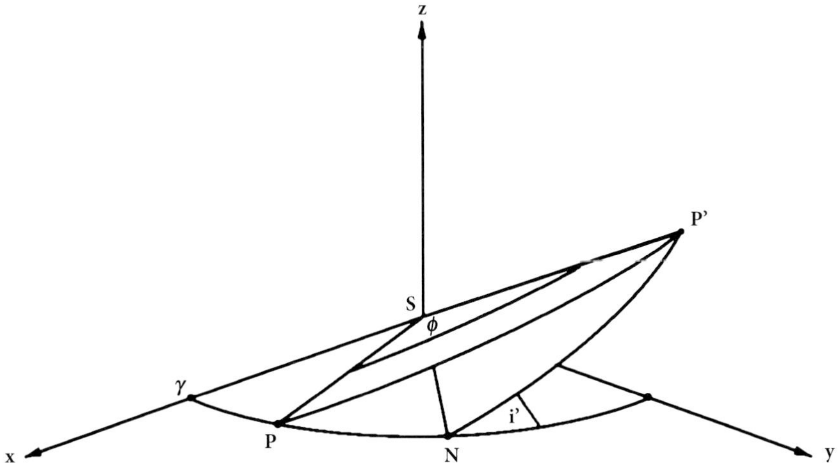


**Figure 8.3** Schematic view of a planet $P$ perturbed by a planet $P'$ in a heliocentric coordinate system. The angle $\phi$ can be determined from the spherical triangle $P N P'$.

the perturbing planet $\phi$ in the spherical triangle $S P P'$.

$$\begin{aligned}
\cos\phi &= \cos\left(\Psi' - \Omega'\right)\cos\left(\Omega' - \Psi'\right) + \sin\left(\Psi' - \Omega'\right)\sin\left(\Omega' - \Psi'\right)\cos i' \\
&= \cos\left(\omega + \nu - \omega' - \nu'\right) - \left(1 - \cos i'\right)\sin\left(\omega + \nu\right)\sin\left(\omega' + \nu'\right) \\
&= \cos\left(\Psi - \Psi'\right) - \frac{1}{2}\left(1 - \cos i'\right)\left[\cos\left(\Psi - \Psi'\right) - \cos\left(\omega + \nu + \omega' + \nu'\right)\right] \\
&= \cos^2\left(\frac{i'}{2}\right)\cos\left(\Psi - \Psi'\right) + \sin^2\left(\frac{i'}{2}\right)\cos\left(\Psi + \Psi' - 2\Omega'\right) \qquad (8.69)
\end{aligned}$$

We now ignore higher orders of $(\sin i)^n = J^n$ from $n = 3$ on; then we can write for the powers of $\phi$ which appear in the Legendre development of $F$

$$\begin{aligned}
\cos\phi &= \left(1 - \frac{J^2}{4}\right)\cos\left(\Psi - \Psi'\right) + \frac{J^2}{4}\cos\left(\Psi - \Psi' - 2\Omega'\right) \\
\cos^2\phi &= \left(\frac{1}{2} - \frac{J^2}{4}\right)\left[1 + \cos\left(2\Psi - 2\Psi'\right)\right] + \frac{J^2}{4}\left[\cos\left(2\Psi - 2\Omega'\right)\right. \\
&\quad \left. + \cos\left(2\Psi' - 2\Omega'\right)\right] \\
\cos^3\phi &= \left(\frac{3}{4} - \frac{9J^2}{16}\right)\cos\left(\Psi - \Psi'\right) + \left(\frac{1}{4} - \frac{3J^2}{16}\right)\cos 3\left(\Psi - \Psi'\right) \\
&\quad + \frac{3J^2}{8}\cos\left(\Psi - \Psi' - 2\Omega'\right) \\
&\quad + \frac{3J^2}{16}\left[\cos\left(\Psi - 3\Psi' + 2\Omega'\right) + \cos\left(3\Psi - \Psi' - 2\Omega'\right)\right] \qquad (8.70)
\end{aligned}$$

To illustrate a planetary theory (only to the first order) we replace $\cos^n\phi$ up to $n = 3$ in the expansion of the perturbing function $F$ (8.52); more precisely we take the first three terms $P_m(\cos\phi)$ for $m = 2, 3, 4$ in the expansion of $F$ into account.

$$F = \frac{k^2 m'}{r'}\left[\frac{\alpha^2}{2}\left(-1+3\cos^2\phi\right)+\frac{\alpha^3}{2}\left(-3\cos\phi+5\cos^3\phi\right)\right.$$
$$\left.+\frac{\alpha^4}{8}\left(3-30\cos^2\phi+35\cos^4\phi\right)+\dots\right] \tag{8.71}$$

We replace the factor $\kappa^2 = k^2(m_0+m') = n'^2a'^3$ in the perturbing function for the planet $m$ perturbed by the planet $m'$. Consequently (8.71) reads

$$F = F_1 + F_2 + F_3 + \dots$$
$$= n'^2a^2\left[\frac{r^2}{a^2}\frac{a'^3}{r'^3}P_2 + \frac{a}{a'}\frac{r^3}{a^3}\frac{a'^4}{r'^4}P_3 + \left(\frac{a}{a'}\right)^2\frac{r^4}{a^4}\frac{a'^5}{r'^5}P_4 + \dots\right]$$

For $F_1$ the terms independent of $J^2$ replacing $\cos^2 J$ is simply

$$F_1 = n'^2a^2\frac{r^2}{a^2}\frac{a'^3}{r'^3}\left[\frac{1}{4}+\frac{3}{4}\cos\left(2\tilde{\omega}-2\tilde{\omega}'+2v-2v'\right)\right] \tag{8.72}$$

Before we can use the expansion of

$$\left(\frac{r}{a}\right)^{\pm k}\sin jv\,;\quad \left(\frac{r}{a}\right)^{\pm k}\cos jv\,;\quad \left(\frac{a}{r}\right)^{\pm k}\sin jv\,;\quad \left(\frac{a}{r}\right)^{\pm k}\cos jv$$

we need to separate the argument of the $\cos v$ and $\sin v$ in (8.72). This can be done with the aid of the classical addition and subtraction theorems of trigonometry[15] which have to be used twice to finally get the appropriate form

$$F = n'^2a'^2\left(\frac{a'}{r'}\right)^{\pm k}\left(\frac{a}{r}\right)^{\pm k} g_1(v)g_2\left(v'\right)g_3\left(\tilde{\omega},\tilde{\omega}',\Omega\right) \tag{8.73}$$

After inserting the expansion and doing different complicated series multiplication we get finally after adding also the terms in $\cos^2\phi$

$$F = n'^2a'^2\left\{\frac{1}{4}+\frac{3}{8}\left(e^2+e'^2\right)-\frac{3}{8}J^2+\left[\frac{3}{4}-\frac{15}{8}\left(e^2+e'^2\right)-\frac{3}{8}J^2\right]\cos\left(2\lambda-2\lambda'\right)\right.$$
$$-\frac{1}{2}e\cos\left(\lambda-\tilde{\omega}\right)-\frac{9}{4}e\cos\left(\lambda-2\lambda'+\tilde{\omega}\right)+\frac{3}{4}e\cos\left(3\lambda-2\lambda'-\tilde{\omega}'\right)$$
$$+\frac{3}{4}e'\cos\left(\lambda'-\tilde{\omega}'\right)-\frac{3}{8}e'\cos\left(2\lambda-\lambda'-\tilde{\omega}'\right)+\frac{21}{8}e'\cos\left(2\lambda-3\lambda'+\tilde{\omega}'\right)$$
$$-\frac{1}{8}e^2\cos\left(2\lambda-2\tilde{\omega}\right)+\frac{15}{8}e^2\cos\left(2\lambda'-2\tilde{\omega}\right)+\frac{3}{4}e^2\cos\left(4\lambda-2\lambda'-2\tilde{\omega}'\right)$$
$$-\frac{3}{4}ee'\cos\left(\lambda-\lambda'-\tilde{\omega}+\tilde{\omega}'\right)+\frac{9}{8}ee'\cos\left(\lambda-\lambda'+\tilde{\omega}-\tilde{\omega}'\right)$$

15) $\sin(\alpha\pm\beta) = \sin\alpha\cos\beta \pm \cos\alpha\sin\beta$ respectively $\cos(\alpha\pm\beta) = \cos\alpha\cos\beta \mp \sin\alpha\sin\beta$.

$$
\begin{aligned}
&-\frac{3}{4}ee'\cos\left(\lambda-\lambda'-\tilde{\omega}+\tilde{\omega}'\right)+\frac{9}{8}ee'\cos\left(\lambda-\lambda'+\tilde{\omega}-\tilde{\omega}'\right)\\
&-\frac{3}{4}ee'\cos\left(\lambda+\lambda'-\tilde{\omega}-\tilde{\omega}'\right)-\frac{3}{8}ee'\cos\left(3\lambda-\lambda'-\tilde{\omega}-\tilde{\omega}'\right)\\
&-\frac{63}{8}ee'\cos\left(\lambda-3\lambda'-\tilde{\omega}+\tilde{\omega}'\right)+\frac{21}{8}ee'\cos\left(3\lambda-3\lambda'-\tilde{\omega}+\tilde{\omega}'\right)\\
&+\frac{9}{8}e'^2\cos\left(2\lambda'-2\tilde{\omega}'\right)+\frac{51}{8}e'^2\cos\left(2\lambda'-4\lambda'+\tilde{\omega}'\right)\\
&+\frac{3}{8}J^2\left[\cos\left(2\lambda-2\Omega\right)+\cos\left(2\lambda'-2\Omega\right)\right]\Bigg\}
\end{aligned}
\tag{8.74}
$$

What we see here is the complete expansion of the perturbing function for the orbit of a planet with the mass $m$ perturbed by another one with mass $m'$ up order 2 in the small parameters inclination[16] and eccentricities.

## 8.5
## Explicit First-Order Planetary Theory for the Osculating Elements

To derive a form where the time $t$ is explicitly in the Fourier expansion we replace the mean longitude $\lambda = M + \omega + \Omega$ and $\lambda' = M' + \omega' + \Omega'$ in (8.74) where the mean anomalies $M = n.t$ and $M' = n'.t$ are also replaced. In the following we denote the involved planets with the masses $m$, the perturbed planet, and $m'$, the perturbing planet. Again, for the detailed derivations, we refer to existing books on this fundamental problem in celestial mechanics like [133]. We thus have an adequate form of $F$ where we are able to build the respective derivatives present in the Lagrange equations and can also accomplish the necessary integration with respect to time.

$$
F = \frac{\kappa_1^4}{2L_1^2} + k^2 m' \sum_{j_1\dots,j_6} C_{j_1,\dots,j_6}\cos\left[\left(j_1 n + j_2 n'\right)t + D_{j_3,\dots,j_6}\right] \tag{8.75}
$$

where

$$
C_{j_1,\dots,j_6} = C_{1,6}(a, a', e, e', i, i')
$$

is a polynomial expression in the action variables of the perturbed and the perturbing planet and

$$
D_{j_3,\dots,j_6} \equiv D_{3,6} = j_3\Omega + j_4\Omega' + j_5\omega + j_6\omega' \,. \tag{8.76}
$$

are phases depending on the remaining angles. Note, that for the purpose of the exposition we also include $\Omega$ to demonstrate a more general case although previously we took as the plane of reference the orbital plane of the perturbed planet. To

16) Note that we took the orbit of the perturbed planet as a reference plane, therefore $i$ is absent.

show an example for the amplitudes $C_{1,6}$ in the series expansion arising as factor we find for the first periodic term of $F_1$

$$C_{2,-2,2,-2,2,-2} = \frac{3}{4} - \left[\frac{15}{8}\left(e^2 + e'^2\right) - \frac{3}{8}J^2\right]$$

for the argument $2\lambda - 2\lambda' = 2M - 2M' + 2\omega - 2\omega' + 2\Omega - 2\Omega$ and consequently $j_1 = j_3 = j_5 = 2$, $j_2 = j_4 = j_6 = -2$ and – according to the d'Alembert rule – the sum of $j_k$ with $k = 1, \ldots, 6$ is zero. Now we can discuss the principles of deriving the perturbations for the osculating elements by introducing the function $F$ into the Lagrange equations

$$F = \frac{\kappa^4}{2L^2} + k^2 m' \sum_{j_1,\ldots,j_6} C_{1,6} \cdot \cos \Phi = F_0 + F_1 \tag{8.77}$$

where we need to replace $\kappa^2 = k^2(m_0 + m) = n^2 a^3$ and $\Phi = \cos[(j_1 \cdot n + j_2 \cdot n')t + D_{3,6}]$. The formulation of the perturbations shows that the mean anomaly $l = M$ is a special case, because we have to take into account the first term in (8.77) when we build the derivatives with respect to the conjugate Delaunay element $L$[17]. Using the canonical equations for the two-body problem it is evident that all the elements are constant with the exception of $l = nt$. For now we concentrate on the other 5 Delaunay elements for which we can directly apply the integration following the derivation to the conjugate Delaunay element to get the perturbations. For every osculating orbital element – we use the greek letter $\chi$ for any of the five elements $L, G, H, g$, and $h$ and $\bar{\chi}$ for their conjugate element – the form of the equations is the following one

$$\frac{d\chi}{dt} = \pm m' \frac{\partial F}{\partial \bar{\chi}} \tag{8.78}$$

Integrating we get the first-order approximation

$$\chi = \pm m' \int_0^t \frac{\partial F}{\partial \chi} dt = \chi_0 + \Delta_1 \chi \tag{8.79}$$

where $\Delta_1 \chi$ stands for the first-order perturbation. For the Delaunay elements $g$ and $h$ and for $L$, $G$ and $H$ we will derive the respective first-order perturbation. To compute $\delta L$ we need to build the derivative in formula (8.79) with respect to the conjugate Delaunay element $l$ (where we use $C_{1,6}$ and $D_{3,6}$). For building the derivative inside the integral expression we can write $(j_1 \cdot n + j_2 \cdot n')t = j_1 l + j_2 l'$ which yields

$$\delta L = -m' \int_0^t \sum_{j_1,\ldots,j_k} C_{1,6} \cdot j_1 \sin\left[\left(j_1 \cdot n + j_2 \cdot n'\right) t + D_{3,6}\right] dt$$

17) Note that $F_0$ is the energy (the Hamiltonian) of the two-body problem. From the Hamiltonian we can derive the so-called velocity relation $v^2 = \kappa^2(\frac{2}{r} - \frac{1}{a})$.

and integration leads to

$$\delta L = m' \sum_{j_1,\ldots,j_k} C_{1,6} \cdot j_1 \frac{\cos\left[(j_1 \cdot n + j_2 \cdot n')\,t + D_{3.6}\right]}{j_1 n + j_2 n'} \tag{8.80}$$

We emphasize that for $\delta\chi'$ (vice versa perturbed by the mass $m$) the same small divisor $j_1 n + j_2 n'$ appears but with a different value of the amplitude (namely $C'_{1,6}$ The Delaunay elements $G$ and $H$ in the series expansion will be replaced by the derivatives of $D_{3,6}$ with respect to the corresponding Delaunay element (e. g. for $H$):

$$\delta H = m' \sum_{j_1,\ldots,j_k} C_{1,6} \cdot \frac{\partial D_{3,6}}{\partial h} \frac{\cos\left[(j_1 \cdot n + j_2 \cdot n')\,t + D_{3,6}\right]}{j_1 n + j_2 n'} \tag{8.81}$$

For the elements $g$ and $h$ the perturbations reads (we show it for $h$)

$$\delta h = -m' \sum_{j_1,\ldots,j_k} \frac{\partial C_{1,6}}{\partial H} \cdot \cos\left[(j_1 \cdot n + j_2 \cdot n')\,t + D_{3,6}\right] \tag{8.82}$$

The separation of the summation for $j_1 = 0$ and $j_2 = 0$ (8.75) leads to a constant term which after integration produces a secular term. This means that any Delaunay element – except $L_i$[18], which we will treat later in detail – has a form

$$\chi = \chi_0 + \chi_1 t + \sum_{j_1 = j_2 \neq 0} \frac{E_{1,6}}{j_1 \cdot n + j_2 \cdot n'} \cos\left[(j_1 \cdot n + j_2 \cdot n')\,t + D_{3,6}\right] \tag{8.83}$$

with $\bar{E}_{1,6}$ the respective amplitude.

### 8.5.1
### Perturbation of the Mean Longitude

To simplify the further discussion we set $D_{3,6}$ to a constant $D_{j_1,j_2}$ since it only contains slowly osculating elements compared to the mean motions. It is necessary to determine separately the perturbations for the Delaunay element $l$ because the perturbations for this osculating element are very large due to a double integration; we show it in more detail in the following.

$$\delta l = \int_0^t \frac{\partial F_1}{\partial L} dt \tag{8.84}$$

In $F$ one has to build the partial derivative with respect to the conjugate variable, which leads to an additional term for $l$ because in the first term in equation (8.78)

18) In the semimajor axes no secular terms arise because for $j_1 = 0$ and $j_2 = 0$ the first-order perturbation disappears.

the conjugate element $L$ is present

$$-\frac{\partial F_0}{\partial L} = \frac{\kappa^4}{L^3}.$$

The variable $L$ has the following form:

$$L = L^{(0)} + m' \int_0^t \frac{\partial F_1}{\partial l} dt = L^{(0)} + \delta L \tag{8.85}$$

and we find

$$\frac{dl}{dt} = -\left(\frac{\partial F_0}{\partial L} + m' \frac{\partial F_1}{\partial L}\right) = \kappa^4 \left(L^{(0)} + \delta L\right)^{-3} - m' \frac{\partial F_1}{\partial L} \tag{8.86}$$

Dividing the quantity $(L^{(0)} + \delta L)^{-3}$ by the small perturbation $\delta L_1$ we develop the expression into a series and just keep the first term[19)]

$$\frac{dl}{dt} = n^{(0)} \left(1 - 3\frac{\delta L}{L^{(0)}}\right) - m' \frac{\partial F_1}{\partial L} \tag{8.87}$$

where we also used the third Kepler law: $n^{(0)} = a^{-3/2}\sqrt{\kappa^2}$. The integration now leads to

$$l = n^{(0)} t - \int_0^t \frac{\partial F_1}{\partial L} dt - 3\frac{n_1^{(0)}}{L^{(0)}} \int_0^t \delta L dt \tag{8.88}$$

and consequently[20)]

$$\begin{aligned} l = &-m' \int_0^t \sum_{j_1=-\infty}^{\infty} \sum_{j_2=-\infty}^{\infty} \frac{\partial C_{j_1,j_2}}{\partial l} j_1 \cos\left[(j_1 n + j_2 \cdot n')\, t + D_{j_1,j_2}\right] dt \\ &- 3\frac{n^{(0)}}{L^{(0)}} \int_0^t \sum_{j_1=-\infty}^{\infty} \sum_{j_2=-\infty}^{\infty} C_{j_1,j_2} j_1 \frac{\cos\left[(j_1 \cdot n + j_2 n')\, t + D_{j_1,j_2}\right]}{j_1 n + j_2 n'} dt \\ &+ n^{(0)} t \end{aligned} \tag{8.89}$$

After integration the $l_1$ has the following form

$$\begin{aligned} l = &-m' \sum_{j_1=-\infty}^{\infty} \sum_{j_2=-\infty}^{\infty} \frac{\partial C_{j_1,j_2}}{\partial L} \frac{\sin\left[(j_1 n + j_2 n')\, t + D_{j_1,j_2}\right]}{(j_1 \cdot n + j_2 n')} \\ &- m' \frac{\partial C_{00}}{\partial L} \cos D_{00} t \\ &- 3\frac{n^{(0)}}{L^{(0)}} \sum_{j_1=-\infty}^{\infty} \sum_{j_2=-\infty}^{\infty} C_{j_1,j_2} j_1 \frac{\sin\left[(j_1 n + j_2 n')\, t + D_{j_1,j_2}\right]}{(j_1 n + j_2 n')^2} + n^{(0)} t \end{aligned} \tag{8.90}$$

19) We make use of the known formula $(1 + x)^{-3} = 1 - 3x + 6x^2 - \ldots$
20) Note that $D_{j_1,j_2}$ can be regarded as constant for $l$.

and finally

$$l = \left(n^{(0)} - m' \frac{\partial C_{00}}{\partial L} \cos D_{00}\right) t + \delta l \left(n^{(0)} + \delta n\right) t + \delta l \tag{8.91}$$

with a similar expression for

$$l = \left(n^{(0)} - m' \frac{\partial C_{00}}{\partial L} \cos D_{00}\right) t + \delta l = \left(n^{(0)} + \delta n\right) t + \delta l \tag{8.92}$$

We can see that also for the mean anomalies we have secular and periodic perturbations. We emphasize that the periodic perturbations – in contrary to the perturbations on the other Delaunay elements – have a denominator $(j_1 n + j_2 n')^2$ which leads to very large amplitudes due to the fact that we have to provide a second integration. These amplitudes in the mean longitude $\lambda = \Omega + \omega + l$ are for Jupiter $\approx 20'$ and for Saturn $\approx 48'$.

A final statement on the validity of the KIII: when we have $\delta n_i$ and $\delta a_i$ – the perturbations in mean motion and in semimajor axis for a planet with mass $m_i$ – then the relation holds for any instant

$$(n^{(0)} + \delta n)(a^{(0)} + \delta a)^{-3/2} = n^{(0)}(a^{(0)})^{-3/2} = \kappa \tag{8.93}$$

This is not true for the mean values of $a$ and $n$ which we derive from observations published in the Nautical Almanac [184] for the published 'mean' elements the third Kepler law is not valid.

## 8.6 Small Divisors

Whenever the ratio of the mean motions $n/n'$ fulfills the condition of being close to commensurability we speak of a *mean motion resonance* (MMR)

$$\frac{n}{n'} \approx -\frac{j_1}{j_2} \tag{8.94}$$

and this leads to divisors close to zero. Consequently the respective perturbation for such planets (or asteroids or comets) in a mean motion resonance are large, because the amplitudes $C_{1,6}$ are divided by a quantity close to zero. For the osculating element $l$ we have shown that the square of these divisors produces especially large amplitudes of the perturbations.

In order to compute the mutual perturbations for a pair of planets one can very well use the so-called Hansen coefficients introduced in a special form by [185]. They have been used e.g in [186] where a development of the inverse distance between the two planets involved is derived in complex variables $z = e \exp \sqrt{-1}\tilde{\omega}$ and $\zeta = \sin i/2 \exp \sqrt{-1}\Omega$; furthermore the results are then used in combination with the Lagrange equations in the complex form given in [187]. Several numerical examples are given there for amplitudes like the great inequality of Jupiter and Saturn or the 13 : 8 MMR between the Earth and Venus.

The appearance of small divisors is important only for small values of $j_1$ and $j_2$ because we have to keep in mind that the mean motions of the planets we are dealing with appear in the ratio $(n/n')^{\nu}$ where $\nu$ is the power up to the value of the expansion with respect to the masses $m, m'$ when taking care of higher orders. In the example from before we see that also products of the masses involved appear $(mm', m^2m', m^2m'^2, \ldots)$ This means that using a development of $F$ to higher orders there will be always integers $j_1$ and $j_2$ which will produce small divisors to any desired value close to zero. Nevertheless we know that for higher orders the amplitudes $C_{1,6}$ are getting small because of the powers of the small parameters masses, inclinations and eccentricities of the planets. This practical limit when using perturbation theory means that from a certain order on the perturbations will increase and blow up the solution. Consequently there is a practical limit for the order of the perturbation theory which we are using to derive an accurate solution. In principal all these series are only semiconvergent which was already known to Poincaré.

Even though in the first-order perturbation theory only in the mean longitude squares of the small divisors appear when proceeding to higher orders in the perturbation theory the second, third, and so on such MMR and consequently small divisors to any power will appear in each solution of the orbital elements.

We will demonstrate the problems arising from small divisors for the perturbations of the eccentricity before integration:

$$\frac{de}{dt} = \sum_{j_1=-\infty}^{\infty} \sum_{j_2=-\infty}^{\infty} C_{1,6} \sin\left[\left(j_1 \cdot n + j_2 \cdot n'\right) t + D_{3,6}\right] \tag{8.95}$$

where we want to keep the nodes and the perihelion as constants $D_{3,6} = \text{const}$. Let us assume that $n/n'$ rational then in the summation in (8.95) we take out $j_1, j_2 = 0$ already before the integration which gives rise to a constant term $C_{0,0} \sin[D_{0,0}]$ which results after integration in a secular term $C_{0,0} \cdot t$ because $D_{0,0}$ is taken as constant. After integrating this series expansion assuming $n/n' \notin \mathbb{R}$ we get

$$\delta e = C_{0,0} t - \sum_{j_1=-\infty}^{\infty} \sum_{j_2=-\infty}^{\infty} \frac{C_{j,k}}{j_1 n + j_2 n'} \cos\left[\left(j_1 \cdot n + j_2 \cdot n'\right) t + D_{3,6}\right]$$

$$(j_1, j_2 \neq 0) \tag{8.96}$$

When $n$ and $n'$ are integers and are close to be commensurable we take out in the summation also the term $j_1 \cdot n + j_2 \cdot n'$ before the integration. As a consequence, in the secular term we also have its contribution, which causes the perturbation of the osculating eccentricity to grow with time. For the periodic terms it means that it is convergent because $j_1 \cdot n + j_2 \cdot n'$ has a lower limit differing from zero in $(n/n')$, which is a rational number; for details we refer to [143].

## 8.7 Long-Term Evolution of Our Planetary System

One may ask whether the appearance of small divisors will lead to big perturbations and later instability of the motion of the planets. From the historical point of view is was Laplace who proved a theorem (around 1770) that up to the second order of the eccentricities and the semimajor axes there appear no secular terms in the perturbation in the mean motions; this means that the semimajor axes of the planets are confined to a ring around the Sun. Later Poisson improved this theorem stating that the stability of the orbits can be assured up to the second order in the masses. In fact it is the representation as Fourier series with respect to the time which seems to make the system unstable because – and we already mentioned it – the series expansions with respect to small parameters are divergent which was shown by Poincaré.

An interesting discussion about the theorem de Poisson is published in a paper by Duriez (see [188]) who could show that the theorem is still valid in the following formulation: "The osculating major axes of the heliocentric orbits of the planets do not contain any secular inequality to the second order in the masses; no secular term appears in these major axes at the second approximation of a classical theory if the expression $n^2a^3$ is a constant[21)], the same for all planets; a secular term of the third order in the masses does appear at the second approximation, if $n^2a^3$ is a function of the planetary masses different from one planet to another".

The only way to give an answer to the question of the stability of the planetary system is to use a combination of analytical methods and numerical integration techniques. This has been undertaken in recent years by [189–192] who on one side could proof that the inner Solar System is chaotic and on the other side – making integration up to billions of years – showed that there is statistically only a small chance that the orbits of the inner planets may be on crossing orbits. Originally his result was that Mercury could achieve such high eccentricities some billion years

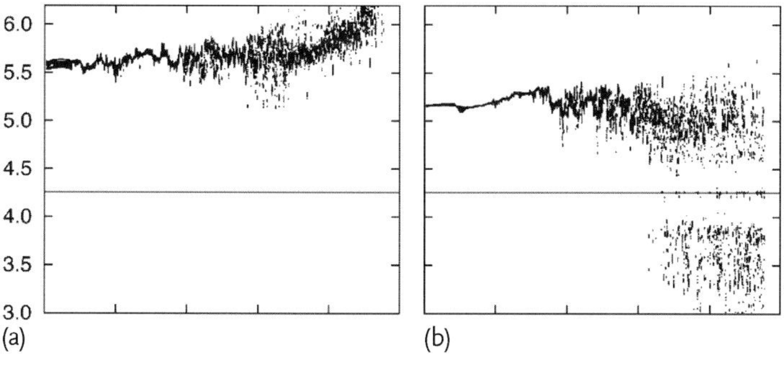


**Figure 8.4** Frequencies of the secular resonances (″/year) on the $y$-axis) versus the eccentricity of Mercury ($x$-axes). $g_5$ is represented as the straight line at $\sim$ 4.6 ″/year, $g_5$ is in resonance ignoring relativity (b) and stays well separated even for large eccentricities (a) (after Laskar).

21) [author comment] Which is in fact not the case, we refer to the correct formulation of KII.

into the future so that it would have a Venus crossing orbit. But including general relativity the results turned out to be different, because in this more realistic model the secular resonances of Mercury and Jupiter (see next chapter) $g_1 \neq g_5$ and the planet Mercury is not trapped into the $g_1 = g_5$ secular resonance. This can be seen in Figure 8.4 where the frequencies of the secular resonance (SR) of Jupiter $g_5$ (straight line) and the SR of Mercury (both on the $y$-axes) are plotted with respect to the eccentricity of Mercury ($x$-axes). In Figure 8.4a (computation with relativity) the two frequencies are far from each other for every eccentricity, while in Figure 8.4b it is obvious that from $e = 0.6$ on $g_1 \sim g_5$. And this causes even larger eccentricities for the orbit of Mercury.

In another recent paper [193] it could be shown that including contributions from general relativity AND the Moon this could even lead to collisions of the orbits of the terrestrial planets (Mercury to Mars) in the inner Solar system in time scales of some gigayears. In fact the development of our planets billions of years into the future is not settled; but then, at any rate, the Sun is expected to develop into a red giant!

# 9
# Resonances

## 9.1
## Mean Motion Resonances in Our Planetary System

When we look at the small divisor problem discussed before in the case of our planetary system we recognize that there are different planets close to such mean motion resonances (see Table 9.1). As an example we discuss the so-called 'great inequality', which is the quasi mean motion resonance of Jupiter $n_{\text{Jup}} = 0.^\circ 08309/\text{day}$ and Saturn $n_{\text{Sat}} = 0.^\circ 03346/\text{day}$. When we develop the approximate ratio of the great inequality (0.4027) we find the sequence for $j_1$ and $j_2$ as shown in Table 9.2.

In planetary theory also multiples of the basic mean motion resonances appear which means that for the couple Jupiter–Saturn also multiples of arguments of the form $2\lambda - 5\lambda'$ appear which contribute also to the size of the amplitude. The value of the small divisor is determined by the order of the resonance which means that for the 5 : 2 MMR the coefficients $C_{1,6}$ are of the order of 3 in the small parameters eccentricity and inclination. This fact makes the multiples of the 5 : 2 resonance as well as the approximation of the great inequality given in Table 9.2 insignificant for the computation of the perturbations. Finally we state the numerical value of the amplitudes of the perturbation in mean longitude for Jupiter (20′) and Saturn 48′. The size of this effect is due to the fact that for the perturbations in $\lambda$ the small divisor is squared, as we have already underlined.

### 9.1.1
### The 13 : 8 Resonance between Venus and Earth

To demonstrate the effect of resonances we take as an example the 13 : 8 MMR between Earth and Venus [194]. Although this is a resonance of order 5 the effect of the mutual perturbations is not negligible even in a simplified three body problem Sun–Venus–Earth. In Figure 9.1 we show the results of an integration of the couple Venus–Earth, where we slightly changed the semimajor axes of the Earth ($x$-axis) in the immediate neighborhood of its actual position as well as the initial eccentricity of the perturbing planet Venus ($y$-axis). As indicator of the perturbations we checked how the eccentricity of the Earth suffers from the presence of

*Celestial Dynamics*, First Edition. R. Dvorak and C. Lhotka.

**Table 9.1** Mean motion resonances in the Solar system (the * denotes exact mean motion resonances).

| System | Resonances | | | |
|---|---|---|---|---|
| Solar system | Jupiter–Saturn | 5 : 2 | Saturn–Uranus | 3 : 1 |
| | Uranus–Neptune | 2 : 1 | Neptune–Pluto* | 3 : 2 |
| Jupiter system | Io–Europa* | 2 : 1 | Europa–Ganymede* | 2 : 1 |
| Saturn system | Mimas–Tethys* | 2 : 1 | Enceladus–Dione* | 2 : 1 |
| | Dione–Rhea | 5 : 3 | Titan–Hyperion* | 4 : 3 |
| Uranus system | Miranda–Umbriel | 3 : 1 | Ariel–Umbriel | 5 : 3 |
| | Umbriel–Titania | 2 : 1 | Titania–Oberon | 3 : 2 |

**Table 9.2** Approximation of the great inequality $j_1 n + j_2 n'$ by rational numbers

| $j_1$ | $j_2$ | ratio | $P_{yrs}$ |
|---|---|---|---|
| 1 | 2 | 0.5 | 61 |
| 2 | 5 | 0.4 | 880 |
| 29 | 72 | 0.4028 | 1 810 |
| 60 | 149 | 0.4027 | 36 000 |

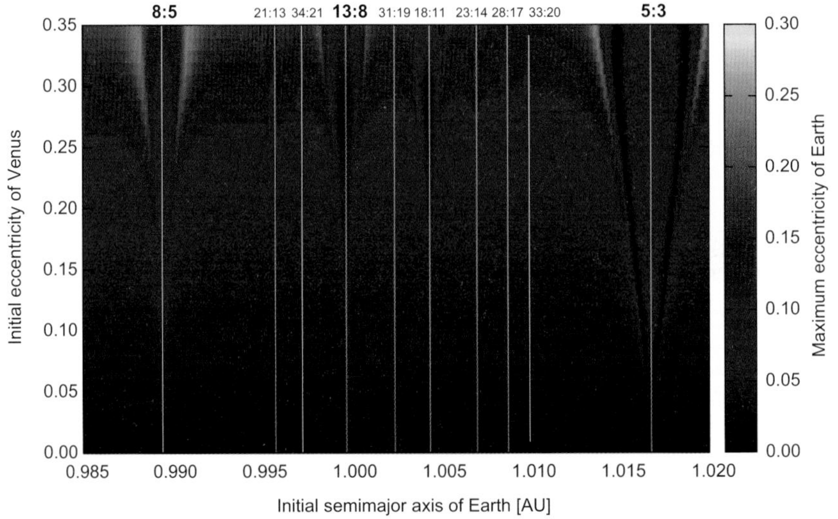


**Figure 9.1** The MMRs between Venus and the Earth; along the x-axis the initial semimajor axis of the Earth is changed, along the y-axis the initial values for the eccentricity of Venus are varied. The color indicates the maximum eccentricity of the Earth, the vertical white lines show the position of the MMR up to high orders.

Venus during its orbital evolution. One can see the typical triangle like structure of the resonance with larger values of the eccentricity on the edge of the triangle and a 'quiet' region in the middle of the resonance. On the left hand of the 13 : 8 MMR of order 5 one sees the 8 : 5 MMR and on the right hand the MMR 5 : 3 (orders 3 respectively 2) where it is evident that the strength of the perturbations increases inverse to the order of the resonance. In Figure 9.2 we checked the region of the 13 : 8 MMR on a finer scale with the aid of a chaos indicator, namely the Fast Liapunov indicator.[1] We see the fine structure of the resonance in detail with chaotic regions down to small eccentricities ($e \sim 0.1$)[2] of Venus on the edges and a very chaotic behavior outside the resonance for $e_{\text{Venus}} > 0.2$. On the side with larger semimajor axes the MMR of order 7 (18 : 11) is well visible with a very similar structure than the 13 : 8 MMR.

Mean motion resonances are of special importance for the asteroids:

- The group of Trojan asteroids shows a 1 : 1 resonance with Jupiter with more than 5000 asteroids around the Lagrange point $L_4$ and about 3400 around the Lagrange point $L_5$.

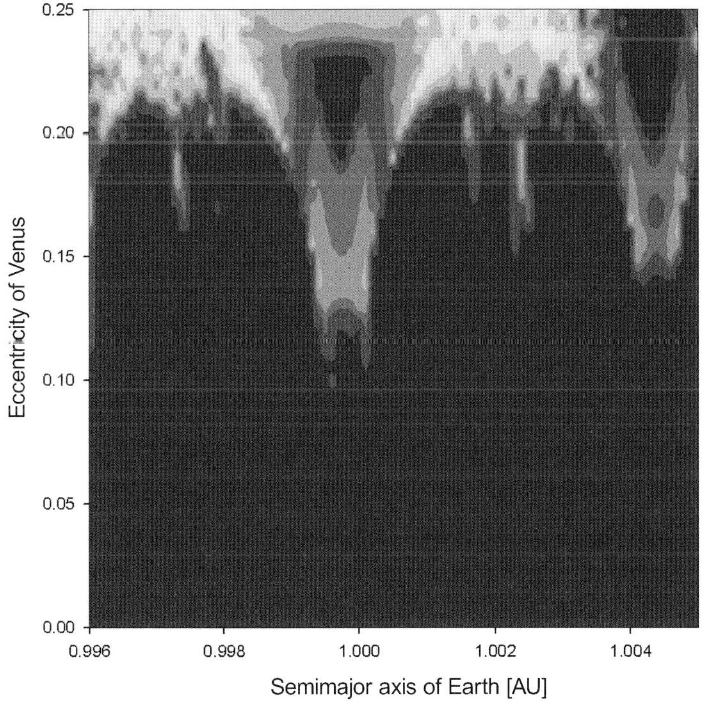


**Figure 9.2** FLI map of the region around the 13 : 8 MMR: semimajor axes of the Earth versus different eccentricities of Venus; black regions indicate stable orbits (after [194]).

1) See Chapter 3.
2) Its actual value of $e \sim 0.007$ is far smaller which means that both planets are moving nowadays in a very stable region.

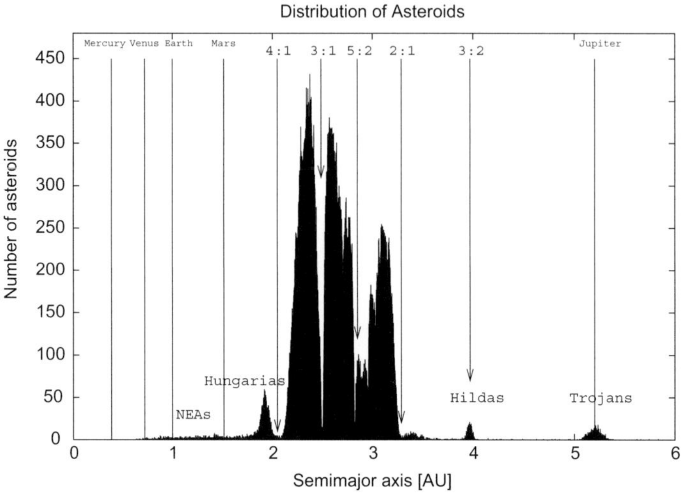


**Figure 9.3** Distribution of the Near Earth Asteroids and the mainbelt of asteroids between Mars and Jupiter. Besides the positions of the planets Mercury to Jupiter the MMR between Jupiter and an asteroid located with a special semimajor axis are marked. Well visible is the depletion for special MMR (the Kirkwood gaps) like for example the 3 : 1 and the 5 : 2 MMR; in contrast an accumulation for the 3 : 2 (the Hilda group) and the 1 : 1 MMR (the Trojans) is also evident.

- The group of the Plutinos are asteroids in the 3 : 2 resonance with Neptune like Pluto itself (up to now we have evidence for several thousands of these bodies[3])
- The Hilda group just at the 3 : 2 MMR with Jupiter consisting of some thousand members.
- The role of MMR is well visible in the structure of the main belt asteroids between Mars and Jupiter (see Figure 9.3) where one can see that for some resonances the number of asteroids is quite small; on the other hand a resonance may protect an asteroid (e.g. visible in the 3 : 2 resonance where the Hilda asteroids reside).

### 9.1.2
### The 1 : 1 Mean Motion Resonance: Trojan Asteroids

A special case of MMR is the motion of small celestial bodies in the same orbit with a large planet. This 1 : 1 MMR is interesting from the point of view of celestial mechanics which was already pointed out and theoretically studied in Chapter 6 (restricted three body problem). In fact therein we discussed the existence of stable equilibrium points which are at the center of the regions of stable motion around

3) More than 1000 with diameters larger than 100 km.

the leading Lagrange point $L_4$ and the trailing one $L_5$. The first Jupiter Trojan[4] was discovered in 1906 (Achilles by Max Wolf in Heidelberg) and this was quite surprising that the equilateral equilibrium points are of astronomical importance.

Since then many Trojan asteroids of Jupiter have been found so that up to now we know of several thousands of objects residing in the $1:1$ mean motion resonance with Jupiter. Theoretical and numerical studies show the symmetry of these two Lagrange points not only in the restricted three body problem [195, 196] but also in the realistic dynamical model of the Solar system consisting of all planets for example [28, 29, 197–199]. To visualize the extent of the stable region of motion for Jupiter besides the former mentioned ones long-term integrations of fictitious bodies have been undertaken for example by [199]. We show one of their results for the $L_4$ region in Figure 9.4 where the libration amplitude $D$ which is related to the semimajor axis $D = \sqrt{3a}$ is plotted versus the eccentricity. The graytone stands for the Liapunov time $T_E$ which is a good measure for the real escape time; the respective number inside the graph show that on the edge (large libration and larger eccentricities) the Trojan escape quite fast ($10^3$–$10^4$ years) whereas in the white region they are stable for the age of the Solar system ($T < 10^9$), Inside this graph the open circles stand for real asteroids which are stable, full circles stand for Trojans in the state of stable chaos [200].

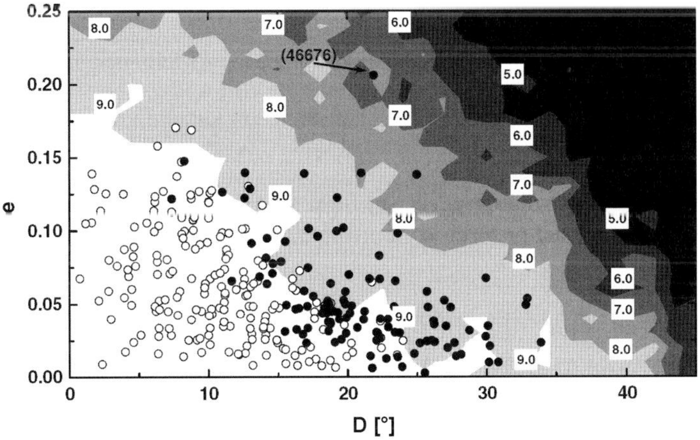


**Figure 9.4** Stability plot for fictitious Trojans in the $L_4$ domain of Jupiter; the initial libration around the equilibrium point is plotted versus the initial eccentricity. Real stable asteroids are plotted as small open circles, unstable ones as full circles (after [200])

4) The asteroids moving in the vicinity of these equilibrium points of Jupiter are named after heroes from the Trojan War of the Greek mythology. According to the Trojan war the two fighting parties reside in different regions on the sky: the asteroids orbiting Jupiter's $L_4$ point are named after the heroes from the Greeks side, and the ones close to $L_5$ are the defenders of Troja. But there are two exceptions: the asteroid 617 named after the Greek Patroclus which is in the middle of the enemies in the Trojan region ($L_5$) and the asteroid 624 with the name of the Troja defending hero Hector ($L_4$).

One of the first authors to study the stability of the Trojan domain for the outer planets with the aid of numerical integration up 20 million years were [201]. [197] have shown in simulations of the dynamical evolution for clouds of hypothetical Trojans around the Lagrange point $L_4$ how fast the depletion of asteroids acts for the gas giants Saturn, Uranus and Neptune. The respective integrations were carried out up 4 Gyr and showed the following results: Saturn Trojans were unstable already after several $10^5$ years, Uranus Trojans were also unstable but after millions of years whereas the Trojans of Neptune survived for the whole integration time. [202] determined the diffusion speed of Trojans of Uranus and Neptune Trojans and computed also the secular frequencies in their region responsible for ejection.

The strange ring structure of the Trojan region of Saturn was already shown in [201] in a numerical study for the Trojans of the outer planets. The very rapid depletion for the Saturn Trojans (without discovering the detailed ring structure) was confirmed by the former mentioned work of [197], and quite recently in a diploma thesis [203] and in [204]. From the last paper we show in Figure 9.5 a stability plot for a grid of fictitious Trojans semimajor axis versus the orbital longitude of the Trojans compared to the Lagrange point $L_4$ where a curious stable ring survived for quite a long time around a small circular domain located just in $L_4$. But for longer integrations all the Trojans escaped!

Confirming the hypothetical results of [197] recently real Trojan asteroids in orbit around $L_4$ and $L_5$ of Neptune were observed. Since this discovery of the first Neptune Trojan [205] – the minor planet 2001 QR322 in the $L_5$ region – many studies have been dedicated to the dynamics of Neptune Trojans. Two more recent ones concerning the Neptune Trojans [206, 207] investigated how the stability depends on the inclination and the eccentricity of the asteroids. Some of the results

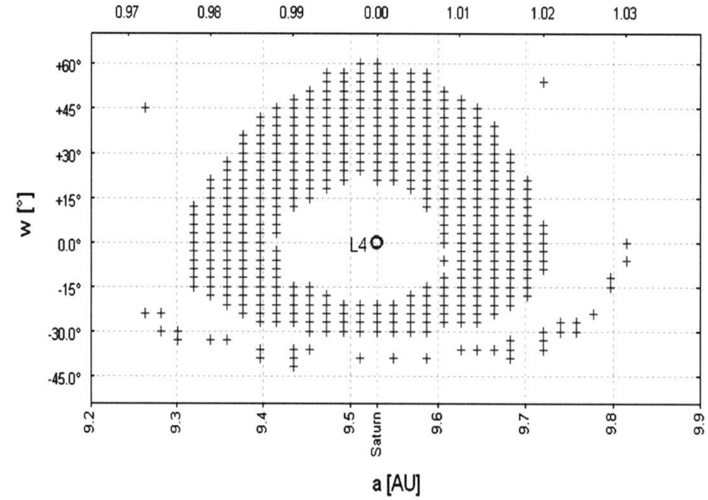


**Figure 9.5** Stability plot for fictitious Trojans in the $L_4$ domain of Saturn after a numerical integration of $10^6$ years; stable orbits are marked with crosses (for details see text)

of these extensive numerical studies of a Trojan cloud of fictitious objects for the whole range of inclinations $0^\circ \leq i \leq 60^\circ$ and different eccentricities are shown in Figures 9.6 and 9.7. The respective plots were derived with the aid of the spectral analysis method (see Chapter 3 and [56]) a measure of the chaoticity of an orbit. From Figure 9.6 on can see that the libration angle $\sigma$ around the equilibrium point does not change with larger inclination. The stable region does not change in extension up to $i = 40^\circ$ and after this gap another stable region up to $i = 60^\circ$ appears. For details concerning the acting secular resonances we refer to [206]. In Figure 9.7 we plotted the stability of the Neptune Trojan region with respect to the initial eccentricity and the semimajor axis of the fictitious objects for a fixed initial inclination $i = 10^\circ$. It is evident that already for small eccentricities of about $e = 0.1$ there are hardly any stable orbits. For details concerning the acting secular resonances we refer to [206, 207]

The region of the Uranus Trojans was also studied in a recent work [208] using the output of long-term numerical integrations with thousands of fictitious Uranus Trojans in inclined orbits. The results confirm the former mentioned ones by [197] that almost none of the Uranus Trojans survive over periods of gigayears. In Figure 9.8 a stability diagram is shown for $10^6$ years where the libration width is depicted on a plot semimajor axes versus inclination. The big gap close to $i \sim 17^\circ$ is due to a secular resonance. For longer integrations all the stable orbits turn out to be unstable, some only after gigayears with the following exception: three Trojans on low inclined orbits survived up to the age of the Solar system of $4.5 \times 10^9$ years!

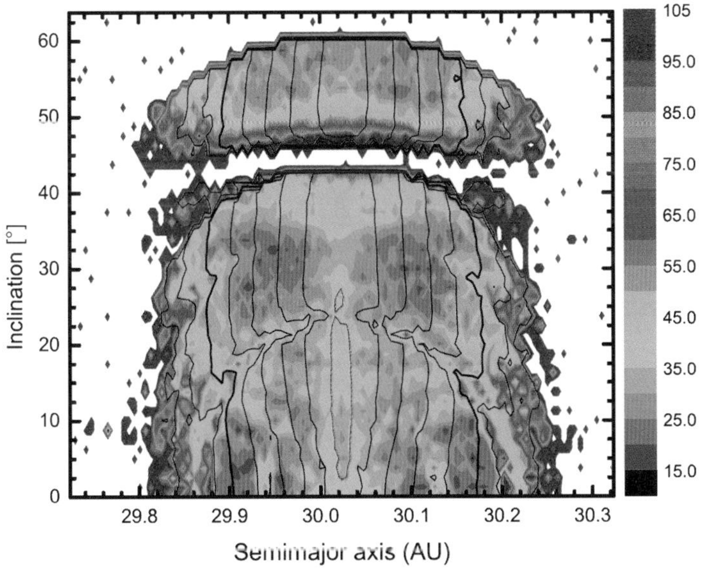


**Figure 9.6** Dynamical map around the $L_4$ point of Neptune where the colors indicate the spectral number from dark green (small spectral number – stable) to blue-red-white (large spectral number – unstable). The contours are for the libration angle $\sigma$ with $\delta\sigma = 10^\circ$ from the center at $a = 30$ AU outwards up to $\sigma = 60^\circ$

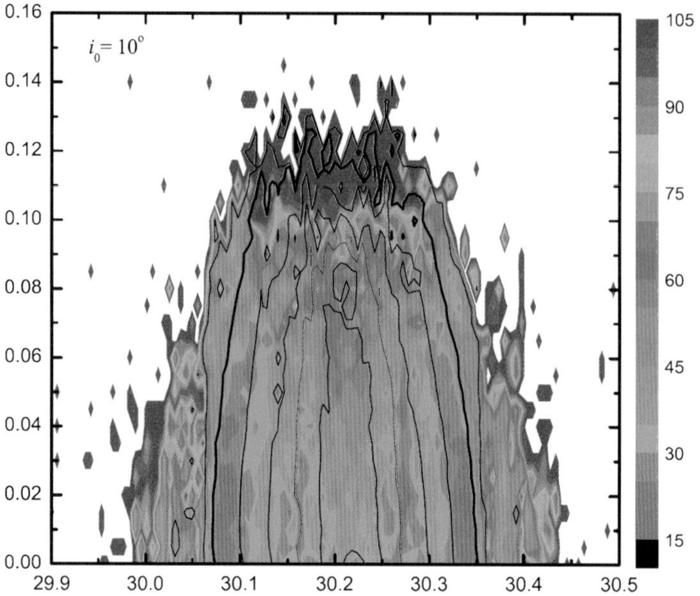


**Figure 9.7** Caption like in Figure 9.6; in this plot the initial eccentricities were increased

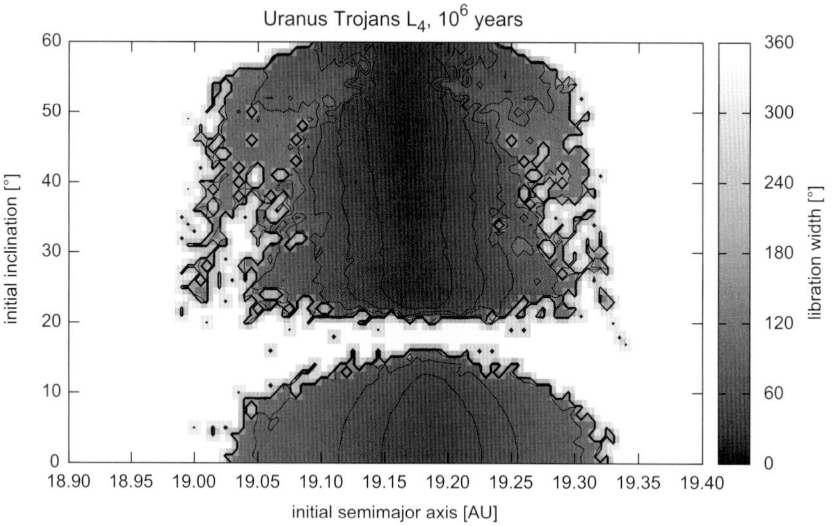


**Figure 9.8** The initial semimajor axes of the fictitious objects ($x$-axis) are plotted versus the initial inclinations of the Neptune Trojans ($y$-axis). The grey-tone visualizes the libration amplitude around the Lagrange point $L_4$; the contour-lines show the limits of the libration amplitudes with a step of 15° (for detail see in the text)

The first real Trojan of the Earth was also discovered recently [209]. In a detailed investigation [210] the dynamical behavior of the asteroid TK7 was carefully checked. At the moment it is in a tadpole motion around $L_4$ but as already shown in [211] this asteroid was jumping from $L_5$ to $L_4$ some thousand years ago. This jumping phenomenon of a Trojan was observed for the first time for the Jupiter Trojan Thersites [212] and seems to be a common dynamical behavior for Trojans on the edge of the stability region. In a stability plot for the whole Trojan region of the Earth one can see in Figure 9.9 how the stable area shrinks with larger inclination. TK7 with an inclination of almost $i = 21°$ is in an unstable region on the edge to a stable one. In fact in the previous mentioned article it was shown that TK7 is in a temporary captured Trojan orbit.

We just mention that also Mars has Trojans – one around $L_4$ and two around $L_5$[5]. A new exciting field for theoretical work regarding motion close to the Lagrange equilibrium points is that in extrasolar planets there could move Trojan planets in the 1 : 1 MMR with a giant planet. This would be especially interesting when the 'hosting' gas giant would move in the so-called habitable zone where water could be liquid (e.g. [213, 214]).

## 9.2 Method of Laplace–Lagrange

In order to be able to see the properties of secular perturbations we introduce the

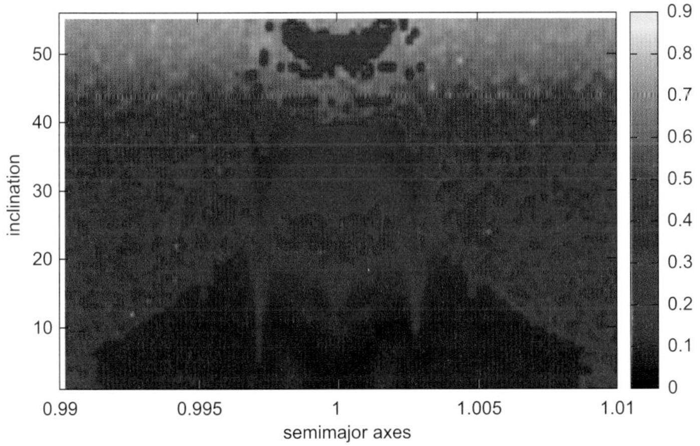


**Figure 9.9** Maximum eccentricities in a stability diagram initial semimajor axes of the fictitious Earth Trojans (*x*-axis) versus its initial inclination (*y*-axis). The colors from blue (small eccentricities – stable) to yellow (large eccentricity – unstable) visualize the stability of an orbit

5) www.minorplanetcenter.net/iau/lists/MarsTrojans.html (last accessed on 18 December 2012).

variables

$$h = e \sin \varpi \,, \qquad p = \sin \frac{i}{2} \sin \Omega$$
$$k = e \cos \varpi \,, \qquad q = \sin \frac{i}{2} \cos \Omega \tag{9.1}$$

for which follows that $e^2 = h^2 + k^2$ and $\sin^2(i/2) = p^2 + q^2$. This choice prevents that neither the eccentricities $e$ nor the inclinations $i$ appear in the denominators. This also makes it easier to identify the variables in the solution because they always keep their polynomial form. We follow closely the paper by Bretagnon [215] in this treatment of determining the secular perturbations. The perturbation function $F$ for a planet $m_1$ perturbed by another one $m_2$ for this new set of elements can be expressed

$$F = \sum S(\alpha) h_1^{r_1} h_2^{r_2} k_1^{s_1} k_2^{s_2} p_1^{t_1} p_2^{t_2} q_1^{u_1} q_2^{u_2} \cos(w_1\lambda_1 + w_2\lambda_2) \tag{9.2}$$

where $\alpha = a_1/a_2$, the ratio of the semimajor axes given in Laplace-coefficients (see Appendix B); the summation over all indices can be chosen according to the desired precision in the small parameters inclination and eccentricity. For these variables the Lagrange equations can be derived, where we use the abbreviation $\hat{e} = \sqrt{1 - e^2}$

$$\begin{aligned}
\frac{dh}{dt} &= \frac{\hat{e}}{na^2} \frac{\partial F}{\partial k} - \frac{h\hat{e}}{na^2(1+\hat{e})} \frac{\partial F}{\partial \lambda} + \frac{kp}{2na^2\hat{e}} \frac{\partial F}{\partial p} + \frac{kq}{2na^2\hat{e}} \frac{\partial F}{\partial q} \\
\frac{dk}{dt} &= -\frac{\hat{e}}{na^2} \frac{\partial F}{\partial h} - \frac{k\hat{e}}{na^2(1+\hat{e})} \frac{\partial F}{\partial \lambda} - \frac{hp}{2na^2\hat{e}} \frac{\partial F}{\partial p} - \frac{hq}{2na^2\hat{e}} \frac{\partial F}{\partial q} \\
\frac{dp}{dt} &= \frac{1}{4na^2\hat{e}} \frac{\partial F}{\partial q} - \frac{p}{2na^2\hat{e}} \frac{\partial F}{\partial \lambda} - \frac{pk}{2na^2\hat{e}} \frac{\partial F}{\partial h} + \frac{ph}{2na^2\hat{e}} \frac{\partial F}{\partial k} \\
\frac{dq}{dt} &= -\frac{1}{4na^2\hat{e}} \frac{\partial F}{\partial p} - \frac{q}{2na^2\hat{e}} \frac{\partial F}{\partial \lambda} - \frac{qk}{2na^2\hat{e}} \frac{\partial F}{\partial h} + \frac{qh}{2na^2\hat{e}} \frac{\partial F}{\partial k}
\end{aligned} \tag{9.3}$$

$$\begin{aligned}
\frac{1}{a}\frac{da}{dt} &= \frac{2}{na^2} \frac{\partial F}{\partial \lambda} \\
\frac{d\lambda}{dt} &= n - \frac{2}{na} \frac{\partial F}{\partial a} + \frac{\hat{e}}{na^2(1+\hat{e})} \left( h \frac{\partial F}{\partial h} + k \frac{\partial F}{\partial k} \right) + \frac{1}{2na^2\hat{e}} \left( p \frac{\partial F}{\partial p} + q \frac{\partial F}{\partial q} \right)
\end{aligned} \tag{9.4}$$

In the following we will concentrate on the perturbations of two planets with the indices 1 and 2. We thus need to introduce the functions

$$F_1 = \kappa^2 \frac{m_1}{a_1} \bar{F}_{1,2} \quad \text{and} \quad F_2 = \kappa^2 \frac{m_2}{a_2} \bar{F}_{1,2} \tag{9.5}$$

with $\bar{F}_{1,2} = a_2/\Delta$,where $\Delta$ is the distance between the two planets[6]. According to the KIII

$$\kappa^2 = \frac{n_1^2 a_1^3}{1 + m_1} = \frac{n_2^2 a_2^3}{1 + m_2} . \tag{9.6}$$

6) See the previous chapter.

For deriving the secular behavior we can neglect (9.4) and take the nonperiodic part of the perturbation function

$$\bar{F}_{1,2} = C_{1,2} + A_{1,2}\left(h_1^2 + k_1^2 + h_2^2 + k_2^2\right) - 4A_{1,2}\left(p_1^2 + q_1^2 + p_2^2 + q_2^2\right)$$
$$+ B_{1,2}\left(k_1 k_2 + h_1 h_2\right) + 8A_{1,2}\left(q_1 q_2 + p_1 p_2\right) \qquad (9.7)$$

In the fourth and fifth terms of this equation the products $k_1 k_2$, $h_1 h_2$, $q_1 q_2$ and $p_1 p_2$ give rise to terms of the form $e_1 e_2 \sin(\tilde{\omega}_1 + \tilde{\omega}_2)$ and $\sin(i_1/2)\sin(\Omega_1 + \Omega_2)$ which are effectively long periodic perturbations; we include them in our considerations. The $A_{1,2}$, $B_{1,2}$ and $C_{1,2}$ of (9.7) are functions of $\alpha = a_1/a_2$; the latter is constant in the here presented first-order perturbation theory. They can be expressed with the aid of Laplace coefficients.[7)]

We see that for the planet $m_1$ perturbed by $m_2$ (and vice versa) the system separates into two subsystems, one dependent on the variables $h$ and $k$ (the system in eccentricity SE) and one dependent on the variables $p$ and $q$ (the system in inclinations SI).

$$[1,2] = \frac{n_1 \alpha m_2}{1 + m_1}\,, \qquad [2,1] = \frac{n_2 m_1}{1 + m_2} \qquad (9.8)$$

Furthermore we introduce 4 quantities $D_{i,j}$, $i, j = 1, 2$ which we will need to derive the solution of the two uncoupled systems

$$D_{11} = [1,2]2A_{1,2}\,, \qquad D_{12} = [1,2]B_{1,2} \qquad (9.9)$$

$$D_{21} = [2,1]B_{1,2}\,, \qquad D_{22} = [2,1]2A_{1,2} \qquad (9.10)$$

Now we can write for the system SE

$$\frac{dh_1}{dt} = D_{11}k_1 + D_{12}k_2\,, \qquad \frac{dk_1}{dt} = -D_{11}h_1 - D_{12}h_2 \qquad (9.11)$$

$$\frac{dh_2}{dt} = D_{21}k_1 + D_{22}k_2\,, \qquad \frac{dk_2}{dt} = -D_{21}h_1 - D_{22}h_2 \qquad (9.12)$$

respectively for the system SI

$$\frac{dp_1}{dt} = -D_{11}q_1 + D_{11}q_2\,, \qquad \frac{dq_1}{dt} = D_{11}p_1 - D_{12}p_2$$
$$\frac{dp_2}{dt} = D_{22}q_1 - D_{22}q_2\,, \qquad \frac{dq_2}{dt} = -D_{21}p_1 - D_{22}p_2 \qquad (9.13)$$

We start to solve the system SE by an ansatz

$$h_1 = M_1 \sin(gt + \beta)\,, \qquad h_2 = M_2 \sin(gt + \beta)$$
$$k_1 = M_1 \cos(gt + \beta)\,, \qquad k_2 = M_2 \cos(gt + \beta) \qquad (9.14)$$

7) The coefficients $b^j_{2/2}$, $j = 1\ldots$ can be computed via an expansion into Fourier series of the direct term in the perturbation function under the assumption that the two orbits are circles ($\alpha = a_1/a_2$): $(1 - 2\alpha\cos\phi + \alpha^2)^{-1/2} = (1/2)b^0_{s/2} + \sum_{j=1}^{\infty} b^j_{s/2}\cos\phi$, for further reference see [143, 180] and Appendix B.

We substitute these solutions in (9.12) and solve the system simultaneously:

$$\begin{aligned} (D_{11} - g) M_1 + D_{12} M_2 &= 0 \\ D_{21} M_1 + (D_{22} - g) M_2 &= 0 \end{aligned} \tag{9.15}$$

It has a solution when the determinant of this system vanishes

$$g^2 - (D_{11} + D_{22}) g + D_{11} D_{22} - D_{12} D_{21} = 0 \tag{9.16}$$

We denote the 'eigenvalues' of the matrix $\mathbf{D}$

$$\mathbf{D} = \begin{pmatrix} D_{11} & D_{12} \\ D_{21} & D_{22} \end{pmatrix}$$

with $g_1$ and $g_2$ which have the respective eigenvectors (not to confuse with the osculating element $\lambda$, the mean longitude)

$$\begin{aligned} \lambda_{11} &= 1 \, , \quad \lambda_{21} = -\frac{D_{21}}{D_{22} - g_1} \\ \lambda_{12} &= -\frac{D_{12}}{D_{11} - g_2} \, , \quad \lambda_{22} = 1 \end{aligned} \tag{9.17}$$

With this notation the solutions for SE read

$$h_1 = \lambda_{11} M_1 \sin(g_1 t + \beta_1) + \lambda_{12} M_2 \sin(g_2 t + \beta_2) \tag{9.18}$$

$$h_2 = \lambda_{21} M_1 \sin(g_1 t + \beta_1) + \lambda_{22} M_2 \sin(g_2 t + \beta_2) \tag{9.19}$$

and for the variables $k$

$$k_1 = \lambda_{11} M_1 \cos(g_1 t + \beta_1) + \lambda_{12} M_2 \cos(g_2 t + \beta_2) \tag{9.20}$$

$$k_2 = \lambda_{21} M_1 \cos(g_1 t + \beta_1) + \lambda_{22} M_2 \cos(g_2 t + \beta_2) \tag{9.21}$$

For initial conditions we take $h_1 = h_1^0 = e_0 \sin \varpi_0$ and so on. The system in inclination SI can be solved like SE with the corresponding initial conditions for $p_1 = p_1^0 = \sin(i_0/2) \sin \Omega_0$ and so on. For SI the particular solution which we use has the corresponding 'eigenvalues' $s_1$ and $s_2$ and the results for $p_1, p_2$ and $q_1, q_2$ are similar to (9.19) and (9.21) but with $\mu_{12}$ instead of $\lambda_{12}$, $N_1$ and $N_2$ instead of $M_1$ and $M_2$ and another phase coefficient $\delta_i$ instead of $\beta_i$ in the solutions.

$$p_1 = \mu_{11} N_1 \sin(s_1 t + \delta_1) + \mu_{12} N_2 \sin(s_2 t + \delta_2) \tag{9.22}$$

$$p_2 = \mu_{21} N_1 \sin(s_1 t + \delta_1) + \mu_{22} N_2 \sin(s_2 t + \delta_2) \tag{9.23}$$

and for the variable $q$

$$q_1 = \mu_{11} N_1 \cos(s_1 t + \delta_1) + \mu_{12} N_2 \cos(s_2 t + \delta_2) \tag{9.24}$$

$$q_2 = \mu_{21} N_1 \cos(s_1 t + \delta_1) + \mu_{22} N_2 \cos(s_2 t + \delta_2) \tag{9.25}$$

To get an adequate solution for the complete solar system with 8 planets we need to take all pairs of planets into account where

$$F_u = \sum_{v<u} \kappa^2 \frac{m_v}{a_u} \bar{F}_{u,v} + \sum_{v>u} \kappa^2 \frac{m_v}{a_v} \bar{F}_{u,v} \tag{9.26}$$

with $\bar{F}_{u,v} = a_v/\Delta$, where $\Delta$ is the distance between the two planets and the parameter

$$\kappa^2 = \frac{n_1^2 a_1^3}{1+m_1} = \frac{n_2^2 a_2^3}{1+m_2} \tag{9.27}$$

For deriving the secular behavior we can neglect (9.4) and take the nonperiodic part of the perturbation function

$$\begin{aligned}\bar{F}_{u,v} = {} & C_{u,v} + A_{u,v}\left(h_u^2 + k_u^2 + h_v^2 + k_v^2\right) - 4A_{u,v}\left(p_u^2 + q_u^2 + p_v^2 + q_v^2\right) \\ & + B_{u,v}\left(k_u k_v + h_u h_v\right) + 8A_{u,v}\left(q_u q_v + p_u p_v\right)\end{aligned} \tag{9.28}$$

with the following factors:

$$\begin{aligned} [u,v] &= \frac{n_u \alpha_{uv} m_v}{1+m_u} \quad \text{for } v > u \\ [u,v] &= \frac{n_u m_v}{1+m_u} \quad \text{for } v < u \end{aligned} \tag{9.29}$$

and the respective equations read

$$\begin{aligned} \frac{dh_u}{dt} &= \sum_{v\neq u}[u,v](2A_{uv}k_u + B_{uv}k_v) \\ \frac{dk_u}{dt} &= -\sum_{v\neq u}[u,v](2A_{uv}h_u + B_{uv}h_v) \\ \frac{dp_u}{dt} &= -\sum_{v\neq u}[u,v](2A_{uv}q_u - 2A_{uv}q_v) \\ \frac{dq_u}{dt} &= \sum_{v\neq u}[u,v](2A_{uv}p_u - 2A_{uv}p_v) \end{aligned} \tag{9.30}$$

With $\mathbf{H} = (h_1, \ldots, h_8)^{\mathrm{T}}$, $\mathbf{K} = (k_1, \ldots, k_8)^{\mathrm{T}}$, $\mathbf{P} = (p_1, \ldots, p_8)^{\mathrm{T}}$ and $\mathbf{Q} = (q_1, \ldots, q_8)^{\mathrm{T}}$[8] we can compute

$$\begin{aligned} \frac{d\mathbf{H}}{dt} &= \mathbf{E}\times\mathbf{K}\,, & \frac{d\mathbf{K}}{dt} &= -\mathbf{E}\times\mathbf{H} \\ \frac{d\mathbf{P}}{dt} &= \mathbf{I}\times\mathbf{Q}\,, & \frac{d\mathbf{Q}}{dt} &= -\mathbf{I}\times\mathbf{P} \end{aligned} \tag{9.31}$$

8) The indices 1 ... 8 stand for the eight planets from Mercury to Neptune.

with the matrix **E** it reads

$$\begin{pmatrix} \sum'_{j\neq 1}[1,j]2A_{1j} & [1,2]B_{12} & [1,3]B_{13} & \dots & [1,8]B_{18} \\ [2,1]B_{21} & \sum'_{j\neq 2}[2,j]2A_{2j} & [2,3]B_{23} & \dots & [2,8]B_{28} \\ [3,1]B_{31} & [3,2]B_{32} & \sum'_{j\neq 3}[3,j]2A_{3j} & \dots & [3,8]B_{38} \\ \dots & \dots & \dots & \dots & \dots \\ [8,1]B_{81} & [8,2]B_{82} & [8,3]B_{83} & \dots & \sum'_{j\neq 8}[8,j]2A_{8j} \end{pmatrix}$$

where $\sum'_{j\neq k} = \sum_{j=1,j\neq k}^{8}$ with $k = 1,\dots,8$. In the matrix **I** all the $B_{12}$ are simply replaced by $-2A_{uv}$ and finally we get the solution

$$\begin{aligned} h_i &= \sum_{j=1,j\neq i}^{8} \lambda_{ij} M_j \sin(g_j t + \beta_j) \\ k_i &= \sum_{j=1,j\neq i}^{8} \lambda_{ij} M_j \cos(g_j t + \beta_j) \\ p_i &= \sum_{j=1,j\neq i}^{8} \mu_{ij} N_j \sin(s_j t + \delta_j) \\ q_i &= \sum_{j=1,j\neq i}^{8} \mu_{ij} N_j \cos(s_j t + \delta_j) \end{aligned} \quad (9.32)$$

We provide the numerical results of the Laplace–Lagrange secular theory in Table 9.3–9.8 as well as Figures 9.10 and 9.11 after A. Süli (2012, personal communication). With the respective values for the different quantities given in the solutions we can compute the secular motion of the planets. We illustrate the difference when one takes into account only the secular perturbations of the pair Jupiter–Saturn in Figure 9.10 compared to the complete system with eight planets.

**Table 9.3** The fundamental frequencies (arcsec/year) and phases (degree) of the Jupiter–Saturn system determined through the Lagrange secular theory

| $k$ | $g_k$ | $\beta_k$ [deg] | $s_k$ | $\delta_k$ |
|---|---|---|---|---|
| 1 | 3.443 034 | 33.590 911 | −0.000 000 | 106.243 103 |
| 2 | 21.683 776 | 129.912 140 | −25.126 810 | 127.484 504 |

**Table 9.4** The coefficients $M_{j,k} \cdot 10^8$ determined through the Lagrange secular theory for the Jupiter–Saturn system

| $j/k$ | 1 | 2 |
|---|---|---|
| 1 | 4 486 013 | −1 519 611 |
| 2 | 3 619 483 | 4 637 743 |

**Table 9.5** The coefficients $N_{j,k} \cdot 10^8$ determined through the Lagrange secular theory for the Jupiter–Saturn system

| $j/k$ | 1 | 2 |
|---|---|---|
| 1 | −315 069 | 1 426 430 |
| 2 | 775 830 | 1 426 430 |

**Table 9.6** The fundamental frequencies (arcsec/year) and phases (degree) of the Solar system determined through the Lagrange secular theory

| k | $g_k$ | $\beta_k$ [deg] | $s_k$ | $\delta_k$ [deg] |
|---|---|---|---|---|
| 1 | 5.457 029 | 89.643 233 | −5.195 513 | 20.041 021 |
| 2 | 7.336 106 | 195.260 924 | −6.562 379 | 318.225 922 |
| 3 | 17.294 160 | 336.107 397 | −18.706 136 | 255.520 208 |
| 4 | 17.974 457 | 318.343 908 | −17.607 500 | 296.712 937 |
| 5 | 3.687 082 | 30.541 533 | −0.000 000 | 107.597 685 |
| 6 | 22.008 854 | 129.178 648 | −25.436 220 | 127.443 805 |
| 7 | 2.706 226 | 100.622 619 | −2.911 734 | 315.660 945 |
| 8 | 0.636 512 | 55.785 622 | −0.680 943 | 202.838 340 |

**Table 9.7** The coefficients $M_{j,k} \cdot 10^8$ determined through the Lagrange secular theory

| $j/k$ | 1 | 2 | 3 | 4 | 5 | 6 | 7 | 8 |
|---|---|---|---|---|---|---|---|---|
| 1 | 18 096 231 | −2 349 881 | 153 881 | −169 064 | 2 428 021 | 11 451 | 65 289 | 918 |
| 2 | 630 069 | 1 930 152 | −1 255 093 | 1 481 565 | 1 650 821 | −58 827 | 64 162 | 1 394 |
| 3 | 404 423 | 1 505 201 | 1 041 991 | −1 482 747 | 1 652 764 | 243 667 | 67 929 | 1 619 |
| 4 | 66 271 | 268 186 | 2 903 884 | 7 410 033 | 1 908 597 | 1 631 198 | 89 996 | 2 586 |
| 5 | −690 | −1059 | −84 | −44 | 4 418 121 | −1 493 365 | 226 862 | 7 520 |
| 6 | −617 | −1 097 | −754 | −898 | 3 475 437 | 4 622 576 | 206 316 | 8 498 |
| 7 | 270 | 270 | 45 | 50 | −4 657 328 | −178 025 | 3 142 667 | 175 505 |
| 8 | 4 | 11 | 3 | 4 | 178 385 | −13 278 | −341 325 | 1 149 411 |

**Table 9.8** The coefficients $N_{j,k} \cdot 10^8$ determined through the Lagrange secular theory

| $j/k$ | 1 | 2 | 3 | 4 | 5 | 6 | 7 | 8 |
|---|---|---|---|---|---|---|---|---|
| 1 | 6 280 126 | −1 790 681 | 204 145 | 59 972 | 1 377 585 | 14 143 | −167 896 | −72 116 |
| 2 | 592 716 | 506 921 | −1 334 201 | −353 229 | 1 377 585 | 4 256 | −96 368 | −65 948 |
| 3 | 426 758 | 408 441 | 1 217 050 | 233 709 | 1 377 585 | 143 680 | −86 990 | −64 632 |
| 4 | 90 887 | 91 220 | −1 818 469 | 2 550 518 | 1 377 585 | 497 403 | −63 029 | −61 210 |
| 5 | −1 038 | −655 | −7 | −81 | 1 377 585 | −314 739 | −47 925 | −58 212 |
| 6 | −1 332 | −929 | −237 | −951 | 1 377 585 | 783 199 | −38 897 | −56 066 |
| 7 | 1 124 | 482 | 20 | 90 | 1 377 585 | −35 399 | 881 473 | 54 860 |
| 8 | 29 | 27 | 2 | 10 | 1 377 585 | −3 927 | −104 013 | 588 602 |

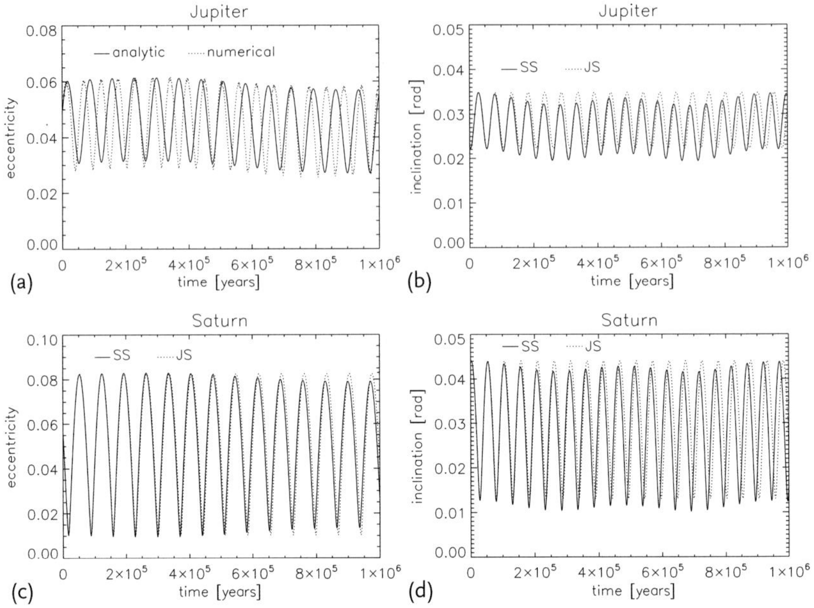


**Figure 9.10** Difference of the solutions for the orbital elements $e$ and $i$ for Jupiter (a,b) and Saturn (c,d) after the theory of Lagrange–Laplace between the pair Jupiter–Saturn (JS) alone and the whole planetary system with eight planets (SS).

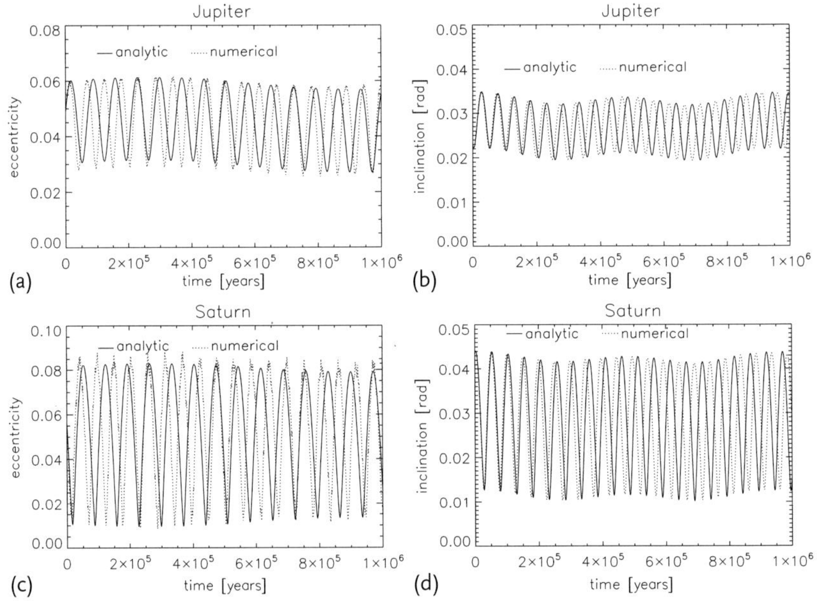


**Figure 9.11** Difference of the solutions for the orbital elements $e$ and $i$ for Jupiter (a,b) and Saturn (c,d) after the theory of Lagrange–Laplace and a numerical integration.

In the evolution of the elements inclination and eccentricity over the time interval of 1 million years it is clear that the perturbations caused by the inner planets is small but is already important – especially for Saturn – after 1 million years. In addition we compare the solution after Lagrange–Laplace with the results of a numerical integration (Figure 9.11). The phase shift is already important after less than $10^5$ years in both osculating elements eccentricity and inclination. Therefore the differences achieve quite large values during one million years of dynamical evolution.

## 9.3 Secular Resonances

### 9.3.1 Asteroids with Small Inclinations and Eccentricities

We have seen in the previous chapter that the osculating elements of the planets (not the semimajor axes) suffer from secular perturbations caused by the other planets. This is not important for the motion of planets with fixed values for the frequencies $g_k$ and $s_k$ (see Table 9.2) but plays an important role in the motion of small bodies like asteroids and comets. We now use canonical variables, the modified Delaunay elements [83]

$$\Lambda_j = L_j \,, \qquad P_j = L_j - G_j \,, \qquad Q_j = G_j - H_j \tag{9.33}$$

and the conjugate ones

$$\lambda_j = l_j + g_j + h_j \,, \qquad p_j = g_j - h_j \,, \qquad q_j = -h_j \tag{9.34}$$

For these variables[9] we can use the secular Hamiltonian where the index $j$ stands for the $j$-th planet

$$\mathcal{H}_{\text{sec}} = \epsilon \mathcal{H}_1 \left( P, Q, p, q, e_j, \tilde{\omega}_j, i_j, \Omega_j \right) \tag{9.35}$$

For the secular behavior (we follow closely the very instructive chapter in the book of Morbidelli [83]) we need to take care of the former determined motions of the perihelion and the node and therefore we introduce the variables $\varpi_k^* = g_k t + \beta_k$ and $\Omega_k^* = s_k t + \delta_k$ and the conjugate variables $\Lambda_{g_k}$ and $\Lambda_{s_k}$ such that (9.35) has the secular normal form

$$\mathcal{H}_2 = \sum_k \left( g_k \Lambda_{g_k} + s_k \Lambda_{s_k} \right) + \epsilon \mathcal{H}_1 \left( P, Q, p, q, \varpi_k^*, \Omega_k^* \right) \tag{9.36}$$

not taking into account the constant term $\mathcal{H}_0(\Lambda, \Lambda_j)$. This form is not integrable and we need to find an integrable approximation with small perturbing parameters which means that we can deal only with eccentricities and inclinations of the

9) We deal with the perturbation of one massive planet (elements with index 'j') on a massless body (elements without an index).

small bodies which are of comparable size as the perturbing planets. In fact in the elements $P$ and $Q$ we expand the factor $\sqrt{(1-e^2)}$ up to the second order and find that $P \sim \sqrt{a}(e^2/2)$ and $Q \sim \sqrt{a}\sin^2(i/2)$. After expanding in power series of $\sqrt{P}$, $\sqrt{Q}$, $e_j$ and $\sin i_j/2$ and taking care of the d'Alembert rule all the odd terms $n$ for $\mathcal{H}_n$ (as well as $\mathcal{H}_0$) vanish and up to the second order in the small parameters we are left with

$$\mathcal{H}_2 = -g_0 P - s_0 Q + \sum_j \Big[ c_j e_j \sqrt{2P}\left(\cos p \cos \tilde{\omega}_j - \sin p \sin \tilde{\omega}_j\right) + d_j \sin \frac{i_j}{2} \sqrt{2Q}\left(\cos q \cos \Omega_j - \sin q \sin \Omega_j\right)\Big] \tag{9.37}$$

We emphasize that $\sqrt{2P}$, $\sqrt{2Q}$, $e_j$ as well $\sin i_j/2$ are of the same order of magnitude. $g_0$, $s_0$, $c_j$, $d_k$ are the respective coefficients of the expansion which depend on the ratio of the semimajor axis of the regarded small body to the planets. If we want to describe the secular behavior of asteroids under the secular perturbation of a big planet we need to introduce in the Hamiltonian explicitly the time $\tilde{\omega}_k = g_j t + \beta_j$ and $\bar{\Omega}_k = s_j t + \delta_j$, we use the results of the Lagrange–Laplace solution and derive

$$\mathcal{H}_2 = -g_0 P - s_0 Q + \sum_j \sum_k \Big[ c_j M_{j,k} \sqrt{2P}\left(\cos p \cos \tilde{\omega}_k - \sin p \sin \tilde{\omega}_k\right) + d_j N_{j,k} \sqrt{2Q}\left(\cos q \cos \Omega^*{}_k - \sin q \sin \Omega^*{}_k\right)\Big] \tag{9.38}$$

We can see that for small values of the eccentricities and the inclinations of the massless body (asteroid) the Hamiltonian $\tilde{\mathcal{H}}$ is integrable

$$\mathcal{H}_{\text{int}} = \sum_k \left(g_k \Lambda_{gk} + s_k \Lambda_{sk}\right) + \mathcal{H}_2 \,. \tag{9.39}$$

After introducing the canonical polynomial variables $x = \sqrt{2P}\sin p, v = \sqrt{2Q}$ $\sin q$ and the conjugate variables $y = \sqrt{2P}\cos p, z = \sqrt{2Q}\cos q$ and taking advantage of the canonical character we just have to integrate the Hamiltonian with respect to the conjugate variables from which the system of eight differential equations follows:

$$\dot{x} = g_0 y + \sum_j \sum_k c_j M_{j,k} \cos \tilde{\omega}^*_k \,, \qquad \dot{y} = g_0 x + \sum_j \sum_k c_j M_{j,k} \sin \tilde{\omega}^*_k \tag{9.40}$$

$$\dot{v} = s_0 z + \sum_j \sum_k d_j N_{j,k} \cos \Omega^*_k \,, \qquad \dot{z} = g_0 v + \sum_j \sum_k d_j N_{j,k} \sin \Omega^*_k \tag{9.41}$$

$$\dot{\Lambda}_{g_k} = \sum_j c_j M_{j,k}\left(y \sin \tilde{\omega}^*_k + x \cos \tilde{\omega}^*_k\right) \,, \qquad \dot{\tilde{\omega}}^*_k = g_k \tag{9.42}$$

$$\dot{\Lambda}_{s_k} = \sum_j c_j N_{j,k}\left(x \sin \Omega^*_k + y \cos \Omega^*_k\right) \,, \qquad \dot{\Omega}^*_k = s_k \tag{9.43}$$

After replacing $\tilde{\omega}_k^* = g_k t + \beta_k$ and $\Omega_k^* = s_k t + \delta_k$ from (9.40) we get the solution of two decoupled harmonic oscillators with a forced term originating from the secular term in $\tilde{\omega}$ and $\tilde{\Omega}$

$$\begin{aligned}
x &= A\sin(-g_0 t + \alpha) + \sum_j \sum_k \frac{c_j M_{j,k}}{g_k - g_0} \sin \tilde{\omega}_k^* \\
y &= A\cos(-g_0 t + \alpha) + \sum_j \sum_k \frac{c_j M_{j,k}}{g_k - g_0} \cos \tilde{\omega}_k^* \\
v &= B\sin(-s_0 t + \beta) + \sum_j \sum_k \frac{c_j N_{j,k}}{s_k - s_0} \sin \Omega_k^* \\
z &= B\cos(-s_0 t + \beta) + \sum_j \sum_k \frac{c_j N_{j,k}}{s_k - s_0} \cos \Omega_k^*
\end{aligned} \tag{9.44}$$

The constants $A$, $B$, $\alpha$ and $\beta$ are derived from the initial conditions; a resonance appears after the integration of (9.40) through the $g_k - g_0$ and $s_k - s_0$ in the denominator whenever the frequencies $g_0$ or $s_0$ of the asteroid are equal (or close) to the respective frequencies of a planet $g_k$ or $s_k$. We then say that the small body is in a *secular resonance of first order*. It is evident that the denominator can cause large perturbations even when the numerator $c_j M_{j,k}$ or $d_j N_{j,k}$ are small quantities. In literature one often finds the notation $\nu_k$ when $g_0 = g_k$ and $\nu_{1k}$ when $s_0 = s_k$; furthermore the indices $k, k = 1, \ldots, 8$ are connected to the number of the planets; so $\nu_5$ is a secular resonance of an asteroid with Jupiter, $\nu_6$ with Saturn.

A final explanation needs to be done for the expression of 'proper elements' for the action variables $e$ and $i$. This kind of very rough mean values can be computed with the aid of the former developed integrable Hamiltonian and new canonical action-angle variables $P'$, $p'$, $Q'$, $q'$ via the 'primed' variables $x'$, $y'$, $v'$, $z'$

$$\begin{aligned}
x' &= x - \sum_j \sum_k \xi_{j,k} \sin \tilde{\omega}_k^* , \qquad & y' &= y + \sum_j \sum_k \xi_{j,k} \cos \tilde{\omega}_k^* \\
v' &= v - \sum_j \sum_k \eta_{j,k} \sin \Omega_k^* , \qquad & z' &= z + \sum_j \sum_k \eta_{j,k} \cos \Omega_k^*
\end{aligned} \tag{9.45}$$

where we use for the possibly large perturbations due to the presence of a resonance the quantities $\xi_{j,k}$ and $\eta_{j,k}$

$$\xi_{j,k} = \frac{c_j M_{j,k}}{g_k - g_0} , \qquad \eta_{j,k} = \frac{d_j N_{j,k}}{s_k - s_0} \tag{9.46}$$

The orbital elements $\varpi_k$ and $\Omega_k$ are the same in the new variables. The solutions in this set of variables can be derived after introducing a new set of canonical variables which are the so-called linear proper Delaunay variables (the primed variables)

$$\begin{aligned}
x' &= \sqrt{2P'} \sin p' , \qquad & v' &= \sqrt{2Q'} \sin q' \\
y' &= \sqrt{2P'} \cos p' , \qquad & z' &= \sqrt{2Q'} \cos q'
\end{aligned} \tag{9.47}$$

for which the Hamiltonian can be integrated because it depends only on the actions.

$$\tilde{\mathcal{H}} = \sum_k \left( g_k \Lambda'_{g_k} + s_k \Lambda_{s_k} \right) - g_0 P' - s_0 Q' \tag{9.48}$$

Finally we find linear proper elements:

$$\begin{aligned} e' &= \sqrt{1 - \left(1 - \frac{P'}{L}\right)}, \qquad \varpi' = -p' \\ i' &= \arccos\left(1 - \frac{Q'}{L - P'}\right), \qquad \Omega' = -q' \end{aligned} \tag{9.49}$$

where $s_0$ and $g_0$ are the linear proper frequencies.

Figure 9.12 shows together with some asteroid families in the main belt the location of the linear secular resonances (= SR) caused by Jupiter and Saturn. $g$ is the asteroid's motion of the $\tilde{\omega}$, $g_5$ and $g_6$ respectively the ones of Jupiter (fifth planet) and Saturn (sixth planet). There are three SR visible which are caused by a resonance between the motion of the perihelia of the asteroid and Jupiter respectively Saturn. Only one resonance where the nodes are involved is visible which is caused by Saturn. The middle line characterizes the exact SR; the two lines aside correspond to a width of $\pm 1''$/yr. The axes are the so-called proper semimajor axes ($x$-axis) and proper inclination ($y$-axis).[10] The location of the SR is modified by the presence of

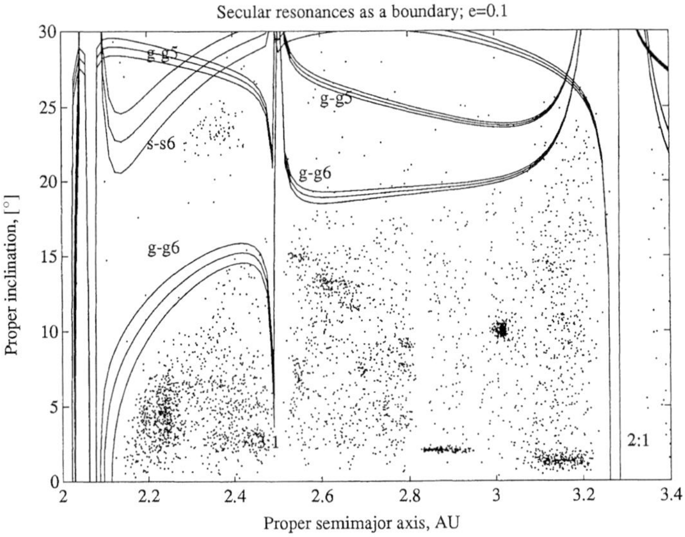


**Figure 9.12** The location of the linear SR of Jupiter and Saturn with asteroids in the inner main belt; one can see (straight vertical lines) how MMR modify the SR. The accumulation of points represent asteroid families (after [216]).

10) The proper orbital elements of a celestial body are constants of motion remaining more or less unchanged over very long timescales. We distinguish the proper semimajor axis ($a_p$), the proper eccentricity ($e_p$) and the proper inclination ($i_p$).

MMR which are represented as straight vertical lines at the 3 : 1 ($a$ = 2.48 AU) and 2 : 1 ($a$ = 3.25 AU). In addition in this figure we can see the accumulation of points for special values $i_p$, $a_p$ which means that a large number of asteroids share almost the same proper elements. These families of asteroids are believed to be the debris of a parent body destroyed by collision with another asteroid long time ago. For a comprehensive review on asteroid families we refer to the work [217]. How can we explain the nature of these proper elements? In the former paragraphs of secular resonances we developed the theory of this kind of 'mean values' of the orbital elements of asteroids. A thoroughly prepared paper by [216] explains – after a short historical introduction – the different definitions and approaches for deriving them. They can also be determined from results of numerical integrations taking the average of the action variables over long time scales ($10^6$ years). It turns out that for most of the asteroids these mean elements are constant and are quasi integrals of motion. To understand the different procedures we refer to the paper mentioned before and an article [218], where the basic steps for deriving them are described more in detail.

In Figure 9.13 we plotted the osculating elements of the main belt asteroids with well established elements according to the MMPCC[11)] semimajor axes versus the inclination. Although here we did not plot the proper elements two asteroid families are well visible: for $i \sim 23°$ and $a < 2$ AU the Hungaria family (with some 5000 members) and for $a \sim 3.93$ AU the Hilda family (with some 2000 members) is well visible. Again one recognizes the sculpting due to the SR and the MMR in the distribution of the main belt asteroids.

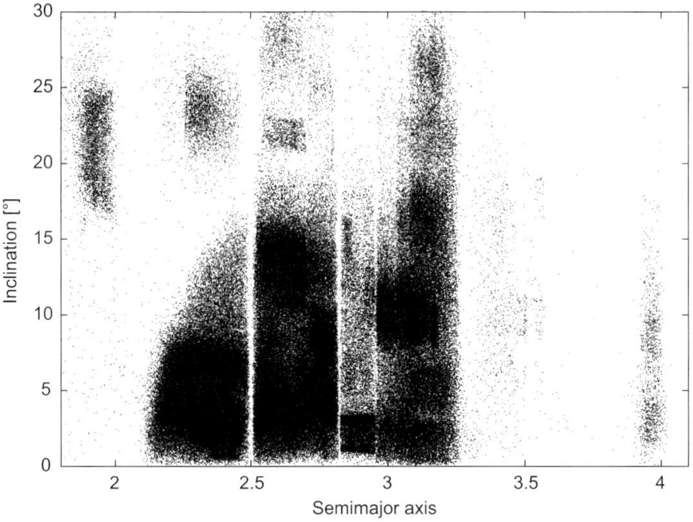


**Figure 9.13** Semimajor axes versus the inclination of main belt asteroids: one can very well see the MMRs (e.g. 3 : 1 MMR at 2.5 AU). The $g_6$ secular resonance is acting from $a = 2.1$ ($i = 0°$) to $a = 2.5$ ($i = 15°$); no asteroids are found here.

11) www.minorplanetcenter.net/iau/MPCStatus.html (last accessed on 20 December 2012).

**Table 9.9** A sample of asteroid families: the upper and lower numbers for the estimated number of members are derived using the different methods of determination namely HCM (upper lines) respectively WAM (lower lines). For details see the text [217].

| 8 Flora | 44 Nysa | 4 Vesta | 24 Themis | 221 Eos | 158 Koronis | 15 Eunomia | 10 Hygia |
|---|---|---|---|---|---|---|---|
| 604 | 381 | 231 | 550 | 477 | 325 | 439 | 103 |
| 575 | 374 | 242 | 517 | 482 | 299 | 393 | 175 |

In Table 9.9 we show some of the most prominent families with some numbers of the membership derived with different methods. We stress that this classification depends very much on the method used. The upper values show the results derived with the HCM (Hierarchical Clustering method) the lower line with the WAM (Wavelet analysis method)[12] In fact the families are regarded to be much more numerous (e.g. the Vesta family is estimated to consist of thousands of members).

## 9.3.2 Comets and Asteroids with Large Inclinations and Eccentricities: the Kozai Resonance

In the former chapter we developed the theory for secular resonances for asteroids with eccentricities and inclinations which are of comparable size with the ones of the planets. The situation is different when dealing with comets and asteroids on high inclined orbits and large eccentricities. It is better to use another expansion of the secular part of the Hamiltonian

$$\mathcal{H}_{\text{sec}} = \sum_k \left(g_k \Lambda_{g_k} + s_k \Lambda_{s_k}\right) + \sum_{n \geq 0} K_n \left(P, Q, p, q, \varpi_k^*, \Omega_k^*\right) \tag{9.50}$$

where $K_n$ is of degree n in the (now not so small) parameters $e_j, \sin(i_j)$. The first term $K_0$ is independent of the eccentricities and the inclinations and consequently also of the angles $\varpi_k$ and $\Omega_k$. Because of the rotational invariance of the Hamiltonian in the expansion of $K_0$ it depends only on the variable $g = \omega$ and is therefore integrable and expressed in Delaunay variables it reads

$$\mathcal{H}_{\text{int}} = \sum_k \left(g_k \Lambda_{g_k} + s_k \Lambda_{s_k}\right) + K_0(G, H, g) \tag{9.51}$$

Because $K_0$ does not contain terms in $h$ the action $H = \sqrt{a(1-e^2)} \cos i$ found with the canonical equation $dH/dt = \partial \mathcal{H}_{\text{int}}/\partial h = 0$ is a constant of motion[13]. This means that the eccentricity and the inclination are coupled via the constant value of $H$.

12) Two of the most used methods are the Hierarchical Clustering Method (HCM) – which means searching for groupings in orbital element space with the aid of smallest distances with respect to the nearest neighbor –, and the Wavelet Analysis Method (WAM) which looks for high densities in the distribution with respect to the orbital elements of the asteroids..

13) Which is the $z$ component of the angular momentum of the asteroid (comet).

To understand the mechanism of the Kozai dynamics (Kozai [219] it is best to use a representation with polar coordinates $h$ and $k$. In the corresponding graph Figure 9.14 in a diagram $h = e \sin \omega$ versus $k = e \cos \omega$ we can draw curves for $K_0 = \text{const}$. Because $H = H(a, e, i)$ we need to fix one of the three parameters and then get a solution for $H = \text{constant}$ varying the two other parameters. We get the value $i_{max} = \arccos(H/\sqrt{a})$ for e=0 and vice-versa the maximum value $e_{max} = \sqrt{1 - H^2/a}$. With such simple considerations we can now plot for specific values of the semimajor axis and a fixed $i_{max}$ curves in the plane $e \sin \omega$ versus $e \cos \omega$. It is interesting to compare the development of the fixed point at the center which changes stability with increasing inclination. For values of $i_{max} < 30$ the equilibrium corresponds to a fixed value of the perihelion $\omega$ of the asteroid with stable curves (librations) around the equilibrium points. For larger values of

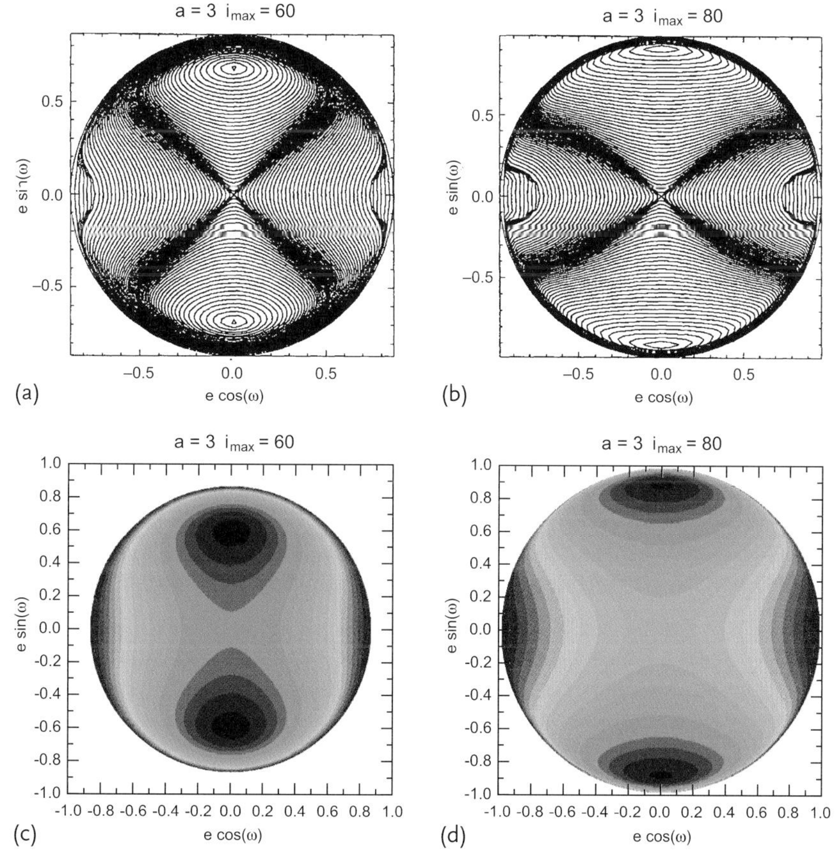


**Figure 9.14** Kozai dynamics of fictitious asteroids with semimajor axes $a = 3$ AU in the Kozai resonance for two different maximum values of the inclination. The closed curves are drawn for a constant Delaunay variable $H$ for $e \sin g$ versus $e \cos g$ after [220] (a,b). Plots after a simplified theory provided by A. Bazso (2012, personal communication) (c,d); for a detailed explanation see the text.

$i_{max} > 30$ it changes its stability, and new stable equilibrium points appear again for $e \cos \omega = 0$ corresponding to $\omega = 90°$ and $\omega = 270°$ with larger and larger eccentricities.

We can now express $K_0$ in the variables $g, h$ for some given value $i_{max}$, which in turn determines the value of $H$. When $i = i_{max} = \arccos\left(H/\sqrt{a}\right)$ then $e = 0$, and for $e = e_{max} = \sqrt{1 - H^2/a}$ it follows $i = 0$. Instead of computing a double integral to obtain the level curves of the Hamiltonian $K_0$ (e.g. [83, 220, 221]), we derived an analytic approximation for $K_0$, valid for any $(e, i)$; but higher powers of $(a/a_k)^2$ have been neglected.

$$K_0 = \sum_k \frac{m_k}{16a_k}\left(\frac{a}{a_k}\right)^2 \left[\left(2 + 3e^2\right)\left(3\cos^2 i - 1\right) + 15e^2 \sin^2 i \cos(2g)\right] \quad (9.52)$$

For instance for Jupiter we have for $a = 3$ and $a_J = 5.2$ a ratio of $(a/a_J)^2 \approx 0.33$. In the following we use the perturbations of the planets of the outer solar system (Jupiter to Neptune) for a fixed value of $a = 3$ AU, and two values of $i_{max} = \{60, 80\}$. First from $(a, i_{max})$ the value of $H = \sqrt{a}\cos i_{max}$ is calculated, from which in turn follows $e_{max} = \sqrt{1 - H^2/a}$. This value of $e_{max}$ defines the boundary curve for the level curves in the figures; it increases with increasing $i_{max}$. Then for a grid in $(e, g)$ (or alternatively $(x, y)$ with $x = e\cos g$, $y = e \sin g$) a suitable $i$ is calculated so as to fulfill the condition of $H(a, e, i) =$ const, and finally the levels of constant

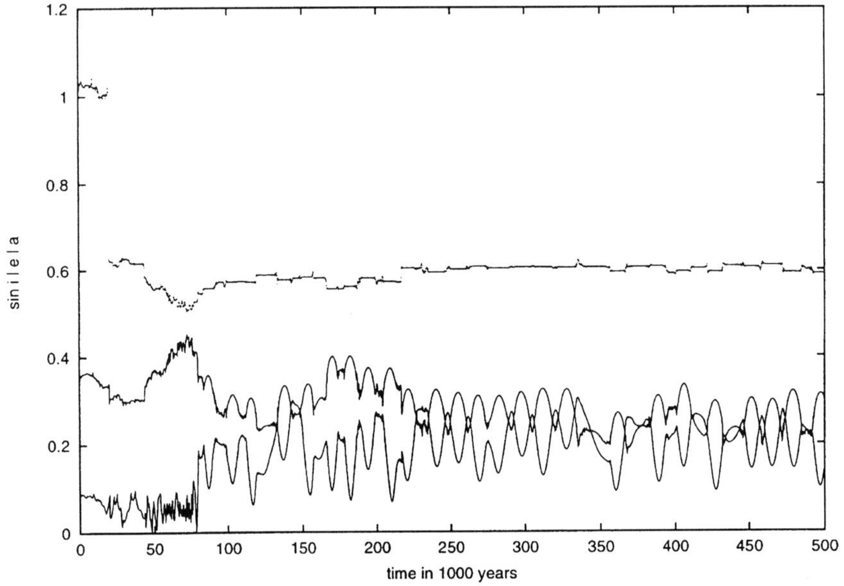


**Figure 9.15** Dynamical evolution of the asteroid (4581) Ascelpius for $5 \times 10^5$ years: top line semimajor axis in AU, middle line eccentricity, low line sin($i$). From $t = 8 \times 10^4$ years on the asteroid is in the Kozai resonance which is visible from the two lower lines with the periodic-opposite variations.

$K_0(a, e, i, g)$ are found from (9.52). The results are shown in the respective Figure 9.14.

To illustrate the Kozai resonance on a real asteroid we show the orbital evolution of the asteroid (4581) Asclepius[14] which is several times in a Kozai resonance (Figure 9.15). It shows the typical dynamics of a small body in Kozai resonance: after some $8 \times 10^4$ years of chaotic evolution caused by multiple encounters with the inner planets it is captured into this secular resonance shown by a more or less constant semimajor axis and increase in inclination causes a decrease in eccentricity according to the constant $H$.

## 9.4
## Three-Body Resonances

Commensurable mean motions between two celestial bodies in our Solar system are a common phenomenon between planets and also between a planet and a small body, an asteroid or a comet. Already these configurations are a severe but more-or-less solvable problem for celestial mechanics, but the difficulties are substantially more challenging when three celestial bodies are in a so-called three-body-resonance (3BR). The general definition is that

$$k_1 \dot{\lambda}_1 + k_2 \dot{\lambda}_2 + k_3 \dot{\lambda}_3 = \sigma \sim 0 \tag{9.53}$$

where $\dot{\lambda}_i, i = 1, 2, 3$ stands for the mean motions of three bodies and the $k_i$ are nonzero integers.

### 9.4.1
### Asteroids in Three-Body Resonances

A special case which is important for the motion of small bodies in the mainbelt of asteroids between Mars and Jupiter and also of asteroids in the Kuiper Belt outside Pluto's orbit was treated in detail in the book by Morbidelli [83] and in some papers on this subject (e.g. [222]). There the locations (values of the semimajor axis of the asteroids) are determined which are in 3BR in combination with Jupiter and Saturn for the region of the main belt of asteroids.

$$k_1 \dot{\lambda}_{\text{asteroid}} + k_2 \dot{\lambda}_{\text{Jupiter}} + k_3 \dot{\lambda}_{\text{Saturn}} \sim 0 \tag{9.54}$$

Some of these resonances together with the most important MMR are shown in Figure 9.16. The complex theory is quite well developed in the previous mentioned book by Morbidelli; he also computed the variation $\Delta a$ in such a 3BR which can be regarded as a measure for the strength of the resonance. In Table 9.10 we compare the values derived from numerical integrations for real asteroids close to these resonances with the theoretical ones. The good agreement is well visible.

14) Originally a member of the Apollo group of Near Earth Asteroids (NEA) with $a = 1.022$ AU, $i = 4.^\circ 9$ and $e = 0.357$.

### 9.4.2
### Three Massive Celestial Bodies in Three-Body Resonances

Generally we can treat the case of three massive planets ($m_1$, $m_2$ and $m_3$) in the following way

$$q n_1 - (p + q) n_2 + p n_3 = 0 \tag{9.55}$$

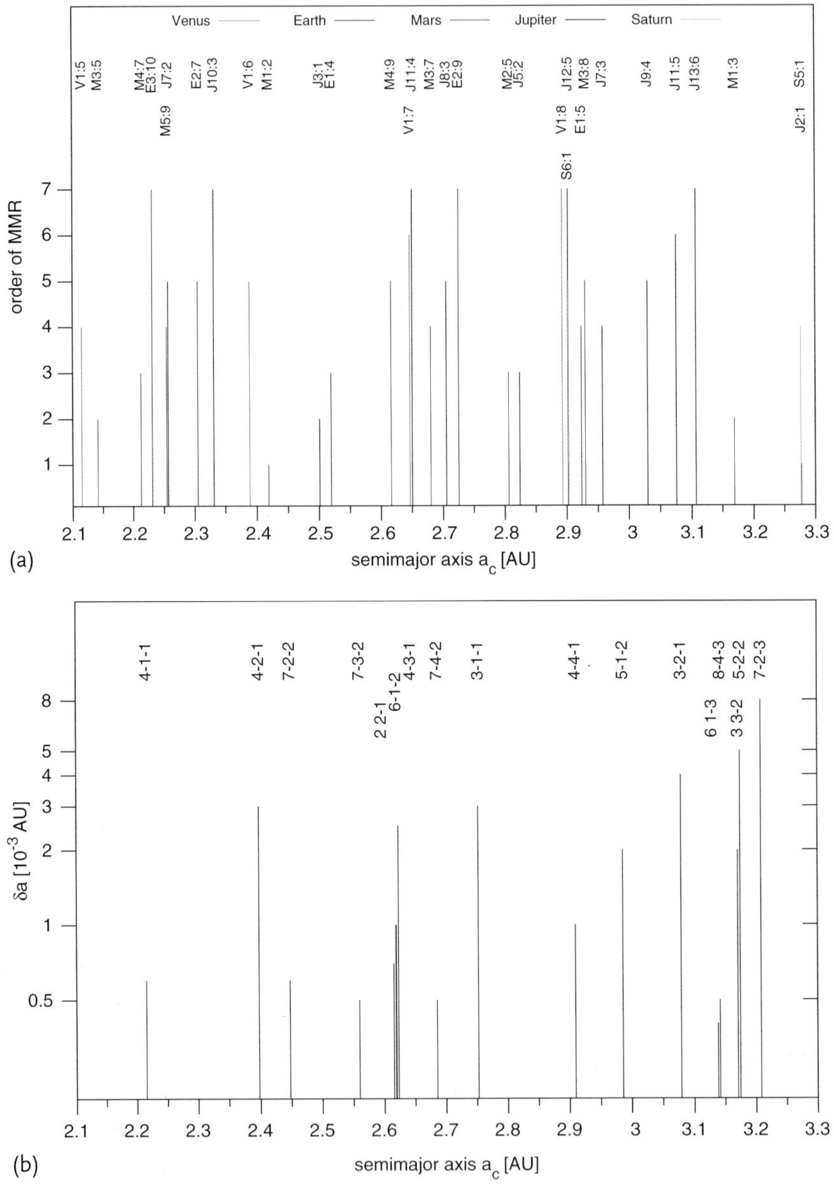


**Figure 9.16** Main MMR in the main belt of asteroids (a) and selected three body resonances of asteroids in the main belt with Jupiter and Saturn semimajor axes versus the variation in semimajor axes $\Delta a$ after [83] (b)

**Table 9.10** Comparison between computed variation in semimajor axes and periods of libration for real asteroids close to this 3BR; $a$ in AU, $\Delta a$ in $10^{-3}$ AU and period of libration in $10^3$ years. The subscripts 'ana' stands for analytical determination and 'int' for the results after numerical integrations (after [222]).

| Resonance | Asteroid | $a_{int}$ | $a_{ana}$ | $\Delta a_{int}$ | $\Delta a_{ana}$ | $P_{int}$ | $P_{ana}$ |
|---|---|---|---|---|---|---|---|
| 4 −1 −1 | 2440 | 2.2155 | 2.2157 | 0.37 | 0.6 | 52 | 50 |
| 4 −2 −1 | 463 | 2.3977 | 2.3978 | 3.0 | 2.4 | 10 | 9.9 |
| 7 −2 −2 | 1966 | 2.4476 | 2.4479 | 0.6 | 0.37 | 30 | 33 |
| 7 −3 −2 | 1430 | 2.5599 | 2.5600 | 0.5 | 0.39 | 30 | 36 |
| 2 2 −1 | 2589 | 2.6155 | 2.6155 | 0.7 | 0.15 | 20 | 200 |
| 6 −1 −2 | 53 | 2.6190 | 2.6192 | 1.0 | 0.25 | 35 | 57 |
| 4 −3 −1 | 792 | 2.6230 | 2.6229 | 2.5 | 0.9 | 25 | 30 |
| 7 −4 −2 | 789 | 2.6857 | 2.6858 | 0.5 | 0.3 | 20 | 49 |
| 3 −1 −1 | 485 | 2.7525 | 2.2527 | 3.0 | 1.9 | 15 | 17.8 |
| 4 −4 −1 | 22 | 2.9095 | 2.9092 | 1.0 | 0.1 | 50 | 370 |
| 5 −1 −2 | 576 | 2.9860 | 2.9864 | 2.0 | 1.1 | 20 | 19 |
| 3 −2 −1 | 2395 | 3.0790 | 3.0794 | 4.0 | 4.5 | 10 | 9.9 |
| 6 1 −3 | 936 | 3.1385 | 3.1389 | 0.4 | 0.12 | 10 | 132 |
| 8 −4 −3 | 10 | 3.1418 | 3.1421 | 0.5 | 0.5 | 30 | 33 |
| 3 3 −2 | 106 | 3.1708 | 3.1705 | 2.0 | 0.13 | – | 180 |
| 7 −2 −3 | 530 | 3.2080 | 3.2100 | <8 | 2.8 | 12 | 5.8 |

or

$$\frac{p}{n_1 - n_2} = \frac{q}{n_2 - n_3} = \frac{p + q}{n_1 - n_3} = \frac{P}{2\pi} \tag{9.56}$$

for $q = 1$ and $p = 2$, where $P$ is the period of time when the three bodies are aligned in a rotating frame. We then speak of the Laplacian three-body resonance which we observe for the Galilean satellites Io, Europa and Ganymede of Jupiter[15]; they are in the 1 : 2 : 4 MMR (see next subsection). For the critical argument one finds

$$\theta = \lambda_1 - 3\lambda_2 + 2\lambda_3 = 180^\circ \tag{9.57}$$

which means whenever Europa and Ganymede are in conjunction ($\lambda_2 = \lambda_3$) the innermost Galilean satellite Io is on the other side of Jupiter. Other examples are the three inner satellites of Uranus Miranda, Ariel and Umbriel and Uranus (1), Neptune (2) and Pluto (3) (the Plutinos) which are also in a 3BR[16], such that $n_1 = 2n_2 = 3n_3$ and $n_1 - 4n_2 + 3n_3 = 0$

$$n_1 : n_2 : n_3 = 1 : \frac{1}{2} : \frac{1}{3} \tag{9.58}$$

As far as the three Jupiter Moons and the Uranus–Neptune–Pluto system concerns the differences $d = n_1 - 2n_2 = n_2 - 2n_3$ respectively $d = n_1 - 2n_2 = 2n_2 - 3n_3$ concerns, are already small which lead to rather small divisors when using

15) With orbital periods of 1.769, 3.551 and 7.155 days.
16) $n_{\text{Uranus}} = 0.^\circ 0117$, $n_{\text{Neptune}} = 0.^\circ 00597$ and $n_{\text{Pluto}} = 0.^\circ 00393$ per day.

perturbation theory. For the theory of motion this means that already in the first order there will be large perturbations present after integration.

We follow the main lines of a very interesting scientific article by Aksnes [223] based on an older paper [224]. We define the angle $\theta$ which describes the libration around an equilibrium point

$$\theta = q\lambda_1 - (p+q)\lambda_2 + p\lambda_3 \tag{9.59}$$

where the $\lambda$'s are the mean longitudes; this angle librates slowly with a small amplitude around a stable configuration with $\theta = 180°$. After a double differentiation with respect to time we get

$$\frac{d^2\theta}{dt^2} = q\frac{d^2\lambda_1}{dt^2} - (p+q)\frac{d^2\lambda_2}{dt^2} + p\frac{d^2\lambda_3}{dt^2} \tag{9.60}$$

The goal is to represent the libration with the equation of a perturbed pendulum

$$\frac{d^2\theta}{dt^2} = c\sin\theta \tag{9.61}$$

This equation has an integral of motion

$$\frac{d\theta}{dt} = \sqrt{2C - 2c\cos\theta}\,. \tag{9.62}$$

where we can determine the constant of Integration $C$ from the initial orbital elements ($t = t_0$)

$$\frac{1}{2}\left(\frac{d\theta}{dt}\right)^2 = \frac{1}{2}\left[qn_1 - (p+q)n_2 + pn_3\right]^2 = C - c\cos\theta_0 \tag{9.63}$$

We see that $\theta$ will librate if $\mid C/c \mid < 1$. The three different possibilities with respect to $c$ are the following ones

- libration $\theta \sim 0°$ for $c < 0$
- libration $\theta \sim 180°$ for $c > 0$
- circulation $|C/c| > 1$

The angle $\theta$ can be expressed with the aid of elliptic integrals; in the case of the 3BR we are primarily interested in the Amplitude $A = \arccos(C/c)$ and the period of libration $P$ for which the following expression holds:

$$P = 4\frac{1}{|\sqrt{c}|}\int_0^{\frac{\pi}{2}} \frac{d\theta}{\sqrt{1 - \sin^2\frac{A}{2}\sin^2\theta}} \tag{9.64}$$

which can be expressed also with the aid of hypergeometric series. When we ignore the inclinations (in most cases this can be done) we can express the longitudes as differences $\alpha_i - \beta_i$ where we also take care of the longitudes of the pericenter

$$\begin{aligned}
\alpha_1^{(k)} &= (p+q-k)\lambda_1 - (p+q)\lambda_2 + k\tilde{\omega}_1 \\
\beta_1^{(k)} &= (p-k)\lambda_1 - p\lambda_3 + k\tilde{\omega}_1 \\
\alpha_2^{(k)} &= q\lambda_1 - (q+k)\lambda_2 + k\tilde{\omega}_2 \\
\beta_2^{(k)} &= (p-k)\lambda_2 - p\lambda_3 + k\tilde{\omega}_2 \\
\alpha_3^{(k)} &= q\lambda_1 - (q+k)\lambda_3 + k\tilde{\omega}_3 \\
\beta_3^{(k)} &= (p+q)\lambda_2 - (p+q+k)\lambda_3 + k\tilde{\omega}_3
\end{aligned} \tag{9.65}$$

and $k = 0, \pm 1, \pm 2 \ldots$ Hence for the perturbing function we can write for one planet $m_1$ perturbed by the two others $(m_2, m_3)$

$$F_1 = k^2 m_2 \left[ \frac{1}{r_{12}} - \frac{(q_1 \cdot q_2)}{r_2^3} \right] + k^2 m_3 \left[ \frac{1}{r_{13}} - \frac{(q_1 \cdot q_3)}{r_3^3} \right] \tag{9.66}$$

which can be written as

$$F_1 = F_1^{(0)} + F_1^{(1)} + \ldots \tag{9.67}$$

and by introducing the respective Laplace coefficients (see Appendix B)

$$F_1^{(0)} = k^2 m_2 \left[ \sum_{k=-\infty}^{\infty} A_{12}^{(k)} \cos(k\lambda_1 - k\lambda_2) \right] + k^2 m_3 \left[ \sum_{k=-\infty}^{\infty} A_{13}^{(k)} \cos(k\lambda_1 - k\lambda_3) \right] \tag{9.68}$$

As already mentioned we write for the gravitational constant $k^2$; the upper index in $F_i^k$ stands for the powers of the eccentricity. This term up to the first order in the eccentricity reads ($j = 2, 3$ and $m_1$ is the innermost planet)

$$\begin{aligned}
F_1^{(1)} = k^2 \sum_{j=2,3} \Bigg\{ m_j \sum_{k=-\infty}^{\infty} & e_1 B_{1j}^{(k)} \cos\left[(k-1)\lambda_1 - k\lambda_j + \tilde{\omega}_1\right] \\
& + e_j C_{1j}^{(k)} \cos\left[k\lambda_1 - (k+1)\lambda_j + \tilde{\omega}_j\right] \Bigg\}
\end{aligned} \tag{9.69}$$

The Laplace coefficients $A_{1j}$, $B_{1j}$ and $C_{1j}$ depend on the respective ratios of the semimajor axis of the couple of planets $a_1/a_2$ and $a_1/a_3$; the respective computation of the Laplace coefficients will be given in Appendix B.
Without going into the details for this derivation (see [223]) one can determine the coefficient $c$ in (9.61) as being composed of terms with increasing power in the eccentricity

$$c = c^{(0)} + c^{(1)} + c^{(2)} + \ldots \tag{9.70}$$

We will not go further than up to the first order; higher-order terms $c^m$, $m > 1$ are truncated. Using the Lagrange equations for $a$ and $\lambda$ just for the first term in the perturbing function $F_1^{(0)}$ and thus for the moment ignoring higher orders in $e$ taking care of the relation between the mean motion and the mean longitude $\lambda_1 = \int n_1 t + \epsilon_1$ we get[17)]

$$\begin{aligned} \frac{da_1}{dt} &= \frac{2}{n_1 a_1}\frac{\partial F_1^{(0)}}{\partial \lambda_1} \\ \frac{d\lambda_1}{dt} &= n_1 + \frac{d\epsilon_1}{dt} = n_1 - \frac{2}{n_1 a_1}\frac{\partial F_1^{(0)}}{\partial a_1} \end{aligned} \tag{9.71}$$

After building the respective derivatives of the perturbing function we get

$$\begin{aligned} \frac{da_1}{dt} &= \sum_j \sum_{k=-\infty}^{\infty} m_j a_{1j}^{(k)} \sin(k\lambda_1 - k\lambda_j) \\ \frac{d\lambda_1}{dt} &= n_{1,0} + \sum_j \sum_{k=-\infty}^{\infty} m_j \lambda_{1j}^{(k)} \cos(k\lambda_1 - k\lambda_j) \end{aligned} \tag{9.72}$$

where we need to remember that $n_{1,0}$ is the constant part of the osculating $n_1$

$$n_1 = n_{1,0} + \frac{\partial n_i}{\partial a_i}\Delta a_1 = n_{1,0} - \frac{3n_1}{2a_1}\int \frac{da_1}{dt}dt \tag{9.73}$$

From these expressions we find with the aid of the Laplace coefficients

$$\begin{aligned} a_{1,j}^{(k)} &= -2n_1 a_1^2 k A_{1j}^{(k)} \\ \epsilon_{1,j}^{(k)} &= -2n_1 a_1^2 \frac{\partial A_{1j}^{(k)}}{\partial a_1} \\ \lambda_{1,j}^{(k)} &= \epsilon_{1,j}^{(k)} + \frac{3n_1 a_{1,j}^{(k)}}{2a_1 k(n_{(1,0)} - n_{(j,0)})} \end{aligned} \tag{9.74}$$

Now we can write the first term in (9.59) using (9.71)

$$\frac{d^2\lambda_1}{dt^2} = \frac{dn_1}{dt} + \frac{d^2\epsilon_1}{dt^2} \tag{9.75}$$

In the next steps we express the two terms in (9.75) and insert the results of a first order perturbations for $\Delta a_1$ and $\Delta\lambda_1$ where we find out – after rather lengthy computations (we refer again to [223]) that $c^{(0)}$ consists in fact of three terms each of them being of second order with respect to the masses

$$c^{(0)} = m_2 m_3 c_1^{(0)}(p, q) + m_3 m_1 c_2^{(0)}(q, -p-q) + m_1 m_2 c_3^{(0)}(-p-q, p) \tag{9.76}$$

17) Note that in the following we concentrate on the perturbation on the mass $m_1$.

To find $c_1$ we have to use $F_1^{(1)}$ given in (9.74) and need to use the Lagrange equations for $e$ and $\tilde{\omega}$

$$\begin{aligned}
\frac{de_1}{dt} &= -\frac{\partial F_1^{(1)}}{\partial \tilde{\omega}} \left(n_1 a_1^2 e_1\right)^{-1} \\
&= n_1 a_1 \sum_j \sum_{k=-\infty}^{\infty} m_j B_{1,j}^{(k)} \sin\left((k-1)\lambda_1 - k\lambda_j + \tilde{\omega}_1\right) \\
e_1 \frac{d\tilde{\omega}_1}{dt} &= \frac{\partial F_1^{(1)}}{\partial e_1} \left(n_1 a_1^2\right)^{-1} \\
&= n_1 a_1 \sum_j \sum_{k=-\infty}^{\infty} m_j B_{1,j}^{(k)} \cos\left((k-1)\lambda_1 - k\lambda_j + \tilde{\omega}_1\right) \qquad (9.77)
\end{aligned}$$

Again use the first-order perturbations for $\Delta e_1$ and $\Delta\tilde{\omega}_1$

$$\begin{aligned}
\Delta e_1 &= -n_1 a_1 \sum_j \sum_{k=-\infty}^{\infty} \frac{m_j B_{1,j}^{(k)}}{(k-1)n_1 - k n_j} \cos\left((k-1)\lambda_1 - k\lambda_j + \tilde{\omega}_1\right) \\
e_1 \Delta\tilde{\omega}_1 &= n_1 a_1 \sum_j \sum_{k=-\infty}^{\infty} \frac{m_j B_{1,j}^{(k)}}{(k-1)n_1 - k n_j} \sin\left((k-1)\lambda_1 - k\lambda_j + \tilde{\omega}_1\right)
\end{aligned} \qquad (9.78)$$

From these expressions we find out – after rather lengthy computations involving the expansion of the two osculating elements $e$ and $\tilde{\omega}$ – that also $c^{(1)}$ consists in fact of three terms each of them being of second order with respect to the masses like $c^{(0)}$.

$$c^{(1)} = m_2 m_3 c_1^{(1)}(p, q) + m_3 m_1 c_2^{(1)}(q, -p-q) + m_1 m_2 c_3^{(1)}(-p-q, p) \qquad (9.79)$$

Thus we have the value of the amplitude $c$ up to the second order in the masses determined and as a consequence we can determine – in principle – the libration around the equilibrium point in the case of three planets in the 3BR.

### 9.4.3
### Application to the Galilean Satellites

As mentioned the three Galilean satellites Io, Europa and Ganymede are in fact in a 3BR of the form

$$n_1 - 2n_2 = n_2 - 2n_3 = 0^\circ.007/\text{day} \qquad (9.80)$$

which means they are in the commensurability $n_1 - 3n_2 + 2n_3 = 0$. For $q = 1$ and $p = 2$ in (9.55) we find $\theta = \lambda_1 - 3\lambda_2 + 2\lambda_3$ and the motion of this angle can be described by a 'simple' perturbed pendulum (9.61). When we determine

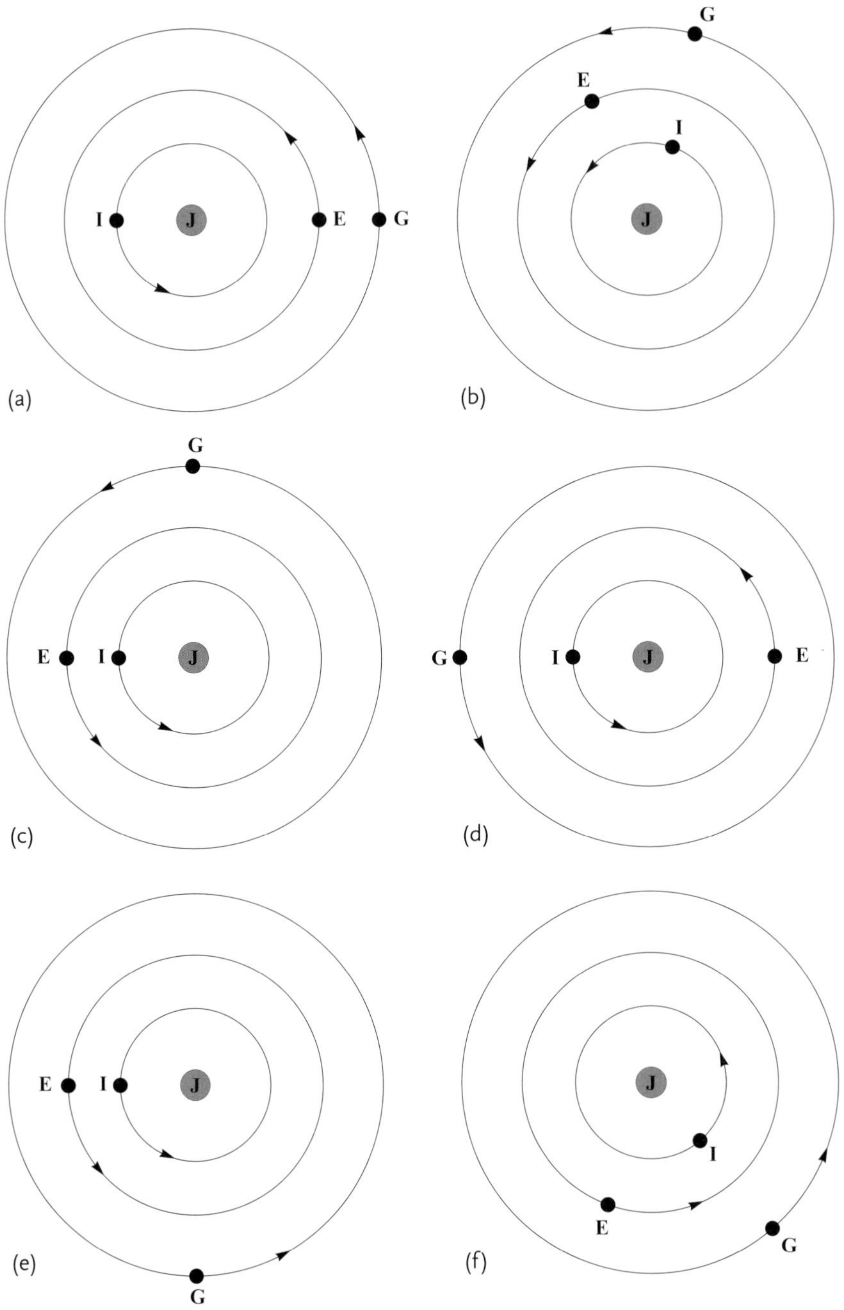


**Figure 9.17** The configurations of Galilean satellites for different phases during their motion, for details we refer to the text

the corresponding quantities $c^{(0)}$ and $c^{(1)}$ for the parameters $p$ and $q$, which can be computed with the aid of the Laplace coefficients, we find (after [223]) that

$$c = (3.31 \cdot m_2 m_3 + 11.93 \cdot m_3 m_1 + 2.05 \cdot m_1 m_2)\, 10^6 \text{ deg/day}^2 \qquad (9.81)$$

which means a period $P = 2260$ days. The amplitude of libration around 180° is only 0.°064!

In Figure 9.17 we can see that the three satellites are never aligned on one side of Jupiter. In the following we mean by (a)–(g) the 6 different configurations of the satellites in this figure.

- (a) the innermost satellite Io is on the left side, Europa and Ganymede are on the other side of Jupiter, all three are aligned.
- (b) after P/6 all three moons are well separated on one side of Jupiter forming an almost equilateral triangle; Io and Ganymede are aligned.
- (c) after P/4 Io is in the same position as in (a) on the left side, where also Europa is located, these two moon are on a connecting line with Jupiter. Ganymede is in its orbit 90° behind.
- (d) after P/2 it is clear that the three bodies are again aligned: the fast moving Io is again on the left side, Ganymede in the same position to the left, but Europa is on the right side
- (e) after 3 P/4 Io and Europa are closed together on the left side (aligned with Jupiter), but Ganymede is 90° in front.
- (f) after 5P/6 the constellation is quite the same as (b), Io and Ganymede are aligned and Europe is behind the two other moons.

Although we expect the realization of a 3BR is quite rare, there may be three planets in extrasolar planetary system in such a stable configuration if one considers the great number of such systems in our galaxy (and in our universe!)

# 10
# Lunar Theory

The theory of the motion of the Moon is connected to the names of the greatest mathematicians and astronomers like Pierre-Simeon Laplace (1749–1827), Leonhard Euler (1707–1783), Charles-Eugene Delaunay (1816–1872), Fraincois Félix Tisserand (1845–1996), George William Hill (1838–1914) and Ernest William Brown (1866–1936) to name a few.[1] Our neighbor in space the Moon is only a bit more distance to the Earth than 30 times the diameter of our planet, by far the closest of all larger bodies in the Solar System. The twin planet Venus, with almost the same mass as the Earth is never closer to the Earth than about thirty times farer away than the Moon and the Sun about 400 times the Moon's distance to the Earth. So it doesn't wonder that its position on the sky was and is now observed with the largest accuracy[2] since ancient times. During the eighteenth century there were Euler and d'Alembert who worked on a Lunar theory which was replaced in the nineteenth century by Laplace's theory and tables. Then, from the middle of the nineteenth century up to the introduction of Brown's tables in 1923 the Lunar theory of Hansen – originally derived for a theory of the motion of the planets – was successfully used. At the same time Delaunay developed a theory where he, in contrary to Hansen who adopted numerical values for the constants, kept them as algebraic quantities. The disadvantage was the slow convergence of his series in the determination of the motion of the node and the perigee. Hill extended the idea but used for the ratio $n/n'$[3] the already well known values from observations. Brown completed this work which is now known as the Hill–Brown Lunar theory. Newcomb found some strange small oscillations which is, as he found out, not a lack of the theory but is due to the irregularities of the rotation of the Earth. Nowadays we have excellent theories by J. Chapront and M. Chapront-Touze [229] adapted to observations and also LLR data (Lunar theory ELP2000, see [230, 231] and review on modern literature at the end of the chapter).

In the following we describe the two main theories of the orbital motion of the Moon on the basis of (i) Hill's Lunar Theory (after [138]) and the (ii) the theory of the

1) A collection of their most important contributions to celestial mechanics can be found for example in [225–228].
2) The Lunar Laser ranging (LLR) gives a precision of its position with an accuracy in the order of 0.1 cm.
3) The ratio of the mean motion $n$ of the Earth to the mean motion of the Moon $n'$.

*Celestial Dynamics*, First Edition. R. Dvorak and C. Lhotka.

Moon after Brown (after [144] and [143]). Both theories give a better understanding of the main problem to be understood and may serve as the basis for more sophisticated models of the motion of the Moon around the Earth. While the former can be seen as a continuation of the formal treatment of the restricted three-body problem (see Chapter 6) the latter resembles more the methodology developed for the general planetary theory (see Chapter 8).

## 10.1 Hill's Lunar Theory

The following assumptions are made by Hill (see [232]) to simplify the problem of the motion of a satellite (the Moon) close to another body (the Earth) perturbed by the motion of a third body (the Sun):

1. The motion of all three bodies lies in the same plane. In case of the Earth–Sun–Moon system the motion of the Moon lies in the ecliptic – the plane in which the Earth moves around the Sun.
2. The orbit of the third body is circular. In case of the Moon the Earth and thus the Sun moves on a circular orbit.
3. The perturbation due to the third body does not depend on the orbital phase of the satellite around the Earth – thus the direction of the perturbation is always parallel to the direction of the Sun–Earth distance.
4. The mass of the satellite can be neglected and is set to zero. The Moon does not influence the motion of the Earth around the Sun.

Let $m_1$ be the mass of the Sun, $m_2$ be the mass of the Earth and $m = 0$ be the mass of the third body (the Moon). We center the origin around the position of the mass $m_2$ and denote by $(x, y)$ the synodic variables of the third body as shown in Figure 10.1. The synodic variables are related to the sidereal coordinates $(\xi, \eta)$ by[4]:

$$\begin{aligned} \xi &= x\cos(nt) - y\sin(nt) \\ \eta &= x\sin(nt) + y\cos(nt) \end{aligned}$$

Here $n$ is the mean motion of the Sun around the Earth which is related to the unit of distance $a$ by Kepler's third law $n^2 a^3 = k^2(m_1 + m_2)$, where $k$ is the gravitational constant. In this setting we are able to start with the equations of motion of the restricted three body problem

$$\begin{aligned} \ddot{x} - 2n\dot{y} &= \frac{\partial\Omega}{\partial x} \\ \ddot{y} + 2n\dot{x} &= \frac{\partial\Omega}{\partial y} \end{aligned} \tag{10.1}$$

4) See also the transformation (6.1).

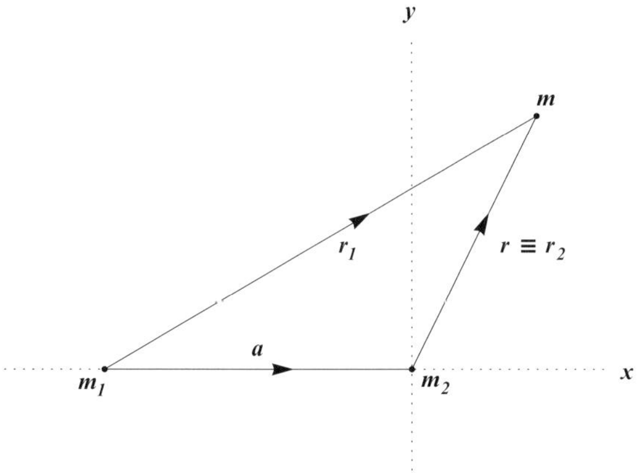


**Figure 10.1** Geometry of the Hill's problem. The axes $(x, y)$ rotate with the mean motion of $m_2$ around $m_1$. In addition to the geometry of the restricted three body problem the origin is set to coincide with the location of $m_2$.

where the potential $\Omega$ is defined as:

$$\Omega(x, y) = U + \frac{n^2}{2}\left(x^2 + y^2\right)$$

with the potential $U = U(x, y)$ defined by:

$$U(x, y) = k^2\left(\frac{m_1}{r_1} + \frac{m_2}{r_2}\right)$$

Here $r_1, r_2$ are the distances from $m_1, m_2$, respectively (see Figure 10.1). Under the above assumptions the potential $\Omega = \Omega(x, y)$ takes the form (compare also with (6.3)):

$$\Omega(x, y) = \frac{n^2}{2}\left[\left(x + \frac{m_1}{m_1 + m_2}a\right)^2 + y^2\right] + \frac{m_1}{r_1} + \frac{m_2}{r_2}$$

with

$$r_1^2 = (x + a)^2 + y^2 = a^2\left[1 + 2\frac{x}{a} + \left(\frac{r}{a}\right)^2\right]$$
$$r^2 \equiv r_2^2 = x^2 + y^2.$$

We aim to find a unit of time such that $n = 1$ and one revolution period of $m_1$ around $m_2$ takes $T = 2\pi$. In contrast to the classical restricted problem we set the mass of the perturber $m_2 = 1$ and choose the unit of distance $a^3 = m_1 + m_2 = 1 + m_1$ such that $k = 1$. In case of the Sun–Earth–Moon system $a^3 = 330\,000$ and thus $a \simeq 69$ and $r \simeq 69/390 = 0.177$ since the distance Sun–Earth (equal 69) is about 390 Sun–Moon distances.

The potential reads:

$$\Omega(x,y) = \frac{1}{2}\left[\left(x + \frac{m_1}{a^2}\right)^2 + y^2\right] + \frac{m_1}{r_1} + \frac{1}{r}$$
$$= m_1\left(\frac{1}{r_1} + \frac{x}{a^2}\right) + \frac{1}{r} + \frac{r^2}{2} + \frac{m_1{}^2}{2a^4}$$

and since the Taylor series expansion of $a/r_1$ is given by:

$$\frac{a}{r_1} = \left(1 + 2\frac{x}{a} + \frac{r^2}{a^2}\right)^{-\frac{1}{2}} = 1 - \frac{x}{a} - \frac{r^2}{2a^2} + \frac{3}{2}\left(\frac{x}{a}\right)^2 + \ldots$$

we find in powers of $a/r_1$:

$$\Omega(x,y) = \frac{m_1}{a}\left(1 - \frac{x}{a} - \frac{r^2}{2a^2} + \frac{3}{2}\frac{x^2}{a^2} + \ldots\right) + m_1\frac{x}{a^2} + \frac{1}{r} + \frac{r^2}{2} + \frac{m_1{}^2}{2a^4}$$

If we moreover neglect the constant terms which do not contribute to the partial derivatives in (10.1) we have:

$$\Omega(x,y) = \frac{3}{2}m_1\frac{x^2}{a^3} + \frac{r^2}{2}\left(1 - \frac{m_1}{a^3}\right) + \frac{1}{r} + \ldots$$
$$= \frac{x^2}{2a^3}\left(2m_1 + a^3\right) + \frac{y^2}{2}\left(1 - \frac{m_1}{a^3}\right) + \frac{1}{r} + \ldots$$

Since $1 + m_1 = a^3$ we also have $2m_1 + a^3 = 3a^2 - 2$ and $1 - m_1 a^{-3} = a^{-3}$ and we get:

$$\Omega(x,y) = \frac{3}{2}x^2 - \frac{2x^2 - y^2}{2a^3} + \frac{1}{r} + \ldots$$

In the above function we have $x, y, r, m_1 a^{-3}$ of order zero but $a^{-1}$ of order one thus terms proportional to $a^{-3}$ are of order three. In a first approximation, may neglect the term proportional to $a^{-3}$. We finally get the approximate potential of Hill's lunar problem:

$$\Omega(x,y) = \frac{3}{2}x^2 + \frac{1}{r} \tag{10.2}$$

The equations of motion reduce to

$$\begin{aligned} \ddot{x} - 2\dot{y} &= \frac{\partial\Omega}{\partial x} = \left(3 - \frac{1}{r^3}\right)x \\ \ddot{y} + 2\dot{x} &= \frac{\partial\Omega}{\partial y} = -\frac{y}{r^3} \end{aligned} \tag{10.3}$$

with

$$r^2 = x^2 + y^2$$

The structure of the potential $\Omega(x, y)$ is shown in Figure 10.2. To be more precise, we show the equipotential lines as the borders between different colors together with Hill's version of the zero velocity curves: if we multiply the first of (10.3) with $\dot{x}$ the second with $\dot{y}$ we get by integrating the sum

$$C(x, y, \dot{x}, \dot{y}) = \frac{2}{r} + 3x^2 - \left(\dot{x}^2 + \dot{y}^2\right) = \text{const} \tag{10.4}$$

which is the Jacobian constant of the motion of the Hill's Lunar problem. From the Jacobi constant (10.4) we can deduce the zero velocity or Hill's curve by setting the sum of the square of the velocities equal zero:

$$r = x^2 + y^2 = \frac{2}{C - 3x^2}$$

Since $r > 0$ we find $x^2 < C/3$ and the boundary situated along the lines $x = \pm\sqrt{C/3}$. Since $r$ does depend on $x^2$ and $y^2$ only it is symmetric with respect to the $x$- and $y$-axes too. The intersection of the curves with the $y$-axis is situated at $y_{1,2} = \pm 2/C$ the intersection with the $x$-axis is determined by the cubic equations:

$$x^3 - \frac{1}{3}Cx + \frac{2}{3} = 0$$

The only solution of interest for $x$ is given by the formula

$$x = \sqrt[3]{-\left(\frac{1}{3} - \sqrt{D}\right)} + \sqrt[3]{-\left(\frac{1}{3} + \sqrt{D}\right)} \quad \text{with } D = \frac{1}{9}\left(1 - \frac{C^3}{81}\right)$$

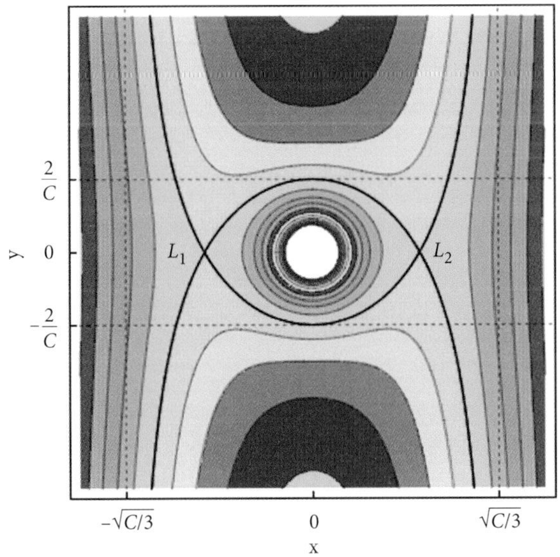


**Figure 10.2** Potential $\Omega(x, y)$ in Hill's problem. The borders between different colors define the curves of the equipotential. The geometry is symmetric with respect to the $x = 0$ and $y = 0$ axes.

which gives from $D > 0$ the condition $C^3 < 81$ or

$$C < 3\sqrt[3]{3}$$

Thus for $C < 3\sqrt[3]{3}$ we get no intersection with $y = 0$ and from the limiting case $C = 3\sqrt[3]{3}$ we get from

$$x^3 - 3x + 2 = (x - 1)^2(x + 2)^2 = 0$$

that $y = 0$ for $x_{1,2} = \pm 3^{-1/3}$. It can be shown that $x_{1,2}$ correspond to the Lagrange points $L_1$ and $L_2$ of the standard restricted problem. See also again Figure 10.2, where we indicate different values of $C$ and show the limiting case $C = 3 \cdot 3^{1/3}$ marked by the black line. We also provide two cases of typical motion found in the Hill's problem in Figure 10.3. In the rotating coordinate system an orbit which reaches a zero-velocity curve follows into the opposite direction until it reaches its zero-velocity curve again. If the curve is closed motion is bounded forever (see Figure 10.3a) if the curve is open it still restricts the area of allowed motion but the orbit may eventually leave the system (see Figure 10.3b. See also Figures 6.7–6.10 for comparison.

Following the description of [138] the formal approximations done in this section are consistent with the assumptions at the beginning in the following way: we fix the location of the Sun, $(x_S, y_S)$ at $y_S = 0$ and let $x_S \to -\infty$ but increase its mass $m_1$ such that $m_1 a^{-3} \simeq (m_1 + 1)\, a^{-3}$ for $a \gg 1$. With this we ensure that $a^3 = 1 + m_1$ with $n = 1$ for $a \to \infty$. Thus the perturbation of the motion of the Moon due to the Sun is replaced by a mean perturbation, which is still proportional to the mass $m_1$ and $r_1^{-3}$ but since we assume $a \simeq r_1$ and moreover neglect the term proportional to $a^{-3}$ the mass $m_1$ and distance $r_1$ disappear from (10.3) completely. This is equivalent to the assumption 3), that is the fact that the perturbation due to the Sun is fixed and does not depend on the orbital phase of the Moon anymore.

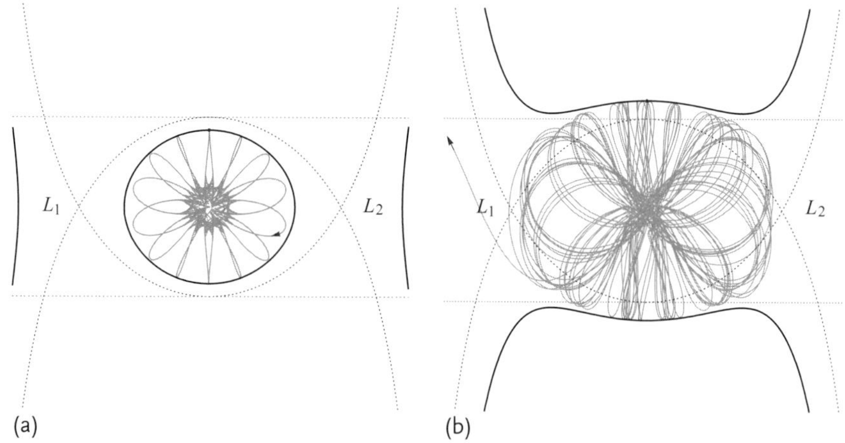


**Figure 10.3** Typical motion found in the Hill's problem. (a) motion is trapped within the zero velocity curve indicated by the solid black line. (b) motion is still restricted by the zero-velocity curves but escapes since the curves are not closed.

### 10.1.1 Periodic Motion

The form of Hill's equations have the advantage that orbits are congruent with respect to the axes $x = 0$ and $y = 0$. As a consequence the solution of the problem with given period $T = 2\pi/\nu = 2\pi m$ has the very simple form

$$\begin{aligned} x(t) &= A_1 \cos(\nu t) + A_3 \cos(3\nu t) + A_5 \cos(5\nu t) + \ldots \\ y(t) &= B_1 \sin(\nu t) + B_3 \sin(3\nu t) + B_5 \sin(5\nu t) + \ldots \end{aligned} \tag{10.5}$$

where $A_j, B_j$ with $j = 1, 3, 5, \ldots$ are the amplitudes of the system. For satellites with small periods $T$ we have small $m^{-1} = \nu$ and the solution (10.5) reduces to

$$x(t) = R\cos(\nu t)\,, \quad y(t) = R\sin(\nu t)$$

Since

$$r(t) = \sqrt{x(t)^2 + y(t)^2} = R_0 + R_2 \cos(2\nu t) + \ldots$$

and for $m \to 0$ we find from Kepler's third law

$$r(t) = R_0 = a = (\nu + 1)^{-\frac{2}{3}} = m^{\frac{2}{3}}\left(1 - \frac{2}{3}m + \ldots\right)$$

The idea of Hill is to simplify (10.3) by making use of the presence of the Jacobi integral. If one multiplies (10.3) by $-yx$, respective $yx$ and by summation we have:

$$\begin{aligned} &x\ddot{y} - y\ddot{x} + 2\,(x\dot{x} + y\dot{y}) + 3xy = 0 \\ &x\ddot{x} + y\ddot{y} - 2\,(x\dot{y} - y\dot{x}) - 3x^2 + \frac{1}{r} = 0 \end{aligned}$$

By making use of (10.4) one eliminates $1/r$ from the second to get instead:

$$x\ddot{x} + y\ddot{y} - 2\,(x\dot{y} - y\dot{x}) + \frac{1}{2}\left(\dot{x}^2 + \dot{y}^2\right) - \frac{9}{2}x^2 + \frac{1}{2}C = 0$$

The final system is of second order and does not depend on $r$ anymore. Furthermore, Hill introduces complex conjugated variables

$$s = x - iy\,, \quad u = x + iy$$

which transforms the system into

$$\begin{aligned} u\ddot{s} - s\ddot{u} - 2i\,(u\dot{s} + s\dot{u}) + \frac{3}{2}\left(s^2 - u^2\right) = 0 \\ u\ddot{s} + s\ddot{u} - 2i\,(u\dot{s} - s\dot{u}) + \dot{u}\dot{s} - \frac{9}{4}(s+u)^2 + C = 0 \end{aligned}$$

The system becomes even more symmetric by adding and subtracting half of its sum to get:

$$u\ddot{s} - 2iu\dot{s} + \frac{1}{2}\dot{u}\dot{s} - \frac{9}{4}us = \frac{3}{8}\left(s^2 + 5u^2\right) - \frac{C}{2}$$
$$s\ddot{u} + 2is\dot{u} + \frac{1}{2}\dot{s}\dot{u} - \frac{9}{4}su = \frac{3}{8}\left(u^2 + 5s^2\right) - \frac{C}{2}$$

To remove $i$ from the preceding equations one replaces the time $t$ by the relation $\zeta = e^{i\nu t}$. If we define the differential operator:

$$D_f \equiv D(f) = \zeta \frac{df}{d\zeta}$$

we get

$$\dot{u} = \frac{du}{d\zeta}\dot{\zeta} = i\nu\zeta\frac{du}{d\zeta} = i\nu D_u$$
$$\ddot{u} = i\nu D\left(i\nu D_u\right) = -\nu^2 D_{u,u}$$

since $\dot{\zeta} = i\nu\zeta$ and where we used the notation $D_{u,u} = D^2(u) = D(D(u))$. In this setting the system of equations becomes:

$$-\nu^2 u D_{s,s} + 2\nu u D_s - \frac{1}{2}\nu^2 D_u D_s - \frac{9}{4}us = \frac{3}{8}\left(s^2 + 5u^2\right) - \frac{C}{2}$$
$$-\nu^2 s D_{u,u} - 2\nu s D_u - \frac{1}{2}\nu^2 D_s D_u - \frac{9}{4}su = \frac{3}{8}\left(u^2 + 5s^2\right) - \frac{C}{2}$$

which we multiply by $-\nu^{-2} = -m^2$ to get:

$$\begin{aligned} u D_{s,s} - 2mu D_s + \frac{1}{2}D_u D_s + \frac{9}{4}m^2 us &= -\frac{3}{8}m^2\left(s^2 + 5u^2\right) + a_0^2 K \\ s D_{u,u} + 2ms D_u + \frac{1}{2}D_s D_u + \frac{9}{4}m^2 su &= -\frac{3}{8}m^2\left(u^2 + 5s^2\right) + a_0^2 K \end{aligned} \qquad (10.6)$$

where we introduced a new constant $K$ and a scaling factor $a_0$ related to the Jacobi constant $C$ by:

$$m^2 C = 2a_0 K$$

The unknown functions $s, u$ are determined by (10.6) in terms of the complex time $\zeta$ and depending on the parameters $m, a_0$ and $K$. Setting $k = 2n + 1$ they are related to the Fourier modes (10.5) by the simple relation

$$\cos(k\nu t) = \frac{1}{2}\left[\left(e^{ki\nu t}\right) + \left(e^{-ki\nu t}\right)\right] = \frac{1}{2}\left(\zeta^k + \zeta^{-k}\right)$$
$$\sin(k\nu t) = \frac{1}{2i}\left[\left(e^{ki\nu t}\right) - \left(e^{-ki\nu t}\right)\right] = -\frac{i}{2}\left(\zeta^k - \zeta^{-k}\right)$$

From

$$u = x + iy = \frac{1}{2}\left[\sum_{n\geq 0}(A_k + B_k)\,\zeta^k + \sum_{n\geq 0}(A_k - B_k)\,\zeta^{-k}\right]$$

(and similar for $s = x - iy$) by setting

$$a_n = \frac{1}{2}(A_k + B_k) , \quad a_{-k} = \frac{1}{2}(A_k - B_k) \tag{10.7}$$

we get

$$u = \zeta U , \quad s = \zeta^{-1} S$$

with

$$U = \sum_{n=-\infty}^{\infty} a_n \zeta^{2n} , \quad S = \sum_{n=-\infty}^{\infty} a_{-n} \zeta^{2n}$$

and

$$A_{2n+1} = a_n + a_{-n-1} , \quad B_{2n+1} = a_n - a_{-n-1}$$

For $m \to 0$ we find $A_1, B_1 \to a$ and from

$$a_0 = \frac{1}{2}(A_1 + B_1)$$

also $a_0 \to a$. If we set $a_0 = 1$ in (10.6) we are able to express $a_n$ in units of $a_0$. If we moreover (this goes back to Poincaré) replace on the left hand sides of (10.6) $m$ with $n$ (which will later be replaced back) but leave on the right hand side $m^2$ we deduce the following form of the unknown $U, S$:

$$U = \sum_{j\geq 0} u_{2j} m^{2j} , \quad S = \sum_{j\geq 0} s_{2j} m^{2j}$$

and also $K, C$:

$$K = \sum_{j\geq 0} k_{2j} m^{2j} , \quad C = \sum_{j\geq 0} c_{2j} m^{2j}$$

In the next step we want to express (10.6) in terms of $U, S$ rather than in terms of $u, s$. Since

$$D\zeta^n = \zeta\left(\frac{d\zeta^n}{d\zeta}\right) = n\zeta^n$$

we find

$$D_u = D(\zeta U) = \zeta(D_U + U) , \quad D_{u,u} = \zeta(D_{U,U} + 2D_U + U)$$
$$D_s = D(\zeta^{-1} S) = \zeta^{-1}(D_S - S) , \quad D_{s,s} = \zeta^{-1}(D_{S,S} - 2D_S + S)$$

and

$$S D_{U,U} + f S D_U + \frac{1}{2}(D_U + U) S_S + g U S = -\frac{3}{8} m^2 (U^2\zeta^2 + 5S^2\zeta^{-2}) + K$$
$$U D_{S,S} - f U D_S + \frac{1}{2}(D_S - S) D_U + g S U = -\frac{3}{8} m^2 (S^2\zeta^{-2} + 5U^2\zeta^2) + K \tag{10.8}$$

with the constants defined:

$$f = \frac{3}{2} + 2p\,, \quad g = \frac{1}{2} + 2p + \frac{9}{4}p^2$$

Moreover we have

$$D_U = m^2 D_{u_1} + m^4 D_{u_2} + \dots\,, \qquad D_{U,U} = m^2 D_{u_1,u_1} + m^4 D_{u_2,u_2} + \dots$$

(and similar for $D_S$ and $D_{S,S}$). Since $a_0 = 1$ we have

$$u_0 = s_0 = 1$$

and moreover $D_{u_0} = D_{s_0} = 0$. We find at zeroth order ($m = 0$) from (10.8):

$$k_0 = g = \frac{1}{2} + 2p + \frac{9}{4}p^2$$

A comparison of terms in (10.8) of order $m^2$ gives:

$$D_{u_1,u_1} + f D_{u_1} + \frac{1}{2} D_{s_1} + g(u_1 + s_1) = -\frac{3}{8}\left(\zeta^2 + 5\zeta^{-2}\right) + k_1$$
$$D_{s_1,s_1} - f D_{s_1} - \frac{1}{2} D_{u_1} + g(s_1 + u_1) = -\frac{3}{8}\left(5\zeta^2 + \zeta^{-2}\right) + k_1$$

It can be shown that the first of the above follows from the second if one replaces $\zeta$ by $\zeta^{-1}$ since $U, S$ (and $u, s$ and also $u_j, s_j$ with $j = 0, 2, \dots$) are complex conjugated functions:

$$u_1 = \alpha_2\zeta^2 + \alpha_{-2}\zeta^{-2}\,, \quad s_1 = \alpha_{-2}\zeta^2 + \alpha_2\zeta^{-2}$$

If we compare equal orders in $\zeta, \zeta^{-1}$ we find the relations:

$$\alpha_2(4 + 2f + g) + \alpha_{-2}(1 + g) = -\frac{3}{8}$$
$$-\alpha_2(1 - g) + \alpha_{-2}(4 - 2f + g) = -\frac{15}{8}$$

with the solution

$$\alpha_2 = \frac{3}{8}\frac{1 + 2f + 4g}{15 - 4f^2 + 8g} = \frac{9}{16}\frac{2 + 4m + 3m^2}{6 - 4m + m^2} = \frac{3}{16} + \frac{1}{2}m + \frac{7}{12}m^2 + \dots$$
$$\alpha_{-2} = -\frac{3}{8}\frac{21 + 2f + 4g}{15 - 4f^2 + 8g} = -\frac{3}{16}\frac{38 + 28m + 9m^2}{6 - 4m + m^2} = -\frac{19}{16} - \frac{5}{3}m - \frac{43}{36}m^2 + \dots$$

The method can be generalized to higher orders in $m$ in a straightforward way. It can be shown that at order $m^{2n}$ we get a linear differential equation of the form

$$D_{u_n,u_n} + f D_{u_n} + \frac{1}{2} D_{s_n} + g(u_n + s_n) = \Phi_n + k_n$$

where $\Phi_n = \Phi_n\left(\zeta, \zeta^{-1}\right)$ is a known function depending on

$$\zeta^{2n}, \zeta^{2n-4}, \ldots, \zeta^{-2n+4}, \zeta^{-2n}$$

If we set $\alpha_2^{(1)} = \alpha_2$ and $\alpha_{-2}^{(2)} = \alpha_{-2}$ and moreover assume

$$\begin{aligned}
u_1 &= \alpha_2^{(1)}\zeta^2 + \alpha_{-2}^{(1)}\zeta^{-2}, \quad & s_1 &= \alpha_{-2}^{(1)}\zeta^2 + \alpha_2^{(1)}\zeta^{-2} \\
u_2 &= \alpha_4^{(2)}\zeta^4 + \alpha_0^{(2)} + \alpha_{-4}^{(2)}\zeta^{-4}, \quad & s_2 &= \alpha_{-4}^{(2)}\zeta^4 + \alpha_0^{(2)} + \alpha_4^{(2)}\zeta^{-4} \\
&\vdots
\end{aligned}$$

we get a system of linear equations which determine the $\alpha_j^{(n)}$ and therefore $u_n, s_n$ with $n \geq 1$. From

$$U = \sum_{n=-\infty}^{\infty} a_n \zeta^{2n} = \sum_{j \geq 0} u_{2j} m^{2j}$$

(and similar for $S$) one obtains $a_n$ in terms of $\alpha_j^{(n)}$ by comparing equal orders in $m$. Up to $m^4$ we get:

$$\begin{aligned}
u_1 &= \frac{3}{16}m^2 + \frac{1}{2}m^3 + \frac{7}{12}m^4 + \ldots \\
a_{-1} &= -\frac{19}{16}m^2 - \frac{5}{3}m^3 - \frac{43}{36}m^4 + \ldots \\
a_2 &= \frac{25}{256}m^4 \\
a_{-2} &= 0
\end{aligned}$$

and from

$$\begin{aligned}
K &= \alpha_0^{(0)} + \alpha_0^{(2)} m^4 + \ldots \\
&= \frac{1}{2}\left(1 + 4m + \frac{9}{2}m^2 - \frac{1147}{128}m^4 + \ldots\right)
\end{aligned}$$

as well as

$$C = \frac{2K}{m^2} = m^{-2}\left(1 + 4m + \frac{9}{2}m^2 - \frac{1147}{128}m^4 + \ldots\right)$$

We obtain the scaling factor $\sigma \equiv a_0$ (which we set to unity in the above calculations) from (10.3). Denoting by $x', y', r'$ the solution found for $a_0 = 1$ we need to set $x = \sigma x'$, $y = \sigma y'$, $r = \sigma r'$ to obtain the absolute one in terms of $x, y, r$. It follows for all times:

$$\begin{aligned}
\sigma^3\left(\ddot{x}' - 2\dot{y}' - 3x'\right) &= -\frac{x'}{r'^3} \\
\sigma^3\left(\ddot{y}' + 2\dot{x}'\right) &= -\frac{y'}{r'^3}
\end{aligned}$$

and thus

$$\sigma^3 = \frac{x'}{r'^3\,(3x' + 2\dot{y}' - \ddot{x}')} = -\frac{y'}{r'^3\,(2\dot{x}' + \ddot{y}')}$$

Since $a_0 \to m^{2/3}$ for close orbits it can be expanded in terms of $m$:

$$\sigma = m^{\frac{2}{3}}\,(1 + \sigma_1 m + \sigma_2 m^2 + \ldots)$$

The value of $\sigma = a_0$ is independent of time $t$. Setting $t = 0$ we find $y'(0) = \dot{x}'(0) = \ddot{y}'(0) = 0$ and $r'(0) = x'$ which simplifies the equation. Moreover we have $\nu = m^{-1}$, $\cos(2n+1)\nu t = 1$ and therefore

$$\begin{aligned} x'(0) &= A_1 + A_3 + \ldots = 1 + (a_1 + a_{-1}) + (a_2 + a_{-2}) + \ldots \\ m\dot{y}'(0) &= B_1 + 3B_3 + 5B_5 + \ldots = 1 + (3a_1 - a_{-1}) + (5a_2 - 3a_{-2}) + \ldots \\ -m^2\ddot{x}'(0) &= A_1 + 9A_3 + 25A_5 + \ldots = 1 + (9a_1 + a_{-1}) + (25a_2 + 9a_{-2}) + \ldots \end{aligned}$$

and from

$$\begin{aligned} \frac{m^2}{\sigma^3} &= x'(0)^2\left\{-m^2\ddot{x}'(0) + 2m\left[m\dot{y}'(0)\right] + 3m^2x'(0)\right\} \\ &= \left(1 + \sigma_1 m + \sigma_2 m^2 + \ldots\right)^{-3} \end{aligned}$$

Since we already know $a_n$ and inserting in the above equation $x'(0)$, $y'(0)$ and its derivatives we get order by order $\sigma_j$ with $j = 1, \ldots$ which reads:

$$\sigma = m^{\frac{2}{3}}\left(1 - \frac{2}{3}m + \frac{7}{18}m^2 - \frac{4}{81}m^3 + \frac{19\,565}{62\,208}m^4 + \ldots\right)$$

We provide the geometry of periodic orbits with $\nu = 1$ and different $m$ in Figure 10.4. The plot shows the geometry of periodic orbits for $m = 0.08$ (our Moon), $m = 0.25$, $m = 0.33$ and $m = 0.56$, respectively.

Hill's Lunar theory is an important step towards more sophisticated models of the Lunar problem. See also [138] and [233] for further reading.

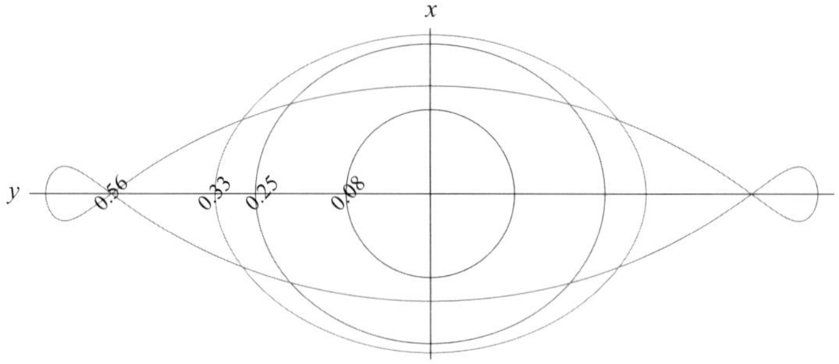


**Figure 10.4** Periodic orbits for $\nu = 1$ and $m$ equal 0.08 (Moon), 0.25, 0.33 and 0.056, respectively.

## 10.2 Classical Lunar Theory

There are two perturbations caused by the planets on the Moon's orbit (see [143, 181]): the *direct planetary perturbations* and the *indirect planetary perturbations*. The second ones are caused by the orbit of the Earth which is not a perfect ellipse. These perturbations are stronger than the direct ones. Due to to the fact that both are significantly smaller than the perturbations of the Sun the first step in a Lunar theory is 'solving' the so-called 'main problem'. In this approximation the planets are ignored and one only solves the problem of the Moon orbiting the Earth perturbed by the Sun. The model is such that (i) all three celestial bodies involved are point masses and (ii) the orbit of the Earth is an unperturbed ellipse around the Sun. In a further step to study the principal inequalities one makes the following simplifications which we will use in the following to derive the principal inequalities:

1. The orbit of the Earth is a circle ($e_{\text{Earth}} \sim 0.016$)
2. Because of the smallness of the Moon's inclination $i \sim 5°$ with respect to the ecliptic high-order terms (from $i^4$ on) will be neglected
3. Because of the smallness of the eccentricity of the Moon's orbit ($e = 0.054$) powers in $e$ from 4 on are neglected

Note that in the following primed quantities are the ones for the Sun. For the perturbing function developed into Legendre polynomials (see (8.67)) for a Lunar theory we need to take only the first term of which is

$$F_1 = \frac{k^2 m'}{r'^3} r^2 \left( -\frac{1}{2} + \frac{3}{2} \cos^2 \phi \right) \tag{10.9}$$

We use Figure 10.5, where the geometry is depicted: the Earth is in the center, its orbit around the Sun is taken as the plane of reference, and $m'$ is now the Sun. We use the same notation for the elements and just replace in (10.9) $\phi^2$ for which we already derived an expression in (8.70). We now get for the motion of the Moon

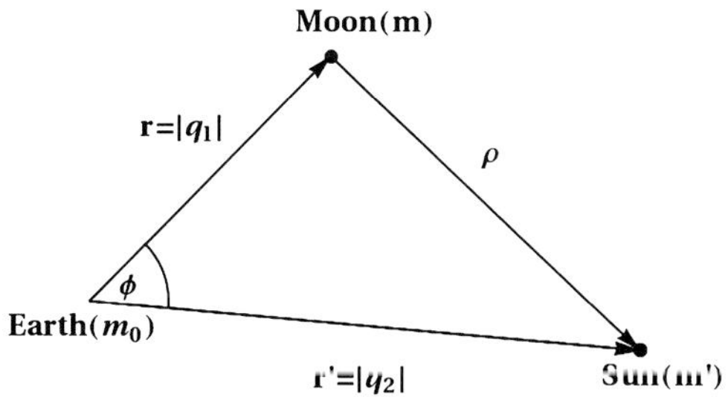


**Figure 10.5** Geometry of the Lunar problem.

perturbed by the Sun (see Figure 8.3)

$$F = \frac{k^2 m'}{r'^3} r^2 \left[ -\frac{1}{2} + \frac{3}{2} \cos^4 \left( \frac{i}{2} \right) \cos^2 (\psi - \psi') \right.$$
$$+3 \cos^2 \frac{i}{2} \cdot \sin^2 \frac{i}{2} \cos (\psi - \psi') \cos (\psi + \psi' - 2\Omega)$$
$$\left. + \frac{3}{2} \sin^4 \frac{i}{2} \cos^2 (\psi + \psi' - 2\Omega) \right] \tag{10.10}$$

We replace the quantities $\psi$ and $\psi'$ and introduce according to the $3^{rd}$ Kepler law $k^2 m' = n'^2 a'^3$

$$F = n'^2 a^2 \left( \frac{a'}{r'} \right)^3 \left( \frac{r}{a} \right)^2 \left\{ -\frac{1}{2} + \frac{3}{4} \cos^4 \frac{i}{2} \cdot [1 + \cos 2 (\omega - \omega' + v - v' + \Omega)] \right.$$
$$+ \frac{3}{8} \sin^2 i \left[ \cos (2\omega + 2v) + \cos (2\omega' + 2v' - 2\Omega) \right]$$
$$\left. + \frac{3}{4} \sin^4 \frac{i}{2} \left[ 1 + \cos 2 (\omega + \omega' + v + v' - \Omega) \right] \right\} \tag{10.11}$$

Taking care of the simplifications we made in point (1) we can immediately replace the true anomaly $v$ of the Earth's orbit by its mean anomaly $M$. Because of simplification (2) for the Moon's eccentricity we neglect terms higher than $e'^2$ and finally because of (3) we may not take into full consideration the inclination of the Moon's orbit and neglect in the power series terms from $i'^4$ on. This mean for the development that we may replace $a'/r' = 1$, in addition we can write for $\sin i$ simply $i$ and as a consequence $\cos^2(i/2)$ by $1 - i^2$ which leads to the expression

$$F = n'^2 a^2 \left( \frac{r}{a} \right)^2 \left[ \frac{1}{4} - \frac{3}{8} e^2 + \frac{3}{4} \left( 1 - \frac{i^2}{2} \right) \cos 2 (\omega + \omega' + v - M' + \Omega) \right]$$
$$+ \frac{3i^2}{8} \left[ \cos (2\omega + 2v) + \cos (2\omega' + 2M' - 2\Omega) \right] \tag{10.12}$$

As next step we have to separate terms of the form

$$\left( \frac{r}{a} \right)^n \cos(mv) \quad \text{and} \quad \left( \frac{r}{a} \right)^n \sin(mv) \tag{10.13}$$

for $n = 0, 2$ and $m = 2$ using known trigonometric formulae. Finally for the Moon we use the following expansion of the perturbing function introducing $\alpha(p) = \cos(2\omega - 2\omega' + pM - 2M' + 2\Omega)$ for $p = 0, 4$ we get

$$F_1 = n'^2 a^2 \left[ \frac{1}{4} + \frac{3}{8} e^2 + \frac{e}{2} \cos M - \frac{e^2}{8} \cos 2M + \frac{15e^2}{8} \alpha(0) \right.$$
$$\left. - \frac{9}{4} e\alpha(1) + \left( \frac{3}{4} - \frac{15e^2}{8} \right) \alpha(2) + \frac{3}{4} e\alpha(3) + \frac{3}{4} e^2 \alpha(4) \right]$$
$$+ (n'ai)^2 \left[ -\frac{3}{8} - \frac{3}{16} \alpha(2) + \frac{3}{8} \cos 2(\omega + M) + \frac{3}{8} \cos 2 (\omega' + M' - \Omega) \right] \tag{10.14}$$

## 10.2.1
## Secular Part: Motion of the Nodes and the Perihelion

To derive the secular part of the Lagrange equations (we follow the books [181, 182]) we simply separate the nonperiodic terms of the perturbing function, which give an estimate for the secular motion of the Moon's node and its perihelion

$$F_{\text{sec}} = n'^2 a^2 \left( \frac{1}{4} + \frac{3}{8} e^2 - \frac{3}{8} i^2 \right) \tag{10.15}$$

To determine an estimate of the secular motion of the Moon's node and its perihelion in the approximated model we use the Lagrange equations (see (8.61)):

$$\begin{aligned} \frac{d\bar{\Omega}}{dt} &= \frac{1}{na^2\sqrt{1-e^2}\sin i} \frac{\partial F_{\text{sec}}}{\partial i} \\ \frac{d\bar{\omega}}{dt} &= \frac{\sqrt{1-e^2}}{na^2 e} \frac{\partial F_{\text{sec}}}{\partial e} - \frac{\cos i}{na^2\sqrt{1-e^2}\sin i} \frac{\partial F_{\text{sec}}}{\partial i} \end{aligned} \tag{10.16}$$

where we need to build the derivatives of the secular perturbation function.[5)]

1. For the motion of the node this is

$$\frac{n'^2 a^2}{na^2 i} \left( 1 + \frac{e^2}{2} \right) \left( -\frac{3i}{4} \right) = -\frac{3n'^2}{4n} \left( 1 + \frac{e^2}{2} \right) \tag{10.17}$$

2. For the motion of the perihelion this is

$$\frac{n'^2 a^2}{na^2 e} \frac{3e}{4} \left( 1 - \frac{e^2}{2} \right) + \frac{n'^2 a^2}{na^2 i} \frac{3i}{4} \left( 1 + \frac{e^2}{2} \right) = \frac{3n'^2}{2n} \tag{10.18}$$

With the introduction of the appropriate numerical values[6)] we find for the retrograde motion of the node $-20.^\circ16$ and for the perihelion $39.^\circ09$ which corresponds to periods of 17.86 and 9.21 years, which is within 5% close to the actual value of about 18.6 and 8.85 years.

Especially the motion of the nodes is quite important for eclipses of the Moon and the Sun. This period of successive passages of the Moon through the node is a *draconitic month*[7)] (27.212 days). It is evident that for an eclipse Sun, Moon and the Earth need to be aligned and therefore this can only be the case when the Moon is full or new. The Sun needs to be near the line of the nodes for an eclipse and therefore the passages of the Sun through the nodes are also essential for an eclipse

5) Note that we use the simple development up to the second order of $\sqrt{1 \pm e^2}$.
6) Mean value of the inclination $5^\circ.14$ varying between $4.^\circ99 < i < 5.^\circ30$; mean eccentricity 0.0549006 varying between $0.026 < e < 0.077$; $n' = 360^\circ$ and $n = 4812^\circ.7$.
7) The word stems from the idea that a dragon eats the Sun at a total eclipse.

to happen; this period is called an eclipse year (also draconitic year). There exists an interesting period of time of 6585 1/3 days which is called the *Saros cycle* consisting of multiples of the synodic month[8], the draconitic month and the draconitic year:

*19 draconitic years ~ 223 synodic month ~ 242 draconitic month ~ 6585 days*

One saros after an eclipse (Moon or Sun), the Sun, Earth, and Moon will have approximately the same geometry and consequently the new eclipse will almost be the same. This period was already known to Babylonian astronomers around 750 BC and served for the prediction of eclipses.

## 10.3 Principal Inequalities

To find out the periodic changes in the position we split into the true longitude and the latitude of the Moon's position (we follow [143]). From observations of the path of the Moon on the sky it was already clear to astronomers in ancient times that its motion is not uniform and does not only suffer from the before-mentioned secular changes of nodes and the perihelion of its orbit[9] but there are periodic changes visible primarely along and perpendicular to its orbit. It is therefore evident that the two values we are interested in are the true longitude of the Moon on the one hand and the latitude on the other hand. In principal this periodic changes can be regarded as the perturbations in polar coordinates, namely $\psi, \beta$ and $r$

$$\begin{aligned} \psi &= \Omega + \omega + v = \Omega + \omega + M + 2e \sin M + \frac{5}{4} e^2 \sin 2M \dots \\ \sin \beta &= \sin i \sin (\psi - \Omega) \\ r &= a\,(1 - e \cos M) \end{aligned} \tag{10.19}$$

One can see that the here as well as the retrograde motion of the node the motion of the perihelion is involved (just replace $M = \lambda - \tilde{\omega}$). To derive the true longitude $\psi$ we make use of the equation of the center and get for $\psi$, when we ignore from the third power in eccentricity on, and additional term $\delta\psi$. The following equations are a description of the elliptic orbit of the Moon[10] which is slightly deformed

$$\begin{aligned} \delta\psi &= \delta\lambda + 2\sin(\lambda - \tilde{\omega})\delta e - 2\cos(\lambda - \tilde{\omega})\, e\delta\tilde{\omega} \\ \delta r &= \delta a - a\cos(\lambda - \tilde{\omega})\,\delta e - a\sin(\lambda - \tilde{\omega})\, e\delta\tilde{\omega} \\ \delta\beta &= \sin(\lambda - \Omega)\delta\gamma - \cos(\lambda - \Omega)\gamma\delta\Omega + \gamma\cos(\lambda - \Omega)\delta\psi \end{aligned} \tag{10.20}$$

where $\gamma = \sin(i)$. To derive the most important inequalities we take care of the first periodic terms in the development of perturbation function up to the first order in $e$

8) From new Moon to the next new Moon.

9) Which are the most important quantities for the determination of the eclipses besides the one we are dealing with here.

10) Remember $e_{\mathrm{Moon}} \sim 0.055$.

(compare with (8.74), where these four periodic terms are shown)

$$F_{\rm i} = n'^2 a^2 \left[ \frac{3}{4} \cos 2\left(\lambda - \lambda'\right) - \frac{1}{2} e \cos(\lambda - \tilde{\omega}) - \frac{9}{4} e \cos\left(\lambda - 2\lambda' + \tilde{\omega}\right) + \frac{3}{4} e \cos\left(3\lambda - 2\lambda' - \tilde{\omega}\right) \right] \tag{10.21}$$

Using the Lagrange equations in the form taking into account the perturbed mean longitude, which is of the form

$$\lambda = \int n dt + \epsilon_1 \tag{10.22}$$

we find the equations of motion:

$$\frac{da}{dt} = \frac{2}{na} \frac{\partial F_i}{\partial \lambda}, \quad \frac{d\epsilon_1}{dt} = -\frac{2}{na} \frac{\partial F_i}{\partial a} \tag{10.23}$$

$$\frac{de}{dt} = -\frac{1}{na^2 e} \frac{\partial F_i}{\tilde{\omega}}, \quad \frac{d\tilde{\omega}}{dt} = \frac{1}{na^2 e} \frac{\partial F_i}{\partial e} \tag{10.24}$$

$$\frac{d\gamma}{dt} = -\frac{1}{na^2 \gamma} \frac{\partial F_i}{\partial \Omega}, \quad \frac{d\Omega}{dt} = \frac{1}{na^2 \gamma} \frac{\partial F_i}{\partial \gamma} \tag{10.25}$$

With $\epsilon_1$ defined above the function $F$ is considered as depending on the osculating elements $a$, $e$, $\gamma$, $\lambda$, $\tilde{\omega}$ and $\Omega$. In what concerns the perturbations $\delta\lambda$ we take care of (10.22) and (10.23) and we get

$$\delta\lambda = -\iint \frac{3}{a^2} \frac{\partial F}{\partial \lambda} dt^2 - \int \frac{2}{na} \frac{\partial F}{\partial a} dt = (\delta\lambda) + \delta\epsilon_1 \tag{10.26}$$

Here we need to proceed as we have done for the planetary theory of first order namely take $a$, $e$, $\gamma$ and also $n^2 = \kappa a^{-3}$ as constants. $\lambda$ on the contrary is a linear function of time as the two orbital elements $\tilde{\omega}$ and also $\Omega$.

### 10.3.1 The Variation

We examine the role of the periodic term

$$\frac{3}{4} n'^2 a^2 \cos\left(2\lambda - 2\lambda'\right) \tag{10.27}$$

Building the respective derivatives of the perturbing function $F$ in the Lagrange equations we get the following results:

$$\frac{da}{dt} = -3 \frac{n'2}{n} a \sin 2\left(\lambda - \lambda'\right)$$

$$\frac{d^2\delta\lambda}{dt^2} = n'^2 \frac{9}{2} a \sin 2\left(\lambda - \lambda'\right)$$

$$\frac{d\epsilon_1}{dt} = -3 \frac{n'2}{n} \cos 2\left(\lambda - \lambda'\right)$$

$$\frac{\delta a}{a} = 3 \frac{n'2}{n\left(2n - 2n'\right)} \cos 2\left(\lambda - \lambda'\right)$$

$$\delta(\lambda) = -\frac{9}{2}\frac{n'2}{2(2n-2n')}\sin 2(\lambda-\lambda')$$
$$\delta\epsilon_1 = -3\frac{n'2}{n(2n-2n')}\sin 2(\lambda-\lambda') \tag{10.28}$$

which simplifies introducing for $n'/n = \hat{n}$ and using for $(1-\hat{n})^{-1} = 1+\hat{n}+\hat{n}^2+\ldots$ – and ignoring high-order terms – to

$$\begin{aligned}\frac{\delta r}{a} = \frac{\delta a}{a} &= \frac{3}{2}\hat{n}^2\cos 2(\lambda-\lambda')\\ \delta\lambda &= -\frac{9}{8}\hat{n}^2\sin 2(\lambda-\lambda')\\ \delta\epsilon_1 &= -\frac{3}{2}\hat{n}^2\sin 2(\lambda-\lambda')\\ \delta\psi = \delta\lambda+\delta\epsilon_1 &= -\frac{21}{8}\hat{n}^2\sin 2(\lambda-\lambda')\end{aligned} \tag{10.29}$$

But there are other terms in the Lagrange equation to take into account[11]: for the third and fourth term in (10.21) we get

$$\begin{aligned}\frac{de}{dt} &= -\frac{9}{4}\frac{n'2}{n}\sin(\lambda-2\lambda'+\tilde{\omega}) - \frac{3}{4}\frac{n'2}{n}\sin(3\lambda-2\lambda'-\tilde{\omega})\\ e\frac{d\tilde{\omega}}{dt} &= -\frac{9}{4}\frac{n'2}{n}\cos(\lambda-2\lambda'+\tilde{\omega}) - \frac{3}{4}\frac{n'2}{n}\cos(3\lambda-2\lambda'-\tilde{\omega})\end{aligned} \tag{10.30}$$

and after integration the perturbation $\delta e$ reads

$$\delta e = \frac{9}{4}\frac{n'2}{n(n-2n')}\cos(\lambda-2\lambda'+\tilde{\omega}) + \frac{3}{4}\frac{n'2}{n(3n-2n')}\cos(3\lambda-2\lambda'-\tilde{\omega}) \tag{10.31}$$

which can be written with the introduction of $\hat{n}$, developing the denominator and retaining only up to $\hat{n}^2$

$$\delta e = \frac{9}{4}\hat{n}^2\cos(\lambda-2\lambda'+\tilde{\omega}) + \frac{1}{4}\hat{n}^2\cos(3\lambda-2\lambda'-\tilde{\omega}) \tag{10.32}$$

In the same way we get

$$e\delta\tilde{\omega} = -\frac{9}{4}\hat{n}^2\sin(\lambda-2\lambda'+\tilde{\omega}) + \frac{1}{4}\hat{n}^2\sin(3\lambda-2\lambda'-\tilde{\omega}) \tag{10.33}$$

For the first-order perturbation of the true longitude $\psi$ and the radius vector $r$ we get

$$\begin{aligned}\delta\psi &= 4\hat{n}^2\sin 2(\lambda-\lambda')\\ \frac{\delta r}{a} &= -\frac{5}{2}\hat{n}^2\cos 2(\lambda-\lambda')\end{aligned} \tag{10.34}$$

11) The second term in (10.21) can be treated in a similar way [143].

and consequently adding the perturbations from all three periodic terms of $F_i$ for the approximation of the variation of $\psi$ and $r$ we get

$$\begin{aligned}\delta\psi &= \frac{11}{8}\hat{n}^2 \sin 2(\lambda - \lambda') \\ \frac{\delta r}{a} &= -\hat{n}^2 \cos 2(\lambda - \lambda')\end{aligned} \tag{10.35}$$

Inserting the numerical value of $\hat{n}$[12] we derive for the amplitude of the longitude $\delta\psi$ a first approximation. To get a better approximation one would need to develop $(1-\hat{n})^{-1}$ to higher orders which leads to a better approximation, namely

$$\delta\psi = \left(\frac{11}{8}\hat{n}^2 + \frac{59}{12}\hat{n}^3 + \frac{893}{72}\hat{n}^4\right) \sin 2(\lambda - \lambda') \tag{10.36}$$

Still the numerical value is some 20% off the precise value. In fact powers in the eccentricities[13] at least up to the second order needs to be included which gives the amplitude of $\delta\psi$ with a precision of about 1%:

$$\delta\psi = \frac{75}{16}e^2\hat{n}^2 + \left(\frac{11}{8} + \frac{1101}{64}e^2\right)\hat{n}^2 + \frac{59}{12}\hat{n}^3 + \frac{893}{72}\hat{n}^4 \sin 2(\lambda - \lambda') \tag{10.37}$$

One should keep in mind that the value of more than half a degree (39′30″) is of the same size of the Moon itself, so why the greeks having observations of Ptolemy did not discover it? Given the period of the perturbations of half of the synodic Moon's period this periodic change was impossible to detect because during the New Moon and Full Moon these inequalities are not visible. Consequently observation of eclipses, which were the major source of their knowledge of the Moon's orbit on the sky could, not unveil it.

### 10.3.2
### The Evection

We have to deal with the term in $F_i$ in which the mean longitude and multiples of it are absent

$$\frac{15}{8}n'^2 a^2 e^2 \cos(2\lambda' - 2\tilde{\omega}) \tag{10.38}$$

We derive for the first-order perturbations in eccentricity $e$ and in longitude of the perihelion $\tilde{\omega}$ of the Moon. After inserting into the Lagrange equations

$$\frac{de}{dt} = -\frac{15}{4}\frac{n'^2}{n} e \sin(2\lambda' - 2\tilde{\omega}) \tag{10.39}$$

and after integration with respect to the time (ignoring the constant of integration) we get

$$\delta e = \frac{15}{4}\frac{n'^2 e}{n(2n - 2n)} \cos(2\lambda' - 2\tilde{\omega}) \tag{10.40}$$

12) $n'/n = 360/4812.7 = 0.0748$.
13) But in principles also in inclinations and in the ratio $a/a'$.

We repeat the calculation for $e\delta\tilde{\omega}$ but now with sin instead of cos to get

$$e\frac{d\tilde{\omega}}{dt} = \frac{15}{4}\frac{n'^2}{n} e \cos\left(2\lambda' - 2\tilde{\omega}\right) \tag{10.41}$$

and

$$e\delta\tilde{\omega} = \frac{15}{4}\frac{n'^2 e}{n(2n - 2n)} \sin\left(2\lambda' - 2\tilde{\omega}\right) \tag{10.42}$$

The amplitudes may be approximated by introducing again $\hat{n} = n'/n$ as $15/8(\hat{n}e)$ and consequently the perturbations on these two orbital elements of the Moon reduce to

$$\begin{aligned} \delta e &= \frac{15}{8}\hat{n}e \cos\left(2\lambda' - 2\tilde{\omega}\right) \\ e\delta\tilde{\omega} &= \frac{15}{8}\hat{n}e \sin\left(2\lambda' - 2\tilde{\omega}\right) \end{aligned} \tag{10.43}$$

Using now (10.20) and replacing the two expression derived above we can see that $\lambda$ is again introduced in the perturbation of the true longitude $\psi$ and the radial component of the Moon with the same amplitudes and periods

$$\begin{aligned} \delta\psi &= \frac{15}{8}\hat{n}e \sin\left(\lambda - 2\lambda' + \tilde{\omega}\right) \\ \frac{\delta r}{a} &= -\frac{15}{8}\hat{n}e \cos\left(\lambda - 2\lambda' + \tilde{\omega}\right) \end{aligned} \tag{10.44}$$

It turns out that this is the largest perturbations in longitude with an amplitude of 1.°27 (in our approximate computation) with a period of 31.27 days. But why is this term so important? The term it comes from does not include the mean longitude $\lambda$. When integrating (10.39) no divisor $n$ appears but rather $n'$ which cancels in part with the enumerator $n'^2$ and consequently that factor $n'/n = \hat{n}$ is present in the amplitude. This inequality was already known by Hipparchos because of such a big displacement of more than two and a half Moon's diameters in its orbit!

### 10.3.3
### Annual Equation, Parallactic Inequality and Principal Perturbation in Latitude

Taking the term

$$\frac{3}{4}n'a^2e' \cos\left(\lambda' - \tilde{\omega}'\right) \tag{10.45}$$

in $F$, and building the derivative gives the following term in $\epsilon_1$:

$$\frac{d\epsilon_1}{dt} = -3\frac{n'^2}{n}e' \cos\left(\lambda' - \tilde{\omega}'\right) \tag{10.46}$$

which after integration can be written as

$$\delta\epsilon_1 = -3\hat{n}e' \sin\left(\lambda' - \tilde{\omega}'\right) \tag{10.47}$$

The same terms arise in the perturbation of the longitude $\delta\psi$. Because in the argument of the sine above there is only $M'$ the period is just one year. Note that for a more precise derivation we need to take care of the terms

$$n'^2 a^2 \left[ -\frac{3}{4} ee' \cos\left(\lambda - \lambda' - \tilde{\omega} + \tilde{\omega}'\right) + \frac{9}{8} ee' \cos\left(\lambda - \lambda' + \tilde{\omega} - \tilde{\omega}'\right) \right] \tag{10.48}$$

In the perturbations $\delta\psi$ and $\delta r/a$ thcy arise in higher orders factors $\hat{n}$, which we don't develop here. The correct Amplitude $A_{\text{ann}}$ up to higher orders is

$$-3e'\hat{n} + \frac{735}{16} e'\hat{n}^3 + \frac{1261}{4} \hat{n}^4 + \ldots \tag{10.49}$$

and gives an exact value of 0.°186 whereas in our – evidently very 'bad' – approximation we get −0.°215 with a negative sign!

Finally we mention the parallactic inequality which is a small perturbation in the Moon's longitude of only 0.°03 with a period of the mean synodic month (29 days). There exist many more of these inequalities, all of them significantly smaller than the one mentioned before.

An exception is the principal perturbation in latitude which stems from the term

$$\frac{3}{8} n'^2 a^2 \gamma^2 \cos\left(2\lambda' - 2\Omega\right) \tag{10.50}$$

which gives after inserting in the Lagrange equations

$$\begin{aligned} \frac{d\gamma}{dt} &= -\frac{3}{4}\frac{n'2}{n} \gamma \sin\left(2\lambda' - 2\Omega\right) \\ \gamma \frac{d\Omega}{dt} &= \frac{3}{4}\frac{n'2}{n} \gamma \cos\left(2\lambda' - 2\Omega\right) \end{aligned} \tag{10.51}$$

and after integration for the first-order perturbation

$$\begin{aligned} \delta\gamma &= \frac{3}{8} \hat{n}\gamma \cos\left(2\lambda' - 2\Omega\right) \\ \gamma\,\delta\Omega &= \frac{3}{8} \hat{n}\gamma \sin\left(2\lambda' - 2\Omega\right) \end{aligned} \tag{10.52}$$

which at the end using (10.20) gives for the latitude $\beta$

$$\delta\beta = \frac{3}{8} \hat{n}\gamma \sin(\lambda - 2\lambda' + \Omega)$$

From this we estimate an amplitude of $A = 0.°143$ (the correct value after the inclusion of high-order terms is – according to Brown's theory – $A = 0.°173$). The corresponding period easily is easily determined from $365.25/(n - 2n') = 32.13$ days. In addition to the aforementioned perturbations there are a total of 13 inequalities in longitude present with amplitudes $A > 100'$ and 46 with amplitudes $1' < A < 100'$.

**Modern Lunar Ephemerides – the ELP2000 solution** The most modern theory of the orbit and rotation of the Moon nowadays is the ELP2000 lunar theory (see [230, 231, 234–237]). The semianalytical theory of the main lunar problem based on Lie-transforms, derived in [238] was compared with a seminumerical one in [239] and with a numerical one in [240] and showed good agreement. The influence of the planetary perturbations (after Bretagnon, [241]) were analyzed in great detail in [242, 243]. Also the relativistic effects of the perturbations were taken into account and investigated in [244]. The analytical aspects of the theory of the motion of the moon was deeply investigated in [245, 246], first purely numerical expressions for the precessions of the mean orbital elements of the moon and the planets were derived in [247]. The outcome of the ELP lunar theory was also compared with the results of observations in [248, 249]. Please see the cited references for further information on the topic.

# 11
# Concluding Remarks

In the present book we have collected many different topics, which have a strong relationship with Celestial Mechanics and Dynamical Systems theory. The choice of them, of course, was motivated not only by our personal interest but also because of the scientific background of our own research work, and was never meant to be complete. As the reader may have noticed important things have been left out: there is nothing said about the important role of tides; the problem of orbit determination and improvement of orbital elements of comets or asteroids (e. g. Near Earth Asteroids) is not even mentioned, the newly discovered extrasolar planets – which are important with respect to the dynamical development – are not discussed, satellite theory, rotational dynamics (like the spin-orbit problem, which is crucial to understand the rotation of the Moon), and the theory of the potential of celestial bodies in general is also not present. Celestial Mechanics is a scientific field which is rich in theories and possible applications, which has been developed by the most famous scientists over the last centuries. A complete treatment of the subject would definitely be beyond the scope of the present book. We think that it would help a lot to mention the books which at least in part discuss these interesting problems that are missing here. We will thus – according to our personal choice – list a variety of important books, starting from the times of the seminal work by Poincaré, and hope that the interested reader will find his own way through this wonderful and interesting subject: Celestial Dynamics.

V. I. Arnold, V. V. Kozlov, A. I. Neishtadt: *Mathematical Aspects of Classical and Celestial Mechanics*, Springer 2010, 3rd edn. reprint. This book emphasizes the mathematical tools for dynamical systems, it presents a review on Lagrangian and Hamiltonian mechanics, with applications to the $n$-body problem. It also includes a detailed treatment of perturbation theory and properly explains KAM theory.

G. Beutler: *Methods of Celestial Mechanics*, Springer 2005, vol. I+II. A textbook for the physical, mathematical and numerical principles of Celestial Mechanics. The equations of motions for the planetary system, Earth's rotation, artificial satellites, and the relativistic equations of motion are discussed. The book includes numerical methods for the solution of ordinary differential equations, with possible applications to dynamical systems found in Celestial Mechanics.

*Celestial Dynamics*, First Edition. R. Dvorak and C. Lhotka.

J. Binney and S. Tremaine: *Galactic Dynamics*, Princeton University Press 1987. The authors cover galactic dynamics in a broad sense; the topics include potential theory applied to the Milky Way, the orbits of stars in various types of potentials, and collision-less and collisional stellar systems.

D. Brouwer and G. M. Clemence: *Methods of Celestial Mechanics*, Academic Press 1961. With the author's own words the book is intended "to provide a comprehensive background for practical applications". The content ranges from all aspects of elliptic motion, to Lunar theory including an extensive part on the subject of planetary perturbation theory.

E. W. Brown: *An Introductionary Treatise on the Lunar Theory*, reprint by Dover 1960 of Cambridge Univ. Press 1896. The book on Brown's Lunar theory gives a deep insight into the main Lunar problem, and includes a concise treatment of the perturbations of the Moon's orbit.

E. W. Brown and C. A. Shook: *Planetary Theory*, reprint by Dover 1964 of Cambridge Univ. Press 1933. The goal of this book is the development of methods for calculating the general orbit of a planet subject to perturbations. It provides the disturbing function, series developments, canonical equations and other tools to finally show the applications on the motion of Trojan asteroids and in the case of resonances.

A. Celletti: *Stability and Chaos in Celestial Mechanics*, Springer 2010. Focusing on the interaction between nonlinear dynamics, computational methods and Celestial Mechanics, this book describes a variety of fields, such as conservative and dissipative systems, rotational dynamics, perturbation theory, KAM and Nekhoroshev's theorem.

C. V. L. Charlier: *Die Mechanik des Himmels* (in German), Veit & Co. 1902–1907, vol. I+II. A compilation of the author's own lectures; he describes the basic properties of differential equations encountered in Celestial Mechanics, gives details on secular perturbations, the convergence of series developments, and formal integrals of motion of the three-body problem.

G. Contopoulos: *Order and Chaos in Dynamical Astronomy*, Springer 2002. Unlike other books in this list, this one puts the main emphasis on order and chaos in general and in galactic dynamics, complemented by applications of the developed tools to problems found in Celestial Mechanics as well as cosmology.

J. M. A. Danby: *Fundamentals of Celestial Mechanics*, Willmann–Bell 1988, revised 2nd edn. Danby covers in his book the fundamental aspects of Celestial Mechanics, starting from an extensive discussion of the two-body problem, continuing with the three- and $n$-body problem. Perturbations and the motion of the Moon as well as the Earth's and Moon's rotation are discussed.

R. Dvorak: *Extrasolar Planets*, Wiley-VCH 2008. Being a collection of contributions of scientists working on extrasolar planets, this book contains chapters on the dynamics of extrasolar planetary systems, planets in double star systems, the impact of stellar evolution on the habitability of exoplanets, the evolution of habitable zones and much more.

R. Dvorak, F. Freistetter, J. Kurths: *Chaos and Stability in Planetary Systems*, Springer 2005. Two main topics govern this book: orbital dynamics and chaotic

Hamiltonian systems on the one hand, and planet formation and dynamics of extrasolar planetary systems on the other hand.

S. Ferraz-Mello: *Canonical Perturbation Theories*, Springer 2007. This book presents the foundations of Hamiltonian perturbation theories used in Celestial Mechanics, starting from Hamilton–Jacobi theory, presenting Lie-series perturbation theory, and applications for resonant or degenerate systems.

Y. Hagihara: *Celestial Mechanics*, MIT Press 1970–1976, vol. I–V. These five volumes intend to recapitulate the results of Celestial Mechanics. Each of the volumes focuses on a special topic, these are: dynamical principles and transformation theory, perturbation theory, differential equations as found in Celestial Mechanics, periodic and quasi-periodic orbits, and the topology of the three-body problem (to name a few).

S. Herrick: *Astrodynamics*, Van Nostrand Reinhold Company 1971/1972, vol. I+II. The author gives an overview and introduction into orbit determination and correction, space navigation, perturbation theory and numerical methods for Celestial Mechanics.

J. Kovalevsky: *Introduction to Celestial Mechanics*, Springer 1967. A concise presentation of topics from Celestial Mechanics, which are useful in calculating the trajectories of objects in space. The main chapters are on perturbation theory and its application to the motion of artificial satellites around the Earth and the Moon.

C. Marchal: *The Three-Body Problem*, Elsevier 1990. The only topic in this book is the three-body problem, and it is presented to such a profoundness, that it is without equal. Among other aspects the invariants in the three-body problem, the Eulerian and Lagrangian solutions and the general three-body problem are discussed.

A. Milani and G. F. Gronchi: *Theory of Orbit Determination*, Cambridge University Press 2010. Devoted to the problem of orbit determination this book includes chapters on the methods of Laplace and Gauss, nongravitational perturbations, and impact monitoring.

A. Morbidelli: *Modern Celestial Mechanics*, Taylor & Francis 2002. The book describes the progress in Solar system dynamics of the last couple of decades, especially in planetary dynamics and the motion of small bodies. The focus is on quasi-integrable and resonant Hamiltonian systems.

F. R. Moulton: *An Introduction to Celestial Mechanics*, reprint by Dover 1970 of Macmillan Company 1914, 2nd revised edn. Starting from the basic principles and fundamentals, Moulton covers the classical subjects of Celestial Mechanics, like the two- and three-body problem, the determination of orbits, and geometrical as well as analytical aspects of perturbation theory.

C. D. Murray and S. F. Dermott: *Solar System Dynamics*, Cambridge University Press 1999. A very complete textbook on Celestial Mechanics, the topics range from two-body and three-body problems, resonant and secular perturbations to tides and rotations. The disturbing function and its expansion are treated in much detail.

H. Poincare: *Leçons de Mécanique Céleste* (in French), Gauthier–Villars 1905–1910, vol. I–III. The three volumes contain the general theory of planetary perturbations, the development of the perturbing function, Lunar theory, and the theory of tides (to name a few).

A. E. Roy: *Orbital Motion*, Institute of Physics Publishing 1988, 3rd edn. This book encompasses orbital motion in a broad sense. Among the indispensable chapters on the two-body problem and orbit determination, there are also sections on artificial satellites, interplanetary trajectories, and on the dynamics of binary and many-body stellar systems.

C. L. Siegel and J. K. Moser: *Lectures on Celestial Mechanics*, Springer 1971. A rather sober but mathematically accurate treatment of the three-body problem. The book includes the discussion of canonical transformations, regularization theory, the stability of solutions in dynamical systems.

W. M. Smart: *Celestial Mechanics*, Longmans 1953. A course on Celestial Mechanics with rather theoretical character. Beginning with expansions for an elliptic orbit, subsequently Lagrange's planetary equations and Delaunay's lunar theory are discussed, followed by the Hill–Brown lunar theory and an overview over precession and nutation.

M. H. Soffel: *Relativity in Astrometry, Celestial Mechanics and Geodesy*, Springer 1989. This book presents various aspects of modern relativistic Celestial Mechanics and astrometry. It contains a description of spatial and temporal reference frames, post-Newtonian equations of motions for artificial and natural satellites.

E. L. Stiefel and G. Scheifele: *Linear and Regular Celestial Mechanics*, Springer 1971. Concentrating on the basic and numerical theory of regularization, this book presents in great detail the Levi–Civita and Kustaanheimo–Stiefel regularization, typical perturbations of the linear equations of motion and numerical aspects of possible applications.

G. Stracke: *Bahnbestimmung der Planeten und Kometen* (in German), Springer 1929. The book deals primarily with the determination of orbits from observations and the computation of ephemerides.

K. Stumpff: *Himmelsmechanik* (in German), VEB 1956–1974, vol. I–III. A very complete treatise on Celestial Mechanics with many detailed computations shown in detail. The book (vol. I) starts with the 2-body problem and explains how one can get the series expansions in the 2-body problem. Then the 3-body problem (vol. II) is introduced and the solutions of Euler and Lagrange are described. Vol. III describes possible applications of perturbation theory.

V. Szebehely: *Theory of Orbits*, Academic Press 1967. Being a monograph on the restricted three-body problem, this book is an extensive introduction to the subject, where the "classical" form as well as modified formulations are presented. That part is followed by a number of regularization techniques, accompanied by a thorough discussion on Lagrange's equilibrium points and periodic solutions.

L. G. Taff: *Celestial Mechanics*, Wiley 1985. The author treats Newtonian gravitation and its effects in Celestial Mechanics. Besides the two- and three- body problems also coordinate and time systems, orbit determination and stellar dynamics are covered.

A. Wintner: *The Analytical Foundations of Celestial Mechanics,* Princeton Univ. Press 1947. A large part of the book is based on the analysis of theoretical dynamics and astronomy from a formal mathematical point of view. Canonical transformations, Hamiltonian and Lagrangian equations are introduced and their application to the restricted problem of three bodies is described.

# Appendix A
# Important Persons in the Field

**Arnold**, Vladimir Igorevich (also Arnol'd); 12 June 1937, Odessa, Soviet Union; 3 June 2010, Paris, France. Russian mathematician; known for his contribution to the Kolmogorov–Arnold–Moser theorem, and giving a partial solution of Hilbert's 13th problem in 1957; developed the Arnold cat map. *Small denominators and stability problems in classical and celestial mechanics, ScD thesis* (1963); *On the classical perturbation theory and stability theory of planetary systems* (1962)

**Bessel**, Friedrich Wilhelm; 22 July 1784, Minden, Germany; 17 March 1846, Königsberg, Germany (today Kaliningrad, Russia). German mathematician and astronomer; accomplished the first parallax measurement on star 61 Cyg (1838), conducted systematic research on Bessel functions. *Fundamenta Astronomiae* (1818); *Tabulae Regiomontanae* (1830)

**Brahe**, Tycho; 14 December 1546, Scania, Denmark (today Scania, Sweden); 24 October 1601, Prague, Czech Republic. Danish astronomer; most precise astrometrical measurements before the invention of the telescope, geoheliocentric model (Tychonic system), (re-)discovery of the variation of the Moon's longitude. *De nova stella (On the new star)* (1573)

**Brown**, Ernest William; 29 November 1866, Hull, England; 23 July 1938, New Haven, USA. British mathematician and astronomer; established precise tables for the motion of the Moon, calculated lunar ephemerides, investigated Trojan asteroid motion. *Tables of the Motion of the Moon* (1919)

**Chirikov**, Valerianovich Boris; 6 June 1928, Oryol, Russia; 12 February 2008. Russian mathematician; interest in general aspects of dynamical systems theory; founder of Hamiltonian chaos, introduced the standard map and invented the Chirikov criterion. *A universal instability of many-dimensional oscillator systems* (1979)

**Delaunay**, Charles-Eugène; 9 April 1816, Lusigny-sur-Barse, France; 5 August 1872, Cherbourg, France. French astronomer and mathematician; contributions to theory of lunar motion and the three body problem. *La Théorie du mouvement de la lune* (1860+1867)

**Euler**, Leonhard; 15 April 1707, Basel, Switzerland; 18 September 1783, Saint Petersburg, Russia. Swiss mathematician and physicist; major contributions to the field of calculus, defined the complex exponential function, developed Eu-

*Celestial Dynamics*, First Edition. R. Dvorak and C. Lhotka.

ler's formula, gave a solution for the two center problem. *Introductio in analysin infinitorum* (1748); *Institutiones calculi differentialis* (1755)

**Fourier**, Jean Baptiste Joseph; 21 March 1768, Auxerre, France; 16 May 1830, Paris, France. French mathematician and physicist; founder of Fourier analysis, Fourier series, applied those methods to the problems of heat transfer and vibrations. *Théorie analytique de la chaleur* (1822)

**Gauss**, Carl Friedrich; 30 April 1777, Braunschweig, Germany; 23 February 1855, Göttingen, Germany. Besides his outstanding mathematical expertise, Gauss contributed important methods of orbit determination (he made the rediscovery of Ceres possible). *Theoria motus corporum coelestium in sectionibus conicis solem ambientium* (1809)

**Hansen**, Peter Andreas; 8 December 1795, Tønder, Denmark; 22 March 1874, Gotha, Germany. Danish astronomer; worked in the field of gravitational astronomy (mutual perturbations of Jupiter and Saturn), contributed to lunar theory. *Tables of the Moon ("Hansen's Lunar Tables")* (1857)

**Hill**, George William; 3 March 1838, New York, USA; 16 April 1914, West Nyack, USA. American astronomer and mathematician; worked on lunar theory and the three-body problem; introduced the notion of Hill sphere. *Researches in Lunar Theory* (1878)

**Jacobi**, Carl Gustav Jacob; 10 December 1804, Potsdam, Germany; 18 February 1851, Berlin, Germany. German mathematician; investigated elliptic functions, laid the fundaments to Hamilton–Jacobi theory, expressed the Jacobi integral in the circular restricted three-body problem. *Fundamenta nova theoriae functionum ellipticarum* (1829); *Sur le movement d'un point et sur un cas particulier du problème des trois corps* (1836)

**Kepler**, Johannes; 27 December 1571, Weil der Stadt, Germany; 15 November 1630, Regensburg, Germany. German mathematician and astronomer; deduced the three Kepler laws, improved the refracting telescope (Keplerian telescope). *Astronomia nova (New Astronomy)* (1609); *Harmonice Mundi (Harmony of the Worlds)* (1619)

**Kolmogorov**, Andrey Nikolaevich; 25 April 1903, Tambov, Russia; 20 Oct 1987, Moscow, Russia. Russian mathematician; known for his contribution to the KAM theorem, worked in many branches of modern mathematics, like harmonic analysis, probability, set- and information theory. *On the conservation of conditionally periodic motions for a small change in the Hamiltonian's function* (1954)

**Lagrange**, Joseph Louis (born Giuseppe Lodovico (Luigi) Lagrangia); 25 January 1736, Torino, Italy; 10 April 1813, Paris, France. Italian-french mathematician and astronomer; creator of variational calculus and Euler–Lagrange equations, discovered the Lagrange points in the circular restricted three-body problem. *Mécanique analytique* (1788/89)

**Laplace**, Pierre Simon, Marquis de; 23 March 1749, Beaumont-en-Auge, France; 5 March 1827, Paris, France. French mathematician and astronomer; worked in the field of Celestial Mechanics, probability theory; inventor of the Laplace transform, and proponent of nebula hypothesis for the Solar system. *Théorie du* *movement et de la figure elliptique des planètes* (1784); *Traité de Mécanique Céleste* (1799)

**Legendre**, Adrien-Marie; 18 September 1752, Paris, France; 10 January 1833, Paris, France. French mathematician; created the Legendre polynomials and Legendre transformation; worked in the field of Celestial Mechanics. *Éléments de géométrie* (1794)

**Le Verrier**, Urbain Jean Joseph; 11 March 1811, Saint-Lô, France; 23 September 1877, Paris, France. French mathematician; predicted the position of Neptune, calculated additional precession of Mercury (perturbations due to additional planet Vulcan as speculated). *Développemens sur différentes points de la théorie des perturbations* (1841)

**Lie**, Sophus; 17 December 1842, Nordfjordeid, Norway; 18 February 1899, Christiania, Norway. Norwegian mathematician; investigated Lie groups, Lie-derivative named after him. *Theorie der Transformationsgruppen* (1888–1893)

**MacMillan**, William Duncan; 24 July 1871, La Crosse, USA; 14 November 1948, St. Paul, USA. American mathematician and astronomer; found a special solution of the restricted three-body problem. *An integrable case in the restricted problem of three bodies* (1911)

**Moser**, Jürgen; 4 July 1928, Königsberg, Germany; 17 December 1999, Zürich, Switzerland. Worked on ordinary and partial differential equations, spectral theory and Celestial Mechanics. The "Moser twist stability theorem" was a major step to the famous KAM theorem. *Stable and Random Motions in Dynamical Systems with Special Emphasis on Celestial Mechanics* (1973, reprint 2001)

**Nekhoroshev**, Nikolay Nikolaevich; 2 October 1946, Russia; 18 October 2008, Milan, Italy. His research interest was Hamiltonian mechanics, perturbation theory, Celestial Mechanics and singularity theory. He was a professor at the Moscow State University and at the University of Milan. Founder of the so-called Nekhoroshev theorem. *Exponential estimates of the stability time of near-integrable Hamiltonian systems* (1977).

**Newton**, Isaac; 4 January 1643, Woolsthorpe-by-Colsterworth, England; 31 March 1727, Kensington, England. English physicist and mathematician; formulated the laws of motion and theory of universal gravitation, contributions to optics. *Philosophiae Naturalis Principia Mathematica* (1687)

**Poincaré**, Jules Henri; 29 April 1854, Nancy, France; 17 July 1912, Paris, France. French mathematician; worked on the three-body problem, laid foundations to chaos theory, formulated the Poincaré conjecture. *Les méthodes nouvelles de la mécanique céleste* (1892–1899)

**Sitnikov**, Kirill Aleksandrowich; 1926, Russia. Russian mathematician; formulated the Sitnikov problem. *The Existence of Oscillatory Motions in the Three-Body Problem* (1961)

**Tisserand**, François Félix; 13 January 1845, Nuits-Saint-Georges, France; 20 October 1896, Paris, France. French astronomer; Tisserand criterion, conducted Venus transit observation, published works on the three-body problem. *Traité de mécanique céleste (Treatise on Celestial Mechanics)* (1889–1896)

# Appendix B
# Formulae

## B.1
## Hansen Coefficients

The Hansen coefficients [250] $X_k^{n,m} = X_k^{n,m}(e)$, which are nowadays mainly used in artificial satellite theory, are given by the relation

$$\left(\frac{r}{a}\right)^n e^{imv} = \sum_k X_k^{n,m}(e) e^{ikM} \tag{B1}$$

where $r/a$ is the ratio of the distance over the semimajor axis and $e$ is the eccentricity of the two body problem, $v$ is the true, $M$ is the mean anomaly and $k, n, m$ are integers. They appear in the development of the perturbing function due to a primary or the presence of a third body and are also related to the so-called eccentricity functions [251]. From Eq. (B1) one finds the integral representation:

$$X_k^{n,m} = \frac{1}{2\pi} \int_{-\pi}^{\pi} \left(\frac{r}{a}\right)^n e^{i(mv-kM)} dM \tag{B2}$$

Using the definition

$$\beta = \frac{e}{1+\sqrt{1-e^2}} , \quad z = e^{iE}$$

where $E$ is the eccentric anomaly, and from the relations

$$\begin{aligned}
\frac{r}{a} &= \left(1+\beta^2\right)^{-1} \left(1-\frac{\beta}{z}\right)(1-\beta z) \\
e^{iv} &= \frac{a}{r}\left(1+\beta^2\right)^{-1} z \left(1-\frac{\beta}{z}\right)^2 \\
e^{-iM} &= z^{-1} e^{\frac{e}{2}\left(z-\frac{1}{z}\right)} \\
dM &= \frac{1}{i}\frac{r}{a} z^{-1} dz
\end{aligned}$$

we get

$$\left(\frac{r}{a}\right)^n e^{i(mv-kM)} dM =$$
$$\frac{1}{i}\left(1+\beta^2\right)^{-n-1}(1-\beta z)^{n-m+1}\left(1-\frac{\beta}{z}\right)^{n+m+1} z^{m-k-1} e^{k\frac{e}{2}\left(z-\frac{1}{z}\right)} dz$$

We expand the binomials $(1-\beta z)^{n-m+1}$ and $(1-\beta/z)^{n+m+1}$ in terms of $\beta$ into a convergent power series since $|z| = 1$ and get:

$$(1-\beta z)^{n-m+1} = \sum_{s=0}^{s_1}\binom{n-m+1}{s}(-\beta)^s z^s$$
$$\left(1-\frac{\beta}{z}\right)^{n+m+1} = \sum_{t=0}^{t_1}\binom{n+m+1}{t}(-\beta)^t z^{-t}$$

where we have defined:

$$s_1 = \begin{cases} n-m+1 & \text{if } n-m+1 \geq 0 \\ \infty & \text{if } n-m+1 < 0 \end{cases}$$

$$t_1 = \begin{cases} n+m+1 & \text{if } n+m+1 \geq 0 \\ \infty & \text{if } n+m+1 < 0 \end{cases}$$

If we substitute the above expressions in (B2) we find the expression:

$$X_k^{n,m} = \frac{1}{2\pi i}\left(1+\beta^2\right)^{-n-1}\sum_{s=0}^{s_1}\sum_{t=0}^{t_1}\binom{n-m+1}{s}\binom{n+m+1}{t}(-\beta)^{s+t}$$
$$\times \oint z^{-1-(k-m-s+t)} e^{k\frac{e}{2}\left(z-\frac{1}{z}\right)} dz$$

Here, the integral on the right is extended in the complex plane along the circle $z = 1$. The definition of the Bessel function of the first kind in terms of $z$ and any number $\alpha$ is

$$J_\lambda(\alpha) = \frac{1}{2\pi i}\oint z^{-1-\lambda} e^{\frac{1}{2}\alpha\left(z-\frac{1}{z}\right)} dz$$

so that we finally get the standard definition of the Hansen coefficients:

$$X_k^{n,m} = \left(1+\beta^2\right)^{-n-1}$$
$$\sum_{s=0}^{s_1}\sum_{t=0}^{t_1}\binom{n-m+1}{s}\binom{n+m+1}{t}(-\beta)^{s+t} J_{k-m-s+t}(ke)$$

For further reading see for example [252].

## B.2
## Laplace Coefficients

The Laplace coefficients [253] denoted by $b_n^{(k)} = b_n^{(k)}(\alpha)$ are defined by the relation

$$(1-\alpha z)^{-n}\left(1-\alpha z^{-1}\right)^{-n} = \frac{1}{2}\sum_k b_n^{(k)}(\alpha) z^k$$

with the basic properties $b_n^{(k)}(-\alpha) = (-1)^k b_n^{(k)}$, $b_n^{(-k)} = b_n^{(k)}$ and $k \geq 0$. In the expansion of the perturbing function of the restricted problem they are needed in the following expansion:

$$\left[1-2\alpha\cos(\psi)+\alpha^2\right]^{-\frac{n}{2}} = \frac{1}{2}\sum_k b_n^{(k)}(\alpha)\cos(k\psi)\,,$$

where the quantity $\alpha = a/a'$ is the ratio of two semimajor axes with $\alpha < 1$ thus $a < a'$ and $\psi$ is the angle between the direction of the semimajor axes $(a, a')$ in the perturbed two body problem, where $a'$ is the perturber. The integral representation of the coefficients is given by:

$$b_n^{(k)}(\alpha) = \frac{2}{\pi}\int_0^{\pi} \frac{\cos(k\psi)}{\left[1-2\alpha\cos(\psi)+\alpha^2\right]^n}\, d\psi$$

which can be solved as

$$\frac{2}{\pi}\int_0^{\pi} \frac{\cos(k\psi)}{\left[1-2\alpha\cos(\psi)+\alpha^2\right]^n}\, d\psi = \frac{(n)_k}{k!}\alpha^k\,{}_1F_2\left(n, n+k, k+1 : \alpha^2\right)$$

where ${}_1F_2$ is the hypergeometric function and we use the notation

$$(n)_k = (-1)^k \binom{-n}{k} k\,!= n(n+1)\cdots(n+k-1) \quad \text{and} \quad (n)_0 = 1$$

which is also called the Pochhammer symbol. The following useful recursive relations between the Laplace coefficients exist:

$$b_{n+1}^{(k)}(\alpha) = \frac{(n+k)}{n}\frac{\left(1+\alpha^2\right)}{\left(1-\alpha^2\right)^2} b_n^{(k)}(\alpha) - \frac{2(k-n+1)}{n}\frac{\alpha}{\left(1-\alpha^2\right)^2} b_n^{(k+1)}(\alpha)$$

$$b_{n+1}^{(k+1)}(\alpha) = \frac{k}{k-n}\left(\alpha+\alpha^{-1}\right) b_{n+1}^{(k)} - \frac{n+k}{k-n} b_{n+1}^{(k-1)}(\alpha)$$

One can generalize the Laplace coefficients (see [254]) as

$$(1-\alpha z)^{-s}\left(1-\alpha z^{-1}\right)^{-r} = \sum_k b_{s,r}^{(k)}(\alpha) z^k$$

with $b_{s,r}^{(k)}(-\alpha) = (-1)^k b_{s,r}^{(k)}(\alpha)$, $b_{s,r}^{(-k)}(\alpha) = b_{r,s}^{(k)}(\alpha)$ and with $k \geq 0$ which turns out to be in terms of hypergeometric function ${}_1F_2$

$$b_{s,r}^{(k)}(\alpha) = \frac{(s)_k}{k!} \alpha^k {}_1F_2\left(r, s+k, k+1 : \alpha^2\right) .$$

Since the classical expansion in terms of Bessel functions takes the form

$$e^{ike\sin(E)} = \sum_n J_n(ke) e^{ikE}$$

it is possible to express the Hansen coefficients in terms of Laplace coefficients:

$$X_k^{n,m} = \frac{1}{2\left(1+\beta^2\right)^{n+1}} \sum_\gamma b_{-n-1+m}^{(k-\gamma-m)}(\beta) J_\gamma(ke)$$

with $\beta$ which is identical to the definition in Section B.1. It is also true that

$$\left(\frac{r}{a}\right)^n e^{imf} = \sum_k Y_k^{n,m} e^{ikf}$$

and

$$\left(\frac{r}{a}\right)^n e^{imf} = \sum_k Z_k^{n,m} e^{ikE}$$

with (after [255])

$$Y_k^{n,m} = \frac{(-1)^{k-m}}{2} \left(1-e^2\right)^n \left(1+\beta^2\right)^n b_n^{(k-m)}(\beta)$$

$$Z_k^{n,m} = \frac{1}{2\left(1+\beta^2\right)^n} b_{-n+m,-n-m}^{(k-m)}(\beta)$$

## B.3
## Bessel Functions

From a mathematical point of view the Bessel functions are the coefficients of the Laurent series of the complex function (after [133])

$$H(z, x) = e^{\frac{x}{2}\left(z-z^{-1}\right)} \tag{B3}$$

which gives:

$$e^{\frac{x}{2}\left(z-z^{-1}\right)} = \ldots J_{-2}(x) z^{-2} + J_{-1}(x) z^{-1} + J_0(x) + J_1(x) z + J_2(x) z^2 + \ldots$$

From the classical Taylor series expansion of the exponential

$$e^z = \sum_{j \geq 0} \frac{1}{j!} z^j$$

we find

$$e^{\frac{x}{2}\left(z-z^{-1}\right)} = \left[\sum_{j\geq 0} \frac{1}{j!}\left(\frac{xz}{2}\right)^j\right]\left[\sum_{k\geq 0} \frac{(-1)^k}{k!}\left(\frac{x}{2z}\right)^k\right]$$
$$= \sum_{j\geq 0}\sum_{k\geq 0} \frac{(-1)^k}{j!k!}\left(\frac{x}{2}\right)^{j+k} z^{j-k}$$

Setting $n = j - k$ and comparing terms of same order in $z$ one finds for $n \geq 0$:

$$J_n(x) = \sum_{k\geq 0} \frac{(-1)^k}{k!(k+n)!}\left(\frac{x}{2}\right)^{2k+n}$$

and

$$J_{-n}(x) = (-1)^n J_n(x)$$
$$J_n(-x) = (-1)^n J_n(x)$$

since $H(z, x) = H\left(-z^{-1}, x\right)$. From the properties above the Laurent series expansion of (B3) reduces to:

$$e^{\frac{x}{2}\left(z-z^{-1}\right)} = J_0(x) + J_1(x)\left(z - z^{-1}\right) + J_2(x)\left(z^2 + z^{-2}\right) + \ldots$$

Their use in celestial mechanics becomes clear when one identifies $z = e^{i\varphi}$ and takes into account the identity:

$$e^{i\varphi} - e^{-i\varphi} = 2i\sin(\varphi)\,.$$

From the expansion of $H\left(e^{i\varphi}, x\right)$ one gets:

$$e^{\frac{x}{2}\left(e^{i\varphi}-e^{-i\varphi}\right)} = e^{ix\sin(\varphi)} = J_0(x) + 2\sum_{k\geq 1} J_{2k}(x)\cos(2k\varphi)$$
$$+ 2i\sum_{k\geq 0} J_{2k+1}(x)\sin\left((2k+1)\varphi\right)$$

and since $e^{i\varphi} = \cos(\varphi) + i\sin(\varphi)$ we have:

$$\cos\left(x\sin(\varphi)\right) = J_0(x) + 2\sum_{k\geq 1} J_{2k}(x)\cos(2k\varphi)$$

$$\sin\left(x\sin(\varphi)\right) = 2\sum_{k\geq 0} J_{2k+1}(x)\sin\left((2k+1)\varphi\right)$$

From the definition of the Fourier expansions of a $2\pi$-periodic function $f(\varphi)$

$$f(\varphi) = a_0 + \sum_{k>0}\left[a_k\cos(k\varphi) + b_k\sin(k\varphi)\right]$$

and the definition of the coefficients $a_k$, $b_k$ from Fourier theory:

$$a_0 = \frac{1}{2\pi}\int_0^{2\pi} f(\varphi)d\varphi$$

$$a_k = \frac{1}{\pi}\int_0^{2\pi} f(\varphi)\cos(k\varphi)d\varphi$$

$$b_k = \frac{1}{\pi}\int_0^{2\pi} f(\varphi)\sin(k\varphi)d\varphi$$

with $k = 0, 1, \ldots$ we also find

$$J_{2n}(x) = \frac{1}{2\pi}\int_0^{2\pi} \cos\left(x\sin(\varphi)\right)\cos(2n\varphi)d\varphi$$

$$J_{2n+1}(x) = \frac{1}{2\pi}\int_0^{2\pi} \sin\left(x\sin(\varphi)\right)\sin\left((2n+1)\varphi\right)d\varphi$$

for $n = 1, 2, \ldots$ From the identity

$$\int_0^{2\pi} \cos(mx)\cos(nx)dx = \int_0^{2\pi} \sin(mx)\sin(nx)dx = 0$$

for integers $m \neq n$ and the fact that $\cos\left(x\sin(\varphi)\right)$ contains only even and $\sin\left(x\sin(\varphi)\right)$ contains only odd multiples of the angle $\varphi$ one can show that

$$\int_0^{2\pi} \sin\left(x\sin(\varphi)\right)\sin(2n\varphi)d\varphi$$

and

$$\int_0^{2\pi} \cos\left(x\sin(\varphi)\right)\cos(2n+1\varphi)d\varphi$$

vanish identically. Together with the above one shows that

$$J_n(x) = \frac{1}{\pi}\int_0^{\pi} \cos\left(n\varphi - x\sin(\varphi)\right)d\varphi \tag{B4}$$

for $n = 0, 1, \ldots$ We aim to use this definition to solve Kepler's equation

$$M = E - e\sin(E)$$

where $M$ is the mean, $E$ is the eccentric anomaly and $e$ is the eccentricity of the orbit of a planet around a central star. From $E(-M) = -E(M)$ one shows that

$$\cos(nE) = a_0 + \sum_{k\geq 1} a_k^{(n)} \cos(kM)$$

and

$$\sin(nE) = \sum_{k\geq 1} b_k^{(n)} \sin(kM)\,.$$

From

$$dM = dE\left[1 - e\cos(E)\right]$$

we find

$$a_0^{(n)} = \frac{1}{2\pi}\int_0^{2\pi} \cos(nE)\, dM = \frac{1}{2\pi}\int_0^{2\pi} \cos(nE)\left[1 - e\cos(E)\right] dE$$

and therefore

$$a_0^{(1)} = -\frac{e}{2}$$

and

$$a_0^{(n)} = 0 \quad \text{for } n > 1$$

We demonstrate the calculation to obtain the coefficient $a_k^{(n)}$:

$$\begin{aligned}
a_k^{(n)} &= \frac{1}{\pi}\int_0^{2\pi} \cos(nE)\cos(kM)dM \\
&= \frac{1}{k\pi}\int_0^{2\pi} \cos(nE)\frac{d\sin(kM)}{dM}dM \\
&= \left[\frac{\cos(nE)\sin(kM)}{k\pi}\right]_0^{2\pi} - \frac{1}{k\pi}\int_0^{2\pi} \frac{d\cos(nE)}{dM}\sin(kM)dM \\
&= -\frac{1}{k\pi}\int_0^{2\pi} \frac{d\cos(nE)}{dE}\sin(nM)dE \\
&= \frac{n}{k\pi}\int_0^{2\pi} \sin(nE)\sin(kM)dE
\end{aligned}$$

$$= \frac{n}{k\pi}\int_0^{2\pi} \sin(nE)\sin(kE - ke\sin(E))dE$$

$$= \frac{n}{2\pi k}\int_0^{2\pi} \left[\cos((k-n)E - ke\sin(E)) - \cos((k+n)E - ke\sin(E))\right]dE$$

$$= \frac{n}{2\pi k}\int_0^{2\pi} \sin((k+n)E - ke\sin(E))dE = \ldots$$

The last integral is of the form (B4) and therefore immediately gives:

$$a_k^{(n)} = \frac{n}{k}\left[J_{k-n}(ke) - J_{k+n}(ke)\right]$$

and we find

$$\cos(nE) = n\sum_{k\geq 1}\left[J_{k-n}(ke) - J_{k+n}(ke)\right]\frac{\cos(kM)}{k}.$$

A similar calculation can be done to get:

$$\sin(nE) = n\sum_{k\geq 1}\left[J_{k-n}(ke) + J_{k+n}(ke)\right]\frac{\sin(kM)}{k}.$$

We conclude with the case $n = 1$. From the properties of Bessel functions (not derived here)

$$\begin{aligned} nJ_n(x) &= \frac{x}{2}\left[J_{n-1}(x) + J_{n+1}(x)\right] \\ \frac{dJ_n}{dx} &= \frac{1}{2}\left[J_{n-1}(x) - J_{n+1}(x)\right] \end{aligned}$$

we have

$$a_k^{(1)} = \frac{1}{k}\left[J_{k-1}(ke) - J_{k+1}(ke)\right] = \frac{2}{k^2}\frac{dJ(ke)}{de}$$

and from

$$r = a\left[1 - e\cos(E)\right]$$

we get

$$\frac{r}{a} = 1 + \frac{1}{2}e^2 - 2e\sum_{k\geq 1}\frac{dJ(ke)}{de}\frac{\cos(kM)}{k^2}$$

which gives by comparison with (B1) if we set $n = 1$, $m = 0$

$$1 + \frac{1}{2}e^2 - 2e\sum_{k\geq 1}\frac{dJ_k(ke)}{de}\frac{\cos(kM)}{k^2} = \sum_k X_k^{1,0}(e)e^{ikM}$$

## B.4
## Expansions in the Two-Body Problem

We collect explicit formulae (from [133]), which are useful for various expansions based on the two-body problem. We start with the expressions

$$\cos(nE) = n \sum_{k=-\infty}^{\infty} J_{-k-n}(ke)\frac{\cos(kM)}{k}$$

$$\sin(nE) = n \sum_{k=-\infty}^{\infty} J_{k-n}(ke)\frac{\sin(kM)}{k}$$

where we exclude $k \neq 0$ in the infinite sum. For $n = 1$ we get:

$$\cos(E) = -\frac{e}{2} + 2\sum_{k=1}^{\infty} \frac{dJ_k(ke)}{de}\frac{\cos(kM)}{k^2}$$

$$\sin(E) = \frac{2}{e}\sum_{k=1}^{\infty} J_k(ke)\frac{\sin(kM)}{k}$$

From

$$r = a\left[1 - e\cos(E)\right]$$

we also get

$$\frac{r}{a} = 1 + \frac{1}{2}e^2 - 2e\sum_{k=1}^{\infty} \frac{dJ_k(ke)}{de}\frac{\cos(kM)}{k^2}$$

In a similar way we obtain from

$$x = a\left[\cos(E) - e\right]$$

$$y = a\sqrt{1-e^2}\sin(E)$$

the expressions

$$\frac{x}{a} = -\frac{3}{2}e + 2\sum_{k=1}^{\infty} \frac{dJ_k(ke)}{de}\frac{\cos(kM)}{k^2}$$

$$\frac{y}{a} = 2\frac{\sqrt{1-e^2}}{e}\sum_{k=1}^{\infty} J_k(ke)\frac{\sin(kM)}{k}$$

From Kepler's equation we get

$$E - M = e\sin(E) = 2\sum_{k=1}^{\infty} J_k(ke)\frac{\sin(kM)}{k}$$

We differentiate with respect to $M$:

$$\frac{dE}{dM} = \frac{1}{1 - e\cos(E)} = \frac{a}{r}$$

from which we finally get

$$\frac{a}{r} = 1 + 2\sum_{k=1}^{\infty} J_k(ke)\cos(kM)$$

Various relations between $\nu$ and $M$ can be derived from

$$\cos(\nu) = \frac{1-e^2}{e}\frac{a}{r} - \frac{1}{e}$$

$$\sin(\nu) = \frac{\sqrt{1-e^2}}{e}\frac{d}{dM}\left(\frac{r}{a}\right)$$

for example we get

$$\cos(\nu) = -e + 2\frac{1-e^2}{e}\sum_{k=1}^{\infty} J_k(ke)\cos(kM)$$

$$\sin(\nu) = 2\sqrt{1-e^2}\sum_{k=1}^{\infty}\frac{dJ_k(ke)}{de}\frac{\sin(kM)}{k}$$

For terms of the form

$$\left(\frac{r}{a}\right)^n \cos(m\nu)$$

$$\left(\frac{r}{a}\right)^n \sin(m\nu)$$

see Section B.1. We just give the results for the special case $n = -2$, $m = 1$ to show the possibility to express them in terms of Bessel functions, which gives

$$\left(\frac{a}{r}\right)^2 \cos(\nu) = 2\sum_{k=1}^{\infty}\frac{dJ_k(ke)}{de}\cos(kM)$$

$$\left(\frac{a}{r}\right)^2 \sin(\nu) = 2\frac{\sqrt{1-e^2}}{e}\sum_{k=1}^{\infty} k\,J_k(ke)\sin(kM)$$

We conclude with the most often used series expansion, which allows to express the true anomaly $\nu$ for given mean anomaly $M$. The difference of the anomalies $\nu - M$ can be obtained from

$$\beta = \frac{1}{e}\left(1 - \sqrt{1-e^2}\right)$$

and

$$\nu - E = 2\sum_{k=1}^{\infty}\frac{1}{k}\sum_{l=1}^{\infty}\beta^l\left[J_{k-l}(ke) + J_{k+l}(ke)\right]\sin(kM)$$

we find from $(\nu - M) = (E - M) + (\nu - E)$ the expansion

$$\begin{aligned}
\nu - M = {} & \sin M \left( 2e - \frac{1}{4}e^3 + \frac{5}{96}e^5 + \frac{107}{4608}e^7 + \ldots \right) \\
& + \sin 2M \left( \frac{5}{4}e^2 - \frac{11}{24}e^4 + \frac{17}{96}e^6 + \ldots \right) \\
& + \sin 3M \left( \frac{13}{12}e^3 - \frac{43}{64}e^5 + \frac{95}{512}e^7 + \ldots \right) \\
& + \sin 4M \left( \frac{103}{96}e^4 - \frac{451}{480}e^6 + \ldots \right) \\
& + \sin 5M \left( \frac{1097}{960}e^5 - \frac{5957}{4608}e^7 + \ldots \right) \\
& + \sin 6M \left( \frac{1223}{960}e^6 - \ldots \right) \\
& + \sin 7M \left( \frac{47273}{32256}e^7 + \ldots \right)
\end{aligned} \tag{B5}$$

# Acknowledgement

R.D. would like to thank all his colleagues in the AstroDynamicsGroup of the Universitätssternwarte of the University of Vienna: H. Baudisch, A. Bazso, S. Eggl, B. Funk, R. Lang, E. Pilat-Lohinger, R. Schwarz, A. Suli, and F. Vrabec.

C.L. thanks A. Celletti and A. Lemaitre for their hospitality during his stays abroad, A. Celletti, C. Efthymiopoulos, and M. Sansottera for proofreading parts of the manuscript, and his parents and his beloved wife Michaela for always supporting his scientific career.

*Celestial Dynamics*, First Edition. R. Dvorak and C. Lhotka.

# References


1 Pedersen, O. (1993) *Early Physics and Astronomy: A Historical Introduction*, Cambridge University Press, Cambridge, UK.

2 Kuypers, F. (2008) *Klassische Mechanik: Mit über 300 Beispielen und Aufgaben mit Lösungen*, Lehrbuch Physik, Wiley-VCH Verlag GmbH.

3 Goldstein, H., Poole, C., and Safko, J. (2002) *Classical Mechanics*, Addison Wesley.

4 Arnold, V. (1989) *Mathematical Methods of Classical Mechanics*, Graduate Texts in Mathematics, Springer.

5 Froeschlé, C. (1992) Mappings in astrodynamics, in *Chaos, Resonance and Collective Dynamical Phenomena in the Solar System* (ed. S. Ferraz-Mello), IAU Symposium 152, 375–389, Kluwer.

6 Hadjidemetriou, J. (1996) Symplectic Mappings, in *Dynamics, Ephemerides and Astrometry of the Solar System* (eds S. Ferraz-Mello *et al.*), IAU Symposium 172, 255–266, Kluwer.

7 Wisdom, J. (1982) The origin of the Kirkwood gaps – A mapping for asteroidal motion near the 3/1 commensurability. *AJ*, **87**, 577–593.

8 Hadjidemetriou, J. (1991) Mapping models for Hamiltonian systems wit application to resonant asteroid motion, in *Predictability, Stability and Chaos in N-Body Dynamical Systems* (ed. A.E. Roy), Kluwer, Dordrecht.

9 Hadjidemetriou, J. (1992) A hyperbolic twist mapping model for the study of asteroid orbits near the 3 : 1 resonance. *ZAMP*, **37**, 776–792.

10 Hadjidemetriou, J. (1999) A symplectic mapping model as a tool to understand the dynamics of 2/1 resonant asteroid motion. *Celest. Mech. Dyn. Astron.*, **73**, 65–76.

11 Ferraz-Mello, S. (2000) The 2/1 and 3/2 resonant asteroid motion: a symplectic mapping approach. *Celest. Mech. Dyn. Astron.*, **78**, 137–150.

12 Ferraz-Mello, S. (1996) A symplectic mapping approach to the study of the stochasticity in asteroidal resonances. *Celest. Mech. Dyn. Astron.*, **65**, 421–437.

13 Efthymiopoulos, C. and Sándor, Z. (2005) Optimized Nekhoroshev stability estimates for the Trojan asteroids with a symplectic mapping model of co orbital motion. *MNRAS*, **364**, 253–271.

14 Lhotka, C., Efthymiopoulos, C., and Dvorak, R. (2008) Nekhoroshev stability at l4 or l5 in the elliptic-restricted three-body problem – application to Trojan asteroids. *MNRAS*, **384**, 1165–1177.

15 various (2012) *Scholarpedia*, www.scholarpedia.org, December 2012.

16 various (2012) *Wolfram MathWorld*, http://mathworld.wolfram.com/, December 2012.

17 Gole, C. (2001) *Symplectic twist maps*, World Scientific, Singapore

18 Lichtenberg, A. and Lieberman, A. (2010) *Regular and Chaotic Dynamics*, Applied Mathematical Sciences, Springer.

19 Celletti, A. and Chierchia, L. (1991) Invariant curves for area-preserving twist maps far from integrable. *J. Stat. Phys.*, **65**, 617–643.


*Celestial Dynamics*, First Edition. R. Dvorak and C. Lhotka.

20 Greene, J. (1979) A method for determining a stochastic transition. *J. Math. Phys.*, **20**, 1183.

21 MacKay, R. and Percival, I. (1985) Transition to chaos for area-preserving maps, in *Lecture Notes in Physics*, vol. 247, Springer, p. 390.

22 Mather, J. (1984) Nonexistence of invariant circles. *Erg. Theory Dyn. Syst.*, **4**, 301.

23 Wisdom, J. (1982) The origin of the Kirkwood gaps: a mapping for asteroidal motion near the 3/1 commensurability. *Astron. J.*, **87**, 577–593.

24 Souchay, J. and Dvorak, R. (eds) (2010) *Dynamics of Small Solar System Bodies and Exoplanets*, Springer, Berlin.

25 Hanslmeier, A. and Dvorak, R. (1984) Numerical integration with Lie series. *A&A*, **132**, 203.

26 Delva, M. (1985) A Lie integrator program and test for the elliptic restricted three body problem. *Astron. Astrophys. Suppl. Ser.*, **60**, 277–284.

27 Lichtenegger, H. (1984) The dynamics of bodies with variable masses. *Celest. Mech.*, **34**, 357–368.

28 Dvorak, R. and Schwarz, R. (2005) On the stability regions of the Trojan asteroids. *Celest. Mech. Dyn. Astron.*, **92**, 19–28.

29 Freistetter, F. (2006) The size of the stability regions of Jupiter Trojans. *A&A*, **453**, 353–361.

30 Schwarz, R., Dvorak, R., Süli, Á., and Érdi, B. (2007) Survey of the stability region of hypothetical habitable Trojan planets. *A&A*, **474**, 1023–1029.

31 Dvorak, R., Pilat-Lohinger, E., Schwarz, R., and Freistetter, F. (2004) Extrasolar Trojan planets close to habitable zones. *A&A*, **426**, L37–L40.

32 Pilat-Lohinger, E., Robutel, P., Süli, Á., and Freistetter, F. (2008) On the stability of Earth-like planets in multi-planet systems. *Celest. Mech. Dyn. Astron.*, **102**, 83–95.

33 Funk, B., Schwarz, R., Pilat-Lohinger, E., Süli, Á., and Dvorak, R. (2009) Stability of inclined orbits of terrestrial planets in habitable zones. *PSS*, **57**, 434–440.

34 Funk, B., Wuchterl, G., Schwarz, R., Pilat-Lohinger, E., and Eggl, S. (2010) The stability of ultra-compact planetary systems. *A&A*, **516**, A82.

35 Gröbner, W. (1967) *Die Lie-Reihen und ihre Anwendungen*, Deutscher Verlag der Wissenschaften, Berlin.

36 Gröbner, W. and Knapp, H. (1967) *Contributions to the Method of Lie-series*, Bibliographisches Institut, Mannheim.

37 Hanslmaier, A. and Dvorak, R. (1984) Numerical integration with Lie series. *A&A*, **132**, 203.

38 Maindl, T.I. and Dvorak, R. (1994) On the dynamics of the relativistic restricted three-body problem. *A&A*, **290**, 335–339.

39 Oseledets, V. (1968) Multiplicative ergodic theorem: Characteristic Lyapunov exponents of dynamical systems. *Mosc. Math. Soc.*, **19**, 197–231.

40 Benettin, G., Galgani, L., Giorgilli, A., and Strelcyn, J.M. (1980) Lyapunov characteristic exponents for smooth dynamical systems: a method for computing all of them. *Meccanica*, **15**, 9–30.

41 Skokos, C. (2010) The Lyapunov Characteristic Exponents and Their Computation, in *Lecture Notes in Physics*, vol. 790, Springer, pp. 63–135.

42 Froeschlé, C. (1984) The Lyapunov characteristic exponents and applications. *J. Méc. Théor. Appl.*, **Numero spécial**, 101–132.

43 Froeschlé, C., Lega, E., and Gonczi, R. (1997) Fast Lyapunov indicators. Application to asteroidal motion. *Celest. Mech. Dyn. Astron.*, **67**, 41–62.

44 Frouard, J., Fouchard, M., and Vienne, A. (2008) Comparison of fast Lyapunov chaos indicators for celestial mechanics, in *SF2A-2008* (eds C. Charbonnel, F. Combes, and R. Samadi), self-published, p. 121.

45 Sándor, Z., Érdi, B., Széll, A., and Funk, B. (2004) The relative Lyapunov indicator: an efficient method of chaos detection. *Celest. Mech. Dyn. Astron.*, **90**, 127–138.

46 Cincotta, P. and Simó, C. (2000) Simple tools to study global dynamics in non-axisymmetric galactic potentials – I. *Astron. Astrophys.*, **147**, 205–228.

47 Cincotta, P., Giordano, C., and Simó, C. (2003) Phase space structure of multidimensional systems by means of

the mean exponential growth factor of nearby orbits. *Physica D*, **182**, 151–178.
48 Goździewski, K., Bois, E., Maciejewski, A.J., and Kiseleva-Eggleton, L. (2001) Global dynamics of planetary systems with the MEGNO criterion. *A&A*, **378**, 569–586.
49 Mestre, M.F., Cincotta, P.M., and Giordano, C.M. (2011) Analytical relation between two chaos indicators: FLI and MEGNO. *MNRAS*, **414**, L100–L103.
50 Skokos, C. (2001) Alignment indeces: a new, simple method to for determining the ordered or chaotic nature of orbits. *J. Phys. A Math. Gen.*, **34**, 10029–10043.
51 Skokos, C., Antonopoulos, C., Bountis, T., and Vrahatis, M. (2004) Detecting order and chaos in in Hamiltonian systems by the SALI method. *J. Phys. A Math. Gen.*, **37**, 6269–6284.
52 Skokos, C., Bountis, T., and Antonopoulos, C. (2007) Geometrical properties of local dynamics in Hamiltonian systems. the Generalized Alignment Index (GALI) method. *Physica D*, **231**, 30–54.
53 Manos, T., Skokos, C., Athanassoula, E., and Bountis, T. (2008) Studying the global dynamics of conservative dynamical systems using the SALI chaos detection method. *Nonlinear Phenom. Complex Syst.*, **11**, 171–176.
54 Bountis, T., Manos, T., and Christodoulidi, H. (2009) Application of the GALI method to localization dynamics in nonlinear systems. *J. Comput. Appl. Math.*, **227**, 17–26.
55 Powell, G. and Percival, I. (1979) A spectral entropy method for distinguishing regular and irregular motion of Hamiltonian systems. *J. Phys. A Math. Gen.*, **12**, 2053–2072.
56 Mitchenko, T. and Ferraz-Mello, S. (1995) Comparative study of the asteroidal motion in the 3:2 and 2:1 resonance with Jupiter. I: Planar model. *A&A*, **303**, 945.
57 Mitchenko, T. and Ferraz-Mello, S. (2001) Resonant structure of the Outer Solar System in the neighborhood of the planets. *AJ*, **122**, 474–481.
58 Voglis, N., Contopoulos, G., and Efthymiopoulos, C. (1999) Detection of ordered and chaotic motion using the dynamical spectra. *Celest. Mech. Dyn. Astron.*, **73**, 211–220.
59 Contopoulos, G. and Voglis, N. (1996) Spectra of stretching numbers and helicity angles in dynamical systems. *Celest. Mech. Dyn. Astron.*, **64**, 1–20.
60 Contopoulos, G. and Voglis, N. (1997) A fast method for distinguishing between ordered and chaotic orbits. *Astron. Astrophys.*, **317**, 73–81.
61 Laskar, J., Froeschlé, C., and Celletti, A. (1992) The measure of chaos by the numerical analysis of the fundamental frequencies. Application to the standard mapping. *Phys. D Nonlinear Phenom.*, **56**, 253–269.
62 Laskar, J. (1993) Frequency analysis of a dynamical system. *Celest. Mech. Dyn. Astron.*, **56**, 191–196.
63 Robutel, P. and Laskar, J. (2001) Frequency map and global dynamics in the Solar System. I: Short period dynamics of massless particles. *Icarus*, **152**, 4–28.
64 Maffione, N., Darriba, L., Cincotta, P., and Giordano, C. (2011) A comparison of different indicators of chaos based on the deviation vectors: application to symplectic mappings. *Celest. Mech. Dyn. Astron.*, **111**, 285–307.
65 Ferraz-Mello, S. (2007) *Canonical Perturbation Theories and Resonance*, Springer, Berlin.
66 Celletti, A. (2009) *Stability and Chaos in Celestial Mechanics*, Springer, Praxis.
67 Hori, G. (1966) Theory of general perturbations with unspecified canonical variables. *Publ. Astron. Soc. Japan*, **18**, 287.
68 Hori, G. (1970) Comparison of Two Perturbation Theories Based on Canonical Transformations. *Publ. Astron. Soc. Japan*, **22**, 191.
69 Hori, G. (1973) *Theory of general perturbations*, Recent Advances in Dynamical Astronomy, Reidel Dodrecht, Tapley, Szebehely.
70 Deprit, A. (1969) Canonical transformations depending on a small parameter. *Celest. Mech.*, **1**, 12.
71 Deprit, A., Henrard, J., Price, J.F., and Rom, A. (1969) Birkhoff's normalization. *Celest. Mech.*, **1**, 222–251.

72 Henrard, J. (1970) On a perturbation theory using Lie transforms. *Celest. Mech.*, **3**, 107–120.
73 Meyer, K.R. (1974) Normal forms for Hamiltonian systems. *Celest. Mech.*, **9**, 517–522.
74 Tupikova, I.V. (1984) Account of additional perturbations in canonical system solutions obtained by Lie transforms. *Celest. Mech.*, **33**, 337–342.
75 Fassò, F. and Benettin, G. (1989) Composition of Lie transforms with rigorous estimates and applications to Hamiltonian perturbation theory. *ZAMP*, **40**, 307–329.
76 Fassò, F. (1989) On a relation among Lie series. *Celest. Mech. Dyn. Astron.*, **46**, 113–118.
77 Fassò, F. (1990) Lie series method for vector fields and Hamiltonian perturbation theory. *ZAMP*, **41**, 843–864.
78 Breiter, S. (1997) On the numerical transformation of variables in perturbation theory. *Celest. Mech. Dyn. Astron.*, **65**, 345–354.
79 Bazzani, A. (1988) Normal forms for symplectic maps in $R^{2n}$. *Celest. Mech. Dyn. Astron.*, **42**, 107–128.
80 Bazzani, A., Marmi, S., and Turchetti, G. (1979) Nekhoroshev estimate for isochronous non resonant symplectic maps. *Celest. Mech. Dyn. Astron.*, **47**, 333–359.
81 Guzzo, M. (2004) A direct proof of the Nekhoroshev theorem for nearly integrable symplectic maps. *Ann. Henri Poincaré*, **5**, 1013–120.
82 Celletti, A. (2010) *Stability and Chaos in Celestial Mechanics*, Springer.
83 Morbidelli, A. (2002) *Modern celestial mechanics: aspects of Solar System dynamics*, Taylor & Francis.
84 Pöschel, J. (2000) A lecture on the classical KAM theorem. *Proc. Symp. Pure Math.*, **69**, 707–732.
85 Chierchia, L. and Mather, J. (2010) Kolmogorov–Arnold–Moser theory. *Scholarpedia*, **5(9)**, 2123.
86 Benettin, G., Henrard, J., and Kuksin, S. (1999) Hamiltonian dynamics – theory and applications, in *Lecture Notes in Mathematics*, Springer.
87 Giorgilli, A. (2002) *Notes on Exponential Stability of Hamiltonian Systems*, Centro di Ricercia Matematics Ennio De Giorgi.
88 Guzzo, M. (2008) An overview on Nekhoroshev theorem, in *Lecture Notes in Physics*, vol. 729, Springer, pp. 1–28.
89 Lyapunov, A. (1892) *The general problem of the stability of motion*, Univ. Kharkov 1892, Engl. Taylor & Francis, London 1992.
90 Arnold, V.E. (1988) *Dynamical Systems III, Encyclopaedia of Mathematical Sciences, vol. 3*, Springer.
91 Contopoulos, G. (1960) A third integral of motion in a galaxy. *Z. Astrophys.*, **49**, 273.
92 Contopoulos, G. and Barbanis, B. (1962) An application of the third integral of motion. *The Observatory*, **82**, 80–82.
93 Contopoulos, G. (1963) On the existence of a third integral of motion. *AJ*, **68**, 1, doi:10.1086/108903.
94 Contopoulos, G. (1963) Some applications of a third integral of motion. *AJ*, **68**, 70, doi:10.1086/109073.
95 Contopoulos, G. (1963) A classification of the integrals of motion. *ApJ*, **138**, 1297, doi:10.1086/147724.
96 Contopoulos, G. (1963) Resonance cases and small divisors in a third integral of motion. I. *AJ*, **68**, 763.
97 Hénon, M. and Heiles, C. (1964) The applicability of the third integral of motion: some numerical experiments. *Astron. J.*, **69**, 73–79.
98 Kolmogorov, A. (1954) On the conservation of conditionally periodic motions for a small change in the Hamiltonian's function. *Dokl. Akad. Nauk. SSSR*, **98**, 527–530.
99 Neishtadt, A. (1967) Scattering by resonances. *Celest. Mech. Dyn. Astron.*, **65**, 1–20.
100 Lichtenberg, A. and Lieberman, M. (1992) *Regular and Chaotic Dynamics*, Springer, Berlin.
101 Chirikov, B. (1960) Resonance processes in magnetic traps. *J. Transl. J. Nucl. Energy Part C: Plasma Phys.*, **1**, 253.
102 Chirikov, B. (1971) Research concerning the theory of nonlinear resonance and stochasticity. *Engl. Trans. CERN Trans.*, **71**, 40.

103 Chirikov, B. (1979) A universal instability of many-dimensional oscillator systems. *Phys. Rep.*, **52**, 263.

104 Contopoulos, G. (2002) *Order and chaos in dynamical astronomy*, Springer, Berlin.

105 Haller, G. (1999) *Chaos near resonance*, Springer.

106 Kolmogorov, A. (1954) On conservation of conditionally periodic motions for a small change in Hamilton's function. *Dokl. Akad. Nauk SSSR*, **98**, 527–530.

107 Arnold, V. (1963) Proof of a theorem of A.N. Kolmogorov on the preservation of conditionally periodic motions under a small perturbation of the Hamiltonian. *Usp. Math. Nauk*, **18**, 13–40.

108 Moser, J. (1962) On invariant curves of area preserving mappings of an annulus. *Nachr. Akad. Wiss. Gött. Math. Phys.*, **K1**, 1–20.

109 Kolmogorov, A. (1953) On dynamical systems with an integral invariant on a torus. *Dokl. Akad. Nauk. SSSR*, **93**, 763–766.

110 Kolmogorov, A. (1957) The general theory of dynamical systems and classical mechanics, in *Proceedings of the International Congress of Mathematicians*, vol. 1, American Mathematical Society, pp. 315–333.

111 Arnold, V.I. (2009) On analytic maps of the circle onto itself, in *Collected Works, Vladimir I. Arnold – Collected Works*, vol. 1 (eds A.B. Givental, B.A. Khesin, J.E. Marsden, A.N. Varchenko, V.A. Vassiliev, O.Y. Viro, and V.M. Zakalyukin), Springer Berlin Heidelberg, pp. 149–151.

112 Arnold, V. and Sinai, Y.G. (1962) Small perturbations of the automorphism of the torus. *Dokl. Akad. Nauk. SSR*, **144**, 693.

113 Arnold, V. (1963) Proof of a theorem by A.N. Kolmogorov on the invariance of quasi-periodic motions under small perturbations of the Hamiltonian. *Russ. Math. Surv.*, **18**, 9–36.

114 Arnold, V. (1961) A new technique for the construction of solutions of nonlinear differential equations. *Proc. Natl. Acad. Sci. USA*, **47**, 24–31.

115 Moser, J. (1967) Convergent series expansions for quasi-periodic motions. *Math. Ann.*, **169**, 136–176.

116 Arnold, V. (1978) *Mathematical Methods of Classical Mechanics*, Springer.

117 Moser, J. (1973) *Stable and Random Motions in Dynamical Systems*, Princeton University Press.

118 Celletti, A. and Chierchia, L. (2007) KAM stability and celestial mechanics. *Mem. Am. Math. Soc.*, **187**, 878.

119 Nekhoroshev, N. (1977) Exponential estimates of the stability time of near-integrable Hamiltonian systems. *Russ. Math. Surv.*, **32**, 1.

120 Nekhoroshev, N. (1979) Exponential estimates of the stability time of near-integrable Hamiltonian systems 2. *Trudy Sem. Petrovos.*, **5**, 5.

121 Benettin, G., Galgani, L., and Giorgilli, A. (1985) A proof of Nekhoroshev's theorem for the stability times in nearly integrable Hamiltonian systems. *Celest. Mech.*, **37**, 1–25.

122 Giorgilli, A. (1988) Rigorous results on the power expansions for the integrals of a Hamiltonian system near an elliptic equilibrium point. *Ann. Inst. H. Poincaré*, **37**, 423–439.

123 Guzzo, M., Fassó, F., and Benettin, G. (1988) On the stability of elliptic equilibria. *Math. Phys. Electron. J.*, **4**, 1–16.

124 Fassó, F., Guzzo, M., and Benettin, G. (1997) Nekhoroshev-stability of elliptic equilibria of Hamiltonian systems. *Commun. Math. Phys.*, **197**, 347–360.

125 Giorgilli, A. and Zehnder, E. (1992) Exponential stability for time dependent potentials. *ZAMP*, **43**, 827–855.

126 Lochak, P. (1992) Canonical perturbation theory via simulataneous approximations. *Usp. Math. Nauk.*, **47**, 59–140.

127 Pöschel, J. (1993) Nekhoroshev estimates for quasi-convex Hamiltonian systems. *Math. Z.*, **213**, 187–216.

128 Morbidelli, A. and Giorgilli, A. (1995) On a connection between KAM and Nekhoroshev's theorems. *Phys. D: Nonlinear Phenom.*, **86**, 514–516.

129 Delshams, A. and Gutiérrez, P. (1995), Effective stability and kam theory.

**130** Niederman, L. (2009) Nekhoroshev theory, in *Encyclopedia of Complexity and Systems Science* (ed. R.A. Meyers), Springer, pp. 5986–5998.

**131** Froeschlé, C. and Lega, E. (2000) On the structure of symplectic mappings. The fast Lyapunov indicator: a very sensitive tool. *Celest. Mech. Dyn. Astron.*, **78**, 167–195.

**132** Froeschlé, C., Guzzo, M., and Lega, E. (2000) Graphical evolution of the Arnold web: From order to chaos. *Science*, **289**, 2108–2110.

**133** Stumpff, K. (1959) *Himmelsmechanik I*, VEB Deutscher Verlag der Wissenschaften.

**134** Stiefel, E. and Scheifele, G. (1971) *Linear and regular celestial mechanics: perturbed two-body motion, numerical methods, canonical theory*, Grundlehren der mathematischen Wissenschaften, Springer.

**135** Davis, J.J., Mortari, D., and Bruccoleri, C. (2010) Sequential solution to Kepler's equation. *Celest. Mech. Dyn. Astron.*, **108**, 59–72.

**136** Danby, J.M.A. (1988) *Fundamentals of celestial mechanics*, Willmann-Bell.

**137** Szebehely, V. (1967) *Theory of orbits. The restricted problem of three bodies*, New York: Academic Press.

**138** Stumpff, K. (1965) *Himmelsmechanik II*, VEB Deutscher Verlag der Wissenschaften.

**139** Marchal, C. (1990) *The Three-Body Problem*, Elsevier.

**140** Conley, C. (1968) Low Energy Transit Orbits in the Restricted Three-Body Problem. *SIAM J. Appl. Math.*, **16**, 732–746.

**141** Conley, C. (1969) On the ultimate behavior of orbits with respect to an unstable critical point. I. Oscillating, asymptotic, and capture orbits. *J. Diff. Eq.*, **5**, 136–158.

**142** Belló, M., Gomez, G., and Masdemont, J. (TT) *Invariant Manifolds, Lagrangian Trajectories and Space Mission Design*, Space Manifold Dynamics (eds E. Perozzi and S. Ferraz–Mello), Springer.

**143** Brouwer, D. and Clemence, G.M. (1961) *Methods of Celestial Mechanics*, Academic Press, New York.

**144** Brown, E. and Shook, C. (1933) *Planetary Theory*, Cambridge University Press.

**145** Szebehely, V. and Giacaglia, G.E.O. (1964) On the elliptic restricted problem of three bodies. *AJ*, **69**, 230.

**146** Erdi, B. (1974) Reduction of the two-dimensional elliptic restricted problem of three bodies to Hill's equation. *AJ*, **79**, 653.

**147** Erdi, B. (1977) An asymptotic solution for the Trojan case of the plane elliptic restricted problem of three bodies. *Celest. Mech.*, **15**, 367–383.

**148** Dvorak, R. (1986) Critical orbits in the elliptic restricted three-body problem. *A&A*, **167**, 379–386.

**149** Sándor, Z. and Érdi, B. (2003) Symplectic mapping for Trojan-type motion in the elliptic restricted three-body problem. *Celest. Mech. Dyn. Astron.*, **86**, 301–319.

**150** Lhotka, C., Efthymiopoulos, C., and Dvorak, R. (2008) Nekhoroshev stability at $L_4$ or $L_5$ in the elliptic-restricted three-body problem – application to Trojan asteroids. *MNRAS*, **384**, 1165–1177.

**151** Hagel, J. and Kallrath, J. (1989) Integration theory for the elliptic restricted three-body problem. *A&A*, **222**, 344–352.

**152** Contopoulos, G. (1967) Integrals of motion in the elliptic restricted three-body problem. *AJ*, **72**, 669.

**153** Beaugé, C. (1996) On a global expansion of the disturbing function in the planar elliptic restricted three-body problem. *Celest. Mech. Dyn. Astron.*, **64**, 313–350.

**154** Palacián, J.F. and Yanguas, P. (2006) From the circular to the spatial elliptic restricted three-body problem. *Celest. Mech. Dyn. Astron.*, **95**, 81–99.

**155** Peale, S. (1993) The effect of the nebula on the Trojan precursors. *Icarus*, **106**, 308–322.

**156** Beaugé, C. and Ferraz-Mello, S. (1992) Corotation solutions in the elliptic asteroidal problem with Stokes drag, in *Chaos, Resonance, and Collective Dynamical Phenomena in the Solar System*, vol. 152 (ed. S. Ferraz-Mello), IAU Symposium, p. 355.

**157** Ferraz-Mello, S. (1992) Averaging the elliptic asteroidal problem with a Stokes drag, in *Interrelations between Physics and*

*Dynamics for Minor Bodies in the Solar System* (eds D. Benest and C. Froeschlé), SFSA, p. 45.

158 Beauge, C. and Ferraz-Mello, S. (1993) Resonance trapping in the primordial solar nebula – The case of a Stokes drag dissipation. *Icarus*, **103**, 301–318.

159 Haghighipour, N. (1999) Dynamical friction and resonance trapping in planetary systems. *MNRAS*, **304**, 185–194.

160 Sitnikov, K. (1960) Existence of oscillatory motion for the three body problem. *Dokl. Akad. Nauk. USSR*, **133**, 303–306.

161 Pavanini, G. (1907) Sopra una nuova categoria di soluzioni periodiche nel problema dei tre corpi. *Ann. Mat. Pura Appl.*, **13**, 179–202.

162 MacMillan, W. (1913) An integrable case in the restricted three body problem. *A.J.*, **27**, 11–13.

163 Soulis, P., Bountis, T., and Dvorak, R. (2007) Stability of motion in the Sitnikov 3 body problem. *Celest. Mech. Dyn. Astron.*, **99**, 129–148.

164 Efthymiopoulos, C., Contopoulos, G., and Voglis, N. (1999) Cantori, islands and asymptotic curves in the stickiness region. *Celest. Mech. Dyn. Astron.*, **73**, 221–230.

165 Alfaro, J.M. and Chiralt, C. (1991) Gaps of regular motion in Sitnikov's problem, in *Positional Astronomy and Celestial Mechanics* (eds A. Lopez Garcia, R.F. Lopez Machi, and A.G. Sokolsky), self-published, pp. 23–28.

166 Sun, Y. (1992) On the Sitnikov problem. *Celest. Mech. Dyn. Astron.*, **49**, 285–302.

167 Hagel, J. (1992) A new analytic solution of the Sitnikov problem. *Celest. Mech. Dyn. Astron.*, **53**, 267–292.

168 Di Ruzza, S. and Lhotka, C. (2011) High order normal form construction near the elliptic orbit of the Sitnikov problem. *Celest. Mech. Dyn. Astron.*, **111**, 449–464.

169 Wodnar, K. (1991) New formulations of the Sitnikov problem, in *Predictability, Stability and Chaos in N-Body Dynamical Systems* (ed A.E. Roy), Plenum Press.

170 Hagel, J. and Trenkler, T. (1993) A Computer-Aided Analysis of the Sitnikov problem. *Celest. Mech. Dyn. Astron.*, **56**, 81–98.

171 Faruque, S. (2003) Solution of the Sitnikov problem. *Celest. Mech. Dyn. Astron.*, **87**, 353–369.

172 Hagel, J. and Lhotka, C. (2005) A High Order Perturbation Analysis of the Sitnikov Problem. *Celest. Mech. Dyn. Astron.*, **93**, 201–228.

173 Lhotka, C. (2005) *Störungsanalyse des Sitnikov Problems für hohe Ordnungen unter Verwendung automatisierter Herleitungsmethoden in Mathematica*, University of Vienna, diploma thesis.

174 Floquet, G. (1883) Sur les équations différentielles linéaires à coefficients périodiques. *Ann. Ecol. Norm. Sup.*, **12**, 47–88.

175 Poincaré, H. (1893) Les Méthodes Nouvelles de la Mécanique Célèste, II. *Dover Publ.*, pp. 123–88.

176 Lindstedt, A. (1882) Sur les équations différentielles linéaires à coefficients périodiques. *Abh. K. Akad. Wiss. St. Petersburg*, **31**.

177 Dvorak, R. (1993) *The Sitnikov problem: A short Review*, Primo Convegno Nazionale di Meccanica Celeste.

178 Lang, R. (2011) Untersuchung des dynamischen Verhaltens des Sitnikov-Problems mittels Alexeev–Moser–Vrabec Mappings. *Diploma thesis, University of Vienna*, p. 102.

179 Moulton, F.R. (1970) *An introduction to celestial mechanics*, Dover, New York.

180 Murray, C.D. and Dermott, S.F. (1999) *Solar system dynamics*, Cambridge University Press.

181 Kovalevsky, J. (1963) *Introduction a la mecanique celeste.*, Paris, Librairie A. Colin.

182 Kovalevsky, J. (1967) *Introduction to celestial mechanics*, Astrophysics and space science library, D. Reidel.

183 Roy, A.E. (1978) *Orbital Motion*, Hilger, Bristol.

184 Seidelmann, P. and Office, U.S.N.O.N.A. (2005) *Explanatory Supplement to the Astronomical Almanac*, University Science Books.

185 Brumberg, V. (1967) A numerical development of a generalized planetary theory. *Bull. Inst. Theor. Astron.*, **2**, 125.

**186** Dvorak, R. (1982) A direct method of computing small divisors in planetary theory. *A&A*, **108**, 14–18.

**187** Chapront, J., Bretagnon, P., and Mehl, M. (1975) A formula for calculating higher-order perturbations in planetary problems. *Celest. Mech.*, **11**, 379–399.

**188** Duriez, L. (1978) Poisson's theorem in heliocentric variables – Conditions for the application of this theorem concerning the invariability of the major axes of planetary orbits to second order in the masses. *A&A*, **68**, 199–216.

**189** Laskar, J. (1989) A numerical experiment on the chaotic behaviour of the solar system. *Nature*, **338**, 237.

**190** Laskar, J. (1990) The chaotic motion of the solar system – A numerical estimate of the size of the chaotic zones. *Icarus*, **88**, 266–291.

**191** Laskar, J. (1994) Large-scale chaos in the solar system. *A&A*, **287**, L9–L12.

**192** Laskar, J. (2008) Chaotic diffusion in the Solar System. *Icarus*, **196**, 1–15.

**193** Laskar, J. and Gastineau, M. (2009) Existence of collisional trajectories of Mercury, Mars and Venus with the Earth. *Nature*, **459**, 817–819.

**194** Bazsó, Á., Dvorak, R., Pilat-Lohinger, E., Eybl, V., and Lhotka, C. (2010) A survey of near-mean-motion resonances between Venus and Earth. *Celest. Mech. Dyn. Astron.*, **107**, 63–76.

**195** Sándor, Z., Érdi, B., and Murray, C.D. (2002) Symplectic Mappings of Co-orbital Motion in the Restricted Problem of Three Bodies. *Celest. Mech. Dyn. Astron.*, **84**, 355–368.

**196** Erdi, B. (1988) Long periodic perturbations of Trojan asteroids. *Celest. Mech.*, **43**, 303–308.

**197** Nesvorný, D. and Dones, L. (2002) How long-lived are the hypothetical Trojan populations of Saturn, Uranus, and Neptune? *Icarus*, **160**, 271–288.

**198** Robutel, P., Gabern, F., and Jorba, A. (2005) The observed Trojans and the global dynamics around the Lagrangian points of the Sun Jupiter system. *Celest. Mech. Dyn. Astron.*, **92**, 53–69.

**199** Tsiganis, K., Varvoglis, H., and Dvorak, R. (2005) Chaotic diffusion and effective stability of Jupiter Trojans. *Celest. Mech. Dyn. Astron.*, **92**, 71–87.

**200** Dvorak, R. and Tsiganis, K. (2000) Why do Trojan ASCs (not) escape? *Celest. Mech. Dyn. Astron.*, **78**, 125–136.

**201** Holman, M.J. and Wisdom, J. (1993) Dynamical stability in the outer solar system and the delivery of short period comets. *AJ*, **105**, 1987–1999.

**202** Marzari, F., Tricarico, P., and Scholl, H. (2003) Stability of Jupiter Trojans investigated using frequency map analysis: the MATROS project. *MNRAS*, **345**, 1091–1100.

**203** Baudisch, H. (2010) *Investigation on the Stability of the Trojans of Saturn around L4 (and L5)*, University of Vienna, diploma thesis.

**204** Baudisch, H.D. and Dvorak, R. (2011) Where are the Saturn Trojans. *Proceedings, Journees 'Systeme de reference spatio-temporels'*, (eds Schuh *et al.*), Vienna University of Technology, p. 225–228.

**205** Chiang, E. (2003) 2001 QR_322. *IAU circular*, 8044, 3.

**206** Zhou, L.Y., Dvorak, R., and Sun, Y.S. (2009) The dynamics of Neptune Trojan – I. The inclined orbits. *MNRAS*, **398**, 1217–1227.

**207** Zhou, L.Y., Dvorak, R., and Sun, Y.S. (2011) The dynamics of Neptune Trojans – II. Eccentric orbits and observed objects. *MNRAS*, **410**, 1849–1860.

**208** Dvorak, R., Bazsó, Á., and Zhou, L.Y. (2010) Where are the Uranus Trojans? *Celest. Mech. Dyn. Astron.*, **107**, 51–62.

**209** Mainzer, A., Bauer, J., Grav, T., Masiero, J., Cutri, R.M., Dailey, J., Eisenhardt, P., McMillan, R.S., Wright, E., Walker, R., Jedicke, R., Spahr, T., Tholen, D., Alles, R., Beck, R., Brandenburg, H., Conrow, T., Evans, T., Fowler, J., Jarrett, T., Marsh, K., Masci, F., McCallon, H., Wheelock, S., Wittman, M., Wyatt, P., DeBaun, E., Elliott, G., Elsbury, D., Gautier, IV, T., Gomillion, S., Leisawitz, D., Maleszewski, C., Micheli, M., and Wilkins, A. (2011) Preliminary Results from NEOWISE: An enhancement to the wide-field infrared survey explorer for Solar System science. *ApJ*, **731**, 53.

210 Dvorak, R., Lhotka, C., and Zhou, L. (2012) The orbit of 2010 TK7. Possible regions of stability for other Earth Trojan asteroids. *ArXiv e-prints.*

211 Connors, M., Wiegert, P., and Veillet, C. (2011) Earth's Trojan asteroid. *Nature*, **475**, 481–483.

212 Tsiganis, K., Dvorak, R., and Pilat-Lohinger, E. (2000) Thersites: a 'jumping' Trojan? *A&A*, **354**, 1091–1100.

213 Schwarz, R., Dvorak, R., Süli, Á., and Érdi, B. (2007) Survey of the stability region of hypothetical habitable Trojan planets. *A&A*, **474**, 1023–1029.

214 Schwarz, R., Dvorak, R., Süli, Á., and Érdi, B. (2007) Stability of fictitious Trojan planets in extrasolar systems. *Astron. Nachr.*, **328**, 785.

215 Bretagnon, P. (1974) Termes a longues periodes dans le systeme solaire. *A&A*, **30**, 141–154.

216 Knezevic, Z., Milani, A., Farinella, P., Froeschlé, Ch., and Froeschlé, Cl. (1991) Secular resonances from 2 to 50 AU. *Icarus*, **93**, 316–330.

217 Cellino, A. and Dell'Oro, A. (2010) Asteroid dynamical families, in *Lecture Notes in Physics* (eds J. Souchay and R. Dvorak), vol. 790, Springer, Berlin, pp. 137–193.

218 Lemaitre, A. (1993) Proper elements – What are they? *Celest. Mech. Dyn. Astron.*, **56**, 103–119.

219 Kozai, Y. (1962) Secular perturbations of asteroids with high inclination and eccentricity. *AJ*, **67**, 591.

220 Thomas, F. and Morbidelli, A. (1996) The Kozai resonance in the Outer Solar System and the dynamics of long-period comets. *Celest. Mech. Dyn. Astron.*, **64**, 209–229.

221 Kinoshita, H. and Nakai, H. (1999) Analytical Solution of the Kozai resonance and its application. *Celest. Mech. Dyn. Astron.*, **75**, 125–147.

222 Nesvorný, D. and Morbidelli, A. (1998) An analytic model of three-body mean motion resonances. *Celest. Mech. Dyn. Astron.*, **71**, 243–271.

223 Aksnes, K. (1988) General formulas for three-body resonances., in *Long-term Dynamical Behaviour of Natural and Artificial N-body Systems* (ed. A.D. Roy), pp. 125–139.

224 Wilkens, A. (1933) Über das Problem der mehrfachen Kommensurabilitäten im Sonnensystem. *Sitzungsber. Math. Naturw. Abt. Bayer. Akad. Wissensch. Heft 1*, **1**, 71–101.

225 Laplace, P. (1825) *Traité de Mécanique Céleste (I–V)*, Paris.

226 Tisserand, F. (1889) *Traité de Mécanique Céleste (I–IV)*, Paris, Gauthier-Villars et fils.

227 Hill, G. (1919) *Tables of the motion of the Moon (I–IV)*, New Have, Yale University Press.

228 Brown, E. (1960) *An Introductory Treatise on the Lunar Theory*, Dover, New York.

229 Chapront, J. and Chapront-Touzé, M. (1996) Lunar motion: Theory and observations. *Celest. Mech. Dyn. Astron.*, **66**, 31–38.

230 Chapront, J., Chapront-Touzé, M., and Francou, G. (1999) Determination of the lunar orbital and rotational parameters and of the ecliptic reference system orientation from LLR measurements and IERS data. *A&A*, **343**, 624–633.

231 Chapront, J., Chapront-Touzé, M., and Francou, G. (2002) A new determination of lunar orbital parameters, precession constant and tidal acceleration from LLR measurements. *A&A*, **387**, 700–709.

232 Hill, E. (1878) Researches in the lunar theory. *Am. J. Math.*, **1**, 129.

233 Stumpff, K. (1966) A new solution of Hill's lunar problem, in *The Theory of Orbits in the Solar System and in Stellar Systems, IAU Symposium*, vol. 25 (ed. G.I. Contopoulos), *IAU Symposium*, vol. 25, pp. 261–265.

234 Chapront-Touze, M. (1980) The ELP solution to the main problem of lunar theory. *A&A*, **83**, 86–94.

235 Chapront-Touze, M. (1982) The ELP solution for the main problem of the Moon and some applications. *Celest. Mech.*, **26**, 63–69.

236 Chapront-Touze, M. and Chapront, J. (1983) The lunar ephemeris ELP 2000. *A&A*, **124**, 50–62.

237 Chapront, J., Chapront-Touzé, M., and Francou, G. (1999) Complements to

Moons'1 lunar libration theory. *Celest. Mech. Dyn. Astron.*, **73**, 317–328.

**238** Chapront-Touze, M. and Henrard, J. (1980) The principal problem of the motion of the Moon – Comparison between two theories. *A&A*, **86**, 221–224.

**239** Chapront-Touzc, M. (1974) Itcrativc construction of a solution of the main problem of the Moon – Influence of small divisors. *A&A*, **36**, 5–16.

**240** Chapront, J. and Chapront-Touze, M. (1981) Comparison of ELP-2000 with a numerical integration at the JPL. *A&A*, **103**, 295–304.

**241** Bretagnon, P. (1980) Second order theory of the inner planets. *A&A*, **84**, 329–341.

**242** Chapront-Touze, M. and Chapront, J. (1980) Planetary perturbations of the Moon. *A&A*, **91**, 233–246.

**243** Chapront, J. and Chapront-Touze, M. (1982) Planetary Perturbations of the Moon in ELP 2000. *Celest. Mech.*, **26**, 83–94.

**244** Lestrade, J.F., Chapront, J., and Chapront-Touze, M. (1982) The relativistic planetary perturbations and the orbital motion of the Moon, in *IAU Colloq. 63: High-Precision Earth Rotation and Earth-Moon Dynamics: Lunar Distances and Related Observations, Astrophysics and Space Science Library*, vol. 94 (ed. O. Calame), *Astrophysics and Space Science Library*, vol. 94, pp. 217–224.

**245** Chapront-Touze, M. (1982) Progress in the analytical theories for the orbital motion of the Moon. *Celest. Mech.*, **26**, 53–62.

**246** Chapront-Touzé, M. and Chapront, J. (1996) La théorie du mouvement de la Lune. *L'Astronomie*, **110**, 118–125.

**247** Simon, J.L., Bretagnon, P., Chapront, J., Chapront-Touze, M., Francou, G., and Laskar, J. (1994) Numerical expressions for precession formulae and mean elements for the Moon and the planets. *A&A*, **282**, 663–683.

**248** Chapront, J. and Chapront-Touzé, M. (1996) Lunar motion: Theory and observations. *Celest. Mech. Dyn. Astron.*, **66**, 31–38.

**249** Chapront, J. and Chapront-Touze, M. (1997) Lunar motion: Theory and observations, in *IAU Colloq. 165: Dynamics and Astrometry of Natural and Artificial Celestial Bodies* (eds I. M. Wytrzyszczak, J. H. Lieske, and R. A. Feldman), Kluwer, p. 31.

**250** Hansen, P. (1855) Abhandlungen der Mathematisch-Physischen Class der Königlich Sächschichen Gessellschaft der Wissenschaften. *Leipzig*, **2**, 181–281.

**251** Kaula, W. (1966) *Theory of satellite geodesy*, Blaisdell.

**252** Giacaglia, G.E.O. (1976) A note on Hansen's coefficients in satellite theory. *Celest. Mech.*, **14**, 515–523.

**253** Laplace, P. (1895) *Théorie de Jupiter et de Saturne*, Mémoire de l'Académie Royale des Sciences, Gauthier-Villars.

**254** Laskar, J. and Robutel, P. (1995) Stability of the planetary three-body problem. I. Expansion of the planetary Hamiltonian. *Celest. Mech. Dyn. Astron.*, **62**, 193–217.

**255** Brumberg, V. (1995) *Analytical Techniques of Celestial Mechanics*, Astronomy and Astrophysics Library, Springer.

# Index